Wearable Sensing and Intelligent Data Analysis for Respiratory Management

Wearable Sensing and Intelligent Data Analysis for Respiratory Management

Edited by

Rui Pedro Paiva
Department of Informatics Engineering, Centre for Informatics and Systems of the University of Coimbra, Coimbra, Portugal

Paulo de Carvalho
Department of Informatics Engineering, Centre for Informatics and Systems of the University of Coimbra, Coimbra, Portugal

Vassilis Kilintzis
Senior Researcher at the Aristotle University of Thessaloniki, Greece

ELSEVIER

ACADEMIC PRESS
An imprint of Elsevier

Academic Press is an imprint of Elsevier
125 London Wall, London EC2Y 5AS, United Kingdom
525 B Street, Suite 1650, San Diego, CA 92101, United States
50 Hampshire Street, 5th Floor, Cambridge, MA 02139, United States
The Boulevard, Langford Lane, Kidlington, Oxford OX5 1GB, United Kingdom

ISBN: 978-0-12-823447-1

For information on all Academic Press publications visit our
website at https://www.elsevier.com/books-and-journals

Publisher: Mara Conner
Acquisitions Editor: Sonnini R. Yura
Editorial Project Manager: Isabella C. Silva
Production Project Manager: Surya Narayanan Jayachandran
Cover Designer: Miles Hitchen
Greek Designer: Costas Platides

Working together
to grow libraries in
developing countries

www.elsevier.com • www.bookaid.org

Typeset by TNQ Technologies

Contents

3. Sensor technologies for mobile and wearable applications in mobile respiratory management

*Josias Wacker, Benjamin Bonnal, Fabian Braun,
Olivier Chételat, Damien Ferrario, Mathieu Lemay,
Michaël Rapin, Philippe Renevey and Gürkan Yilmaz*

4. Textiles and smart materials for wearable monitoring systems

Rita Paradiso and Laura Caldani

Part III
Data analysis and management

5. Automated respiratory sound analysis

Diogo Pessoa, Bruno Machado Rocha, Paulo de Carvalho and Rui Pedro Paiva

6. Respiratory image analysis

Inéz Frerichs, Zhanqi Zhao, Meng Dai, Fabian Braun,
Martin Proença, Michaël Rapin, Josias Wacker, Mathieu Lemay,
Kostas Haris, Georgios Petmezas, Aris Cheimariotis,
Irini Lekka, Nicos Maglaveras, Claas Strodthoff, Barbara Vogt,
Livia Lasarow, Norbert Weiler, Diogo Pessoa, Bruno Machado Rocha,
Paulo de Carvalho, Rui Pedro Paiva and Andy Adler

7. Respiratory data management

Vassilis Kilintzis and Nikolaos Beredimas

Part IV
Current challenges in respiratory management systems

8. The edge-cloud continuum in wearable sensing for respiratory analysis

Anaxagoras Fotopoulos, Pantelis Z. Lappas and Alexis Melitsiotis

11. Integrated care in respiratory function management
Iman Hesso, Reem Kayyali and Shereen Nabhani-Gebara

Contributors

Andy Adler, Systems and Computer Engineering, Carleton University, Ottawa, Canada

Nikolaos Beredimas, Laboratory of Computing, Medical Informatics and Biomedical Imaging Technologies, Medical School, Aristotle University of Thessaloniki, Thessaloniki, Greece

Benjamin Bonnal, Swiss Center for Electronics and Microtechnology (CSEM, Centre Suisse d'Electronique et de Microtechnique), Neuchâtel, Switzerland

Fabian Braun, Swiss Center for Electronics and Microtechnology (CSEM, Centre Suisse d'Electronique et de Microtechnique), Neuchâtel, Switzerland

Laura Caldani, Smartex srl, Prato, PO, Italy

Georgia Chasapidou, Pulmonology Department, "G. Papanikolaou" General Hospital, Thessaloniki, Greece

Aris Cheimariotis, Lab of Computing, Medical Informatics and Biomedical Imaging Technologies, Aristotle University of Thessaloniki, Thessaloniki, Greece

Ioanna Chouvarda, School of Medicine, Aristotle University of Thessaloniki, Thessaloniki, Greece

Olivier Chételat, Swiss Center for Electronics and Microtechnology (CSEM, Centre Suisse d'Electronique et de Microtechnique), Neuchâtel, Switzerland

Meng Dai, Department of Biomedical Engineering, Fourth Military Medical University, Xi'an, China

Paulo de Carvalho, Department of Informatics Engineering, Centre for Informatics and Systems of the University of Coimbra, Coimbra, Portugal

Damien Ferrario, Swiss Center for Electronics and Microtechnology (CSEM, Centre Suisse d'Electronique et de Microtechnique), Neuchâtel, Switzerland

Anaxagoras Fotopoulos, EXUS AI Labs, Athens, Attika, Greece

Inéz Frerichs, Department of Anaesthesiology and Intensive Care Medicine, University Medical Centre Schleswig-Holstein, Campus Kiel, Kiel, Germany

Kostas Haris, Lab of Computing, Medical Informatics and Biomedical Imaging Technologies, Aristotle University of Thessaloniki, Thessaloniki, Greece; Department of Informatics and Computer Engineering, University of West Attica, Athens, Greece

Iman Hesso, Kingston University London, KT, United Kingdom

Laura Romero Jaque, Universitat Politecnica de Valencia, Valencia, Spain

Evangelos Kaimakamis, 1st Intensive Care Unit, "G. Papanikolaou" General Hospital, Thessaloniki, Greece

Reem Kayyali, Kingston University London, KT, United Kingdom

Vassilis Kilintzis, Laboratory of Computing, Medical Informatics and Biomedical Imaging Technologies, Medical School, Aristotle University of Thessaloniki, Thessaloniki, Greece

Pantelis Z. Lappas, EXUS AI Labs, Athens, Attika, Greece

Livia Lasarow, Department of Anaesthesiology and Intensive Care Medicine, University Medical Centre Schleswig-Holstein, Campus Kiel, Kiel, Germany

Irini Lekka, Lab of Computing, Medical Informatics and Biomedical Imaging Technologies, Aristotle University of Thessaloniki, Thessaloniki, Greece

Mathieu Lemay, Swiss Center for Electronics and Microtechnology (CSEM, Centre Suisse d'Electronique et de Microtechnique), Neuchâtel, Switzerland

Bruno Machado Rocha, Department of Informatics Engineering, Centre for Informatics and Systems of the University of Coimbra, Coimbra, Portugal

Nicos Maglaveras, Lab of Computing, Medical Informatics and Biomedical Imaging Technologies, Aristotle University of Thessaloniki, Thessaloniki, Greece

Alda Marques, Lab3R-Respiratory Research and Rehabilitation Laboratory, School of Health Sciences (ESSUA), Institute of Biomedicine (iBiMED), University of Aveiro, Aveiro, Portugal

Alexis Melitsiotis, EXODUS SA, Athens, Attika, Greece

Shereen Nabhani-Gebara, Kingston University London, KT, United Kingdom

Rui Pedro Paiva, Department of Informatics Engineering, Centre for Informatics and Systems of the University of Coimbra, Coimbra, Portugal

Rita Paradiso, Smartex srl, Prato, PO, Italy

Eleni Perantoni, School of Medicine, Aristotle University of Thessaloniki, Thessaloniki, Greece

Diogo Pessoa, Department of Informatics Engineering, Centre for Informatics and Systems of the University of Coimbra, Coimbra, Portugal

Georgios Petmezas, Lab of Computing, Medical Informatics and Biomedical Imaging Technologies, Aristotle University of Thessaloniki, Thessaloniki, Greece

Martin Proença, Swiss Center for Electronics and Microtechnology (CSEM, Centre Suisse d'Electronique et de Microtechnique), Neuchâtel, Switzerland

Michaël Rapin, Swiss Center for Electronics and Microtechnology (CSEM, Centre Suisse d'Electronique et de Microtechnique), Neuchâtel, Switzerland

Philippe Renevey, Swiss Center for Electronics and Microtechnology (CSEM, Centre Suisse d'Electronique et de Microtechnique), Neuchâtel, Switzerland

Vicente Traver Salcedo, ITACA − Universitat Politecnica de Valencia, Valencia, Spain

Sara Souto-Miranda, Lab3R-Respiratory Research and Rehabilitation Laboratory, School of Health Sciences (ESSUA), Institute of Biomedicine (iBiMED), University of Aveiro, Aveiro, Portugal

Paschalis Steiropoulos, Medical School, Democritus University of Thrace, Alexandroupolis, Greece

Claas Strodthoff, Department of Anaesthesiology and Intensive Care Medicine, University Medical Centre Schleswig-Holstein, Campus Kiel, Kiel, Germany

Barbara Vogt, Department of Anaesthesiology and Intensive Care Medicine, University Medical Centre Schleswig-Holstein, Campus Kiel, Kiel, Germany

Josias Wacker, Swiss Center for Electronics and Microtechnology (CSEM, Centre Suisse d'Electronique et de Microtechnique), Neuchâtel, Switzerland

Norbert Weiler, Department of Anaesthesiology and Intensive Care Medicine, University Medical Centre Schleswig-Holstein, Campus Kiel, Kiel, Germany

Gürkan Yilmaz, Swiss Center for Electronics and Microtechnology (CSEM, Centre Suisse d'Electronique et de Microtechnique), Neuchâtel, Switzerland

Zhanqi Zhao, Institute of Technical Medicine, Furtwangen University, Villingen-Schwenningen, Germany; Department of Biomedical Engineering, Fourth Military Medical University, Xi'an, China

Preface

Respiratory diseases, such as chronic obstructive pulmonary disease (COPD), lower respiratory tract infections, or asthma, have significant impact on patient's health-related quality of life, health-care systems, and society in general. Recent studies estimate that, worldwide, around 339 million people suffer from asthma and that, by 2030, COPD will become the third leading cause of death. This poses severe burdens to health-care systems in terms of outpatient and inpatient care, as well as pharmaceutical costs, which are highly correlated with the severity of exacerbation episodes.

In this scenario, the use of wearable sensing and intelligent data analysis algorithms for respiratory management assumes particular relevance, offering several potential clinical benefits. Namely, it allows for the early detection of respiratory exacerbations in patients with chronic respiratory diseases, allowing earlier and, therefore, more effective treatment. Early intervention in exacerbations of these conditions has been shown to decrease hospitalization rates and improve long-term outcomes, including survival.

As such, presently, the problem of continuous, noninvasive, remote, and real-time monitoring of such patients is deserving increasing attention from the scientific community. Wearable and portable systems with sensing technology and automated analysis of respiratory sounds and pulmonary images are some of the problems that are the subject of current research efforts. Such systems have the potential for substantial clinical benefits, promoting the so-called P4 medicine (personalized, participative, predictive, and preventive).

To this end, this book covers the most recent research and development on wearable technologies for respiratory management. The book, organized into 4 parts and 11 chapters, starts with an introductory overview of the process of respiration, its physiology, pathologies, and treatment, followed by the current needs and gaps of respiratory management in daily life. The second part addresses the aspects involved on wearable sensing, namely portable and noninvasive sensor technologies for mobile and wearable applications, and textiles and smart materials. Part III covers the data analysis and management pipeline, from data acquisition, transmission, storage, and representation, to feature engineering and machine learning for respiratory sound and image analysis. Finally, Part

IV addresses the current key challenges of respiratory management systems, namely the edge-cloud continuum in wearable sensing, strategies for long-term patient adherence, decision support systems, and integrated care in respiratory management.

We believe this book offers three main distinctive features: (i) an integrated, unified, and holistic coverage of the main topics and trends in wearable sensing and intelligent data analysis for respiratory management; (ii) an up-to-date review of the current trends and hot topics in the different subfields (e.g., wearable technologies, respiratory sound analysis, and pulmonary image analysis, particularly electrical impedance tomography); (iii) a comprehensive guide for starting researchers, namely, PhD students, offering them the necessary tools to start performing cutting-edge research in their area of interest.

Hence, this book will best suit the needs of researchers, particularly PhD students, working on different aspects of engineering issues for respiratory function management, namely in the areas of biomedical engineering, informatics engineering, electrical engineering, and data science and engineering. It will also work as an integrated and comprehensive entry point for any researcher who needs a holistic overview of the field. Health-care professionals will also benefit from the topics covered in the book, which aim at the active promotion of P4 medicine.

As such, the reader will be able to make use of the book mainly in two ways: (i) as someone with a broad interest in the whole process of technology use for respiratory management, where the whole book will offer the reader a broad and deep understanding of the area; (ii) as a researcher aiming to acquire specific knowledge in some of the identified subtopics, in which case the reader might be interested in the chapters setting the big picture of the whole field and then focusing on a specific subtopic.

Rui Pedro Paiva
Paulo de Carvalho
Vassilis Kilintzis

Acknowledgments

First and foremost, we would like to express our utmost gratitude to all the contributors of this book, who have offered their expertise and generous hard work to make this project a reality. It has been a pleasure working with you throughout the years in different projects that served as the catalyst to this book. In particular, we would like to acknowledge European projects WELCOME (Wearable Sensing and Smart Cloud Computing for Integrated Care to COPD Patients with Comorbidities—FP7-611223) and WELMO (Wearable Electronics for Effective Lung Monitoring—H2020-825572).

We also would like to express our great appreciation to the staff at Academic Press for their guidance and support during this process.

Finally, our gratitude goes to our host institutions, namely the Centre for Informatics and Systems of the University of Coimbra, Portugal, and the Aristotle University of Thessaloniki, Greece.

Rui Pedro Paiva
Paulo de Carvalho
Vassilis Kilintzis

Part I

Respiratory management: overview

Chapter 1

Respiration: physiology, pathology, and treatment

Evangelos Kaimakamis[1], Georgia Chasapidou[2]
[1]*1st Intensive Care Unit, "G. Papanikolaou" General Hospital, Thessaloniki, Greece;*
[2]*Pulmonology Department, "G. Papanikolaou" General Hospital, Thessaloniki, Greece*

Introduction

Human respiration is the physiological process enabling the gases (oxygen and carbon dioxide—O_2 and CO_2) exchange between the atmosphere and the respiratory circulation. It is achieved by means of the respiratory movements of the thoracic cavity and the lungs and is governed by a series of physiological laws and parameters which describe the respiratory mechanics and the gas exchange itself.

Respiration is controlled by superior centers in the brain stem and is affected by numerous internal and external stimuli.

Normal respiratory function is compromised when certain diseases affect the lung parenchyma (lung tissue), the thoracic wall formations, or the vascular network of the lungs. The range of disease states that affect the respiratory function is wide, including diseases of the lung itself or other systematic pathologies having a direct or indirect effect on respiratory functionality. Treatment of such diseases is highly dependent on underlying pathophysiology and the extent of the damage.

This chapter comprises two parts. In Part A, the physiology of respiration is presented and analyzed, covering issues related to the description of lung function, pulmonary circulation, gas exchange (perfusion, diffusion), respiratory mechanics, and neural control of respiration. In Part B, the main respiratory pathological states are presented, with brief description of their effect on lung function and available treatment options.

Part A: physiology of respiration

Overview of respiration

According to current physiological definitions, respiration is the "interchange of gases between an organism and the medium in which it lives" [1]. In the

Wearable Sensing and Intelligent Data Analysis for Respiratory Management
https://doi.org/10.1016/B978-0-12-823447-1.00004-X

human body, respiration can be further classified to external and internal [2]. The external respiration involves the transfer of oxygen (O_2) and carbon dioxide (CO_2) that occurs in the lungs between the atmosphere and the pulmonary circulation. The internal process is an analogous process that occurs at the cellular level. The external respiration comprises of three primary components: ventilation, perfusion, and diffusion. A thorough understanding of each of these components and their potential impairments is essential for researchers to understand the origin and implications of the biosignals that are related to respiration.

The respiratory system (physiology)

The ultimate function of the respiratory system is gas exchange [3]. This gas exchange consists of obtaining O_2 from the atmosphere and removing CO_2 from the blood. The presence of O_2 is necessary for normal cellular metabolism and CO_2 is a waste product of this metabolism. While CO_2 plays a role in acid−base balance, it must be cleared from the body in appropriate levels through ventilation.

Air passes through the nose and mouth into the airways, where it is warmed, humidified, and filtered. From the trachea to the alveoli, there are 23 branching generations of airways. The first 16 (in average) constitute the conducting zone, which is an anatomic dead space (i.e., no gas exchange takes place). The 17−23 generations form the respiratory zone. Each generation of branching increases the total cross-sectional area of the airways but reduces the radius of each airway and the velocity of air flowing through that. The exchange-effective respiratory zone comprises of the respiratory bronchioles, alveolar ducts, and alveolar sacs. At this end of the airways, approximately 300 million alveoli are situated. An adult alveolus has an average diameter of 200 μm, with an increase in diameter during inhalation. The alveoli consist of an epithelial layer and extracellular matrix surrounded by capillaries. Each capillary is in contact with several alveoli, so the capillaries present a sheet of blood to the alveolar air for gas exchange. The total area between pulmonary capillary blood and alveolar air ranges from 70 to 140 m^2 in adult humans (increased during exercise through recruitment of new capillaries in particular in the apical parts of the lungs).

Turbulent flow is the agitated random movement of molecules, which accounts for the sounds heard over the chest during breathing. This flow develops at the branch points of the upper airways even in quiet breathing. Turbulence also develops when constriction, mucus, infection, solid tumors, or foreign bodies decrease the radius of the airways. Vagal (parasympathetic) stimulation (by smoke, dust, cold air, and irritants) leads to airway constriction, whereas sympathetic stimulation dilatates the airways.

Volumes and capacities related to the lung function

The following schema (Fig. 1.1) depicts the volumes and capacities that are related with the breathing pattern in humans. The term Capacity refers to a sum of two or more volumes. In each normal resting respiratory movement, a volume equal to 350–500 mL of air enters the respiratory tract. This is called tidal volume (VT). The volume that remains from the end of a tidal inspiration to the maximum inspiratory effort is the inspiratory reserve volume (IRV). Similarly, the expiratory reserve volume (ERV) is the remaining volume that can be expired after the end of a tidal breath. The residual volume (RV) is the volume that remains in the lungs after a maximum expiratory effort. The sum of all the above volumes is the total lung capacity (TLC), whereas the sum of IRC and VT is the inspiratory capacity and the sum of IRV, VT, and ERV is the vital capacity. Finally, the sum of ERV and RV is the functional residual capacity (FRC) which is a very important measure, since it constitutes the balance point of the lung expansion (i.e., the point where the forces that tend to expand and close the lungs are equal and at a balance under normal conditions).

Two additional metrics are produced during the spirometry test: forced expiratory volume in one second (FEV_1) is the 1-s volume exhaled with forceful pressure from maximal inspiration. FEV_1 is often expressed in relation to the total forced expiratory volume (FEV). These metrics are essential for the study of obstructive and restrictive lung pathologies.

Ventilation

The most crucial component of respiration is the act of breathing, during which the lungs are filled with air through inhaling and CO_2 is removed

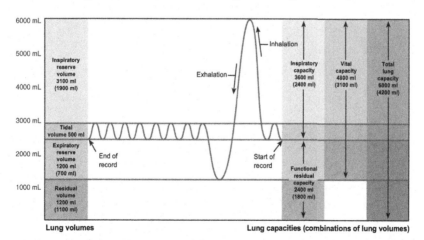

FIGURE 1.1 Diagram showing the volumes and capacities of the lungs (the volumes that are shown in the left column are representative of an average adult male).

through exhalation [1]. This process of moving air into and out of the lungs is known as ventilation [4]. While it may seem a simplistic process, the ability of air to flow into and out of the alveoli is dependent on a number of factors including integrity and compliance of the lung tissue and resistance to airflow within the airways [2]. Compliance (C) is a measure of how much the lung can increase its volume (ΔV) when under an increasing pressure (ΔP) applied on it, so $C = \Delta V/\Delta P$. The inverse property is called elastance (E), thus $E = 1/C$.

The lungs normally have the tendency to return to a volume slightly smaller than the RV due to the elastic forces associated with their structure. In order to maintain a volume larger than this, an expansion pressure must be applied to the lungs. This pressure is considered to equal the alveolar pressure (P_{alv}) minus the pleural pressure (P_{pl}) (the pressure inside the pleural cavity covering the lungs). By plotting the transmural pressure over the lung volume, the static pressure–volume curve is derived, as in Fig. 1.2. The sigmoid form of the curve reflects the fact that the required transmural pressure for the inflow of a specific volume of air into the lungs is significantly lower when the inspiratory effort starts from the level of the functional respiratory capacity compared to when the effort begins after previous hyperinflation (i.e., closer to the TLC).

Normal lungs are very distensible at FRC, but stiffen progressively toward TLC. The falling compliance is caused by an increase in the air–liquid surface tension because the liquid contains tension-reducing molecules (a substance called surfactant) that are spread further and further apart. Thereby the compliance of the lung is reduced. The contribution of surface tension to the overall lung elasticity is more than 50%. Compliance also decreases with age; there are corresponding decreases in lung volumes. The presence of surfactant has multiple beneficiary effects on the lung parenchyma: It reduces the surface

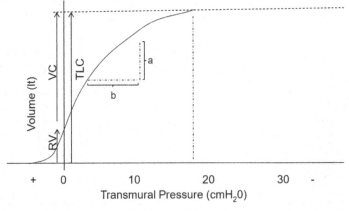

FIGURE 1.2 The static pressure–volume curve of the respiratory system. Lung compliance can be calculated by dividing the difference in volume ($\Delta V = a$) with difference in pressure ($\Delta P = b$).

tension of alveoli, stabilizes the walls of smaller alveoli, reduces the work necessary to distend the alveoli, and finally, prevents the filling of the alveolar space with extracellular fluid, thereby preventing pulmonary edema. Absence of surfactant can induce severe respiratory distress syndrome.

The driving pressure for air to move is $P_{alv}-P_B$, where PB = Barometric Pressure. The driving pressure and airway resistances are studied when air moves into and out of the lungs and the condition is, therefore, called dynamic. The driving pressure for inspiration is a negative alveolar pressure (P_{alv}) relative to barometric pressure.

Respiratory volume is recorded graphically. The VT is plotted against the driving pressure, which is equal to the dynamic alveolar pressure (Fig. 1.3).

Integrating pressure with respect to volume gives the green area corresponding to the elastic work of one inspiration. This is the work needed to overcome the elastic resistance against inspiration. The red area to the right of the diagonal is the extra work of inspiration called the flow-resistive work or alternatively nonelastic work (Fig. 1.3). During expiration, the flow-resistive work is equal to the green area. The inspiratory and expiratory curve forms a so-called hysteresis loop. The lack of coincidence of the curves for inspiration and expiration is known as elastic hysteresis (lag). With deeper and more rapid breathing the hysteresis loop becomes larger, and the nonelastic work relatively higher.

The Dynamic Pressure - Volume Curve

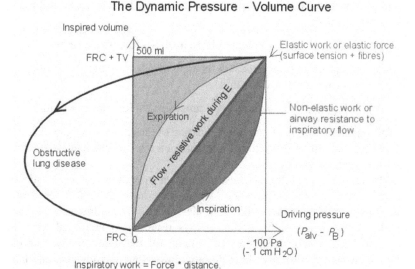

FIGURE 1.3 Tidal volume (TV) and the dynamic transmural pressure ($P_{alv}-P_{barometric}$) in a healthy person during one respiratory cycle: The expiratory curve from a patient with obstructive lung disease is shown to the left.

The airflow into the respiratory system depends on the driving pressure (ΔP) and the airways resistance (R_{AW}) to this flow, therefore: Flow $= V^* = \Delta P/R = (P_{ALV} - P_{ATM})/R_{AW}$. The airways resistance (R_{AW}) is dependent on the viscosity of the inhaled air, the length of the airways, and their diameter (it is proportional to the length and inversely proportional to the fourth power of the radius of the airways). The diameter of the airways depends on the position of the latter in the bronchial tree, the lung volume, the constriction of smooth muscles around the bronchi, presence of mucous, and the transmural pressure.

Regarding the mechanical work required to increase the volume of the lungs above the level of FRC, it can be divided into the elastic and the friction work components. The former refers to the work required to distend the elastic tissues of the lung and expand the thoracic cavity. It is stored in the elastic bands of the lungs to allow for a passive expiratory movement. The latter refers to the work required to overcome the resistance of the airways. Those work components are affected by the lung volume at any given point of time, as smaller lung volumes mean increased friction due to high resistance, whereas larger volumes lead to increased elastic work necessary to further distend the thoracic cavity. The FRC represents the volume in which the elastic recoil forces of the lungs are in balance with the analogous forces of the thoracic wall (point of minimum workload). The pressure that must be applied by the inspiratory muscles plus the pressure applied at the level of airways (or the level of pressure of a ventilator in cases of mechanical ventilation) equals the VT multiplied by the lung elastance plus the resistance of the airways multiplied by the flow of air into the lungs. Therefore, the following equation is created:

$$P_{mus} + P_{vent}(P_{aw}) = (\text{Volume} * \text{Elastance}) + (\text{Resistance} * \text{Flow})$$

Since the elastance is reciprocal to the lung compliance, the above equation can be transformed to the following:

$$P_{mus} + P_{vent}(P_{aw}) = (\text{Volume} / \text{Compliance}) + (\text{Resistance} * \text{Flow})$$

This equation is called the motion equation of the respiratory system and is suitable for explaining the settings and parameters of mechanical ventilation (invasive or noninvasive).

Different lung pathologies cause significant alterations in the elastic or friction forces of the respiratory system and have an effect on FRC. Obstructive lung diseases (like chronic obstructive pulmonary disease (COPD)) cause constriction of the airways resulting in increased work to overcome friction and also increased volume for the minimum work point (FRC). Conversely, restrictive lung diseases lower lung compliance, and lead to increased muscular work for the lung expansion and reduced FRC.

Under normal conditions, the energy expenditure for respiration is minimal: only an oxygen consumption of less than 1 mL/min is required for every L/min of lung ventilation, constituting 1%−2% of the total O_2 consumption. Also, during tidal breathing air is not evenly distributed inside the lungs, since there is a vertical variation in pleural pressure due to anatomical and gravitational reasons. Pleural pressure is higher in the apical and lower in the basal parts of the lungs. Thus, apical alveoli contain larger volumes of air. On the other hand, the alveoli in the basal parts lie at the lower part of the static pressure−volume curve and have greater distensibility while they are also under a higher tension during inspiration because of their proximity to the diaphragm, the main inspiratory muscle.

Respiratory movements

Approximately 12−15 times per minute in an adult, the diaphragm and thoracic muscles receive impulses from the brain signaling them to contract. This contraction moves the diaphragm downward and the rib cage up and out, which increases the volume of the thoracic cavity and creates a negative pressure within the lungs. This causes air from the higher-pressure environment outside the body to flow into the lower-pressure environment in the lungs. This is the active phase of ventilation, known as inhalation [3]. Air continues flowing through the airway openings and into the lungs while equalization of pressure occurs [5]. After full expansion of the lungs, stretch receptors signal the brain stem and inhalation ceases. The passive phase of ventilation, known as exhalation, subsequently takes place. The diaphragm and thoracic muscles relax and the lungs recoil, decreasing the volume and increasing the pressure in the thoracic cavity. The air inside the lungs flows back out to the lower-pressure atmosphere outside the body [3]. Since exhalation is a passive process, it typically takes twice as long as the active process of inhalation [2]. Due to the acoustics in the thoracic cavity, however, the sound of expiration lasts less than that of the inspiration in lung auscultation under normal conditions. In cases when the expiration must take place more rapidly (due to increased metabolic demands for Oxygen) or in lung pathologies (like obstructive pulmonary diseases) the expiratory phase becomes an energetic process, with utilization of various accessory expiratory muscles, as can be seen in Fig. 1.4. The same image also shows the accessory inspiratory muscles.

Through this process of inhalation and exhalation, the average human cycles 5−10 L of air through the lungs each minute [2]. The amount of air taken into the lungs during each breath (approximately 500 mL in an adult) is known as tidal volume (V_T), while the collective volume over the course of a minute (respiratory rate $\times$ V_T) is known as the minute volume (V_E). Due to

INHALATION (Restrictive Conditions)	EXHALATION (Obstructive Conditions)
Scalenes	Rectus Abdominis
Sternocleidomastoid	External Obliques
Pectoralis Major	Internal Obliques
Trapezius	Transverse Abdominis
External Intercostals	Internal Intercostals

FIGURE 1.4   The human accessory inspiratory and expiratory muscles.

the lack of gas exchange that occurs in the conducting airways (from the mouth to the terminal bronchioles), a portion of each breath is ineffective for gas exchange. This anatomical dead space (V_D) is approximately 150 mL in the average adult and must be subtracted from the V_T in order to determine the volume of air that reaches the alveoli (V_A) and can be used for gas exchange.

As previously noted, CO_2 is created within the body, and it is the role of ventilation to remove this by-product. It is also for this reason that ventilation is best evaluated through a measure of CO_2 (partial pressure of CO_2 in arterial blood - $PaCO_2$ or partial pressure of CO_2 in end-tidal volume of air - $P_{ET}CO_2$) [4]. If breathing stops (apnea) or if the V_E decreases (hypoventilation), CO_2 will accumulate within the blood and rapidly reach toxic levels (hypercapnia), resulting in respiratory acidosis. Conversely, if V_E increases (hyperventilation), the excessive elimination of CO_2 (hypocapnia) will result in respiratory alkalosis. It is in this manner that our respiratory system affects the body's pH and can also serve as a compensatory mechanism to offset metabolic derangements (e.g., hyperventilation can compensate for metabolic acidosis due to varied etiology).

Perfusion

The second component of respiration is perfusion. This process involves the circulation of blood through the capillaries, which facilitates nutrient exchange [6]. External respiration requires adequate delivery of blood to the capillary beds of the lungs via the pulmonary circulation. In the absence of this blood supply, there will be no transport mechanism for O_2.

Gases are exchanged between the atmosphere and the alveolar air and diffuse between the alveolar air and the blood flowing through the pulmonary capillary vessels.

Oxygen is transported from the atmosphere, via the alveolar ventilation and then carried by the pulmonary blood flow (equal to the cardiac output), into the cells and their mitochondria for metabolic purposes. Carbon dioxide, the final end-product of metabolism, migrates from the cells to the atmosphere.

A healthy normal person at rest ventilates his lungs with 5 L/min of fresh air (V^*_A). The respiratory quotient (RQ) is a metabolic ratio between the carbon dioxide output $\left(V^*_{CO_2} \right)$ and the oxygen uptake $\left(V^*_{O_2} \right)$ defined for all body cells as a whole. On a diet dominated by carbohydrate the metabolic RQ for all cells of the body is approaching the value 1.

The normal resting carbon dioxide output is 10 mmol or 224 mL/min from an adult person, and the cardiac output is typically 5 L/min. The blood volume of 5 L carries each min about 10 mmol (or 224 mL) of oxygen toward the mitochondria. Following passage of the capillary system, the same amount of CO_2 is carried toward the lungs in the venous blood as long as RQ is 1.

Blood passing the pulmonary capillaries of a healthy person is rapidly equilibrating with the alveolar air. Oxygen from the air diffuses into the blood and binds reversibly with hemoglobin. The normal oxygen capacity is 200 mL per L of blood (150 g hemoglobin per L carrying 1.34 mL/g).

Normally, alveolar ventilation (V^*_A) and perfusion (Q^*) are matched and the total V^*_A/Q^* ratio is between 0.8 and 1.2 with normal alveolar and blood gas tensions. In the normal upright lung, the regional V^*_A/Q^* ratio is approximately 0.6 at the lower and about 3 at the upper lung region. The pulmonary blood flow decreases from the lower to the upper parts of the lung of a resting person. Likewise, the relative ventilation of the lung also decreases linearly from the base to the apex, but at a slower rate. Thus, the regional ventilation—perfusion ratio varies from zero in the lower region, where there is only blood flow and no ventilation to infinity in the upper region, where there is only ventilation and no blood flow. At the lower lung region, regional V^* approaches zero and at the top of the lung regional perfusion approaches zero. In a PO_2-PCO_2 diagram each point on the curve represents partial pressures at which alveolar air and blood can equilibrate at a certain V^*_A/Q^* ratio. Thus, for any practically obtainable point, a single value exists for blood gas concentrations. Lung regions at the base with low V^*_A/Q^* have low P_AO_2 and high P_ACO_2, relative to normal mean values. Upper lung regions with high V^*_A/Q^* have relatively high P_AO_2 and low P_ACO_2.

Normally, up to 5% of the venous return passes directly into the systemic arterial circulation. This shunt-blood includes nutrient blood flow coming from the upper airways and collected by the bronchial veins. The coronary venous blood that drains directly into the left ventricle through the Thebesian veins is also shunt-blood.

Pulmonary vascular resistance (PVR) is minimal compared to that of the systemic circulation. The pulmonary vascular system is basically a low-pressure, low-resistance, highly compliant vessel system with a blood flow sensitive to gravity and to P_AO_2.

The system is meant to accommodate the entire cardiac output and not to meet special metabolic demands as in the case of the systemic circulation. The pressure in the right ventricle is 3.3 kPa systolic and 0.133 kPa diastolic in a healthy, supine person at rest. The pressure in the pulmonary artery is about 3.3 kPa systolic and 1 kPa diastolic, with a mean of 1.7 kPa. The blood flow of the pulmonary capillaries pulsates and its mean pressure is below 1 kPa. The pressure in the left atrium is 0.7 kPa. This value implies a pressure drop across the pulmonary circulation of $(1.7-0.7) = 1$ kPa. This driving pressure is less than 1/10 of the systemic driving pressure.

The walls of the pulmonary vessels are thin, hence their pressure must fall at each inspiration, because the intrapulmonary pressure falls. Change of posture from supine to erect position reduces the pressure toward zero in the apical vessels, whereas it increases the pressure in the basal vessels due to gravity. When the driving pressure in the apical blood vessels approaches zero, the blood flow will also approach zero. Apart from its implication for gas exchange, this phenomenon limits the supply of nutrients. Lung disorders often occur in the apical regions.

The PVR remains low in healthy persons, even when cardiac output increases to 30 L/min, because of distensibility and recruitment of pulmonary vessels. Stretch receptors, found in the left atrium and in the walls of the inlet veins, are believed to be stimulated by distension. Such a distension blocks liberation of vasopressin (antidiuretic hormone) from the posterior pituitary and releases atrial natriuretic factor from the atrial tissue. Hereby, the urine volume increases and the extracellular volume decreases.

Changes in PVR are achieved mainly by passive factors, but also by active modification. Passive factors: the larger arteries and veins are located outside the alveoli (extra-alveolar); they are tethered to the elastic lung parenchyma and are exposed to the pleural pressure. The pulmonary capillaries lie between the alveoli and are exposed to the alveolar pressure.

Alveolar capillary volume: the intra-alveolar vessels are wide open at low alveolar volumes, so that their PVR must be minimal. With increasing alveolar distension these vessels are compressed. This increases the intra-alveolar PVR. However, at low alveolar (lung) volumes, the extra-alveolar vessels are small because of the small transmural vascular pressure gradient, and their PVR is high.

With increasing lung distension, the intrathoracic pressure becomes more subatmospheric. This elevates the transmural vascular gradient and is coupled with the radial traction on these vessels by the surrounding lung parenchyma as it expands. Thus, the extra-alveolar PVR decreases. The greatest cross-sectional area exists in the many intra-alveolar vessels, hence increasing PVR in these vessels leads to decreased extra-alveolar PVR. Thus, total pulmonary vascular resistance is increased at higher alveolar volumes when intra-alveolar PVR is high. PVR is minimal at FRC, where there is air enough to open the extra-alveolar vessels with minimal closure of the intra-alveolar vessels.

Pulmonary artery pressure: A healthy person at rest (at the level of FRC) has approximately half of the pulmonary capillaries open, but with increasing arterial pressure, the previously closed capillaries open (recruitment). As the arterial pressure continues to rise, the capillaries become distended. The net effect is a rise in the total cross-sectional area of the lung capillaries, leading to decreased PVR.

A decrease in P_AO_2 in an occluded region of the lung produces hypoxic vasoconstriction of the vessels in that region. The reduced P_AO_2 causes constriction of the precapillary muscular arteries leading to the hypoxic region. This hypoxic effect is not nerve mediated. The reaction shifts blood away from poorly ventilated alveoli to better-ventilated ones. Nitric oxide (NO) seems to dilatate the vessels of the well-ventilated segments of the lung. Perfusion is hereby matched with ventilation.

Diffusion

Diffusion is another important method of transport within the body and is the third component of respiration. Diffusion involves the movement of a substance in a solution (liquid or air) from higher concentration areas to lower concentration areas [7]. In the case of respiration, diffusion involves the distribution of O_2 from the atmosphere through the epithelial membrane of the alveoli, the basic membrane, and the pulmonary capillary (endothelial) walls and into the bloodstream. At the same time, CO_2 diffuses from the bloodstream into the alveoli following the opposite route. This process of diffusion is dependent on the characteristics of each individual gas, the rate of perfusion, and the integrity of the alveolar–capillary membrane [4].

The six zones of the alveolar–capillary barrier are: (1) a fluid layer containing surfactant; (2) the alveolar epithelium; (3) a fluid-filled interstitial space; (4) the capillary endothelium with basement membrane; (5) the blood plasma; and (6) the erythrocyte membrane. The six zones form an almost ideal gas exchanger for oxygen and carbon dioxide diffusion.

Fick's law of diffusion states that the flux of gas transferred across the alveolar–capillary barrier is related to the solubility of the gas, the diffusion area (A), the length of the diffusion pathway from the alveoli to the blood (L), and the driving pressure (P1 − P2).

The Earth's atmosphere contains approximately 21% O_2. At sea level, under normal conditions, barometric (i.e., atmospheric) pressure is 760 mmHg. According to Dalton's law, this pressure is comprised of the partial pressures of the individual gases that make up our atmosphere: primarily nitrogen (N_2) and O_2. In this situation, the partial pressure of O_2 is 159 mmHg (21% of 760 mmHg). By the time O_2 diffuses into the human circulation, its partial pressure (PaO_2) is reduced to 80–100 mmHg [4]. While the percentage of O_2 in the atmosphere remains constant, the process of diffusion can be enhanced through a combination of supplemental O_2 and altering the airway pressure (e.g., continuous positive airway pressure or high-flow nasal cannula devices) or by altering a combination of barometric pressure and O_2 concentration (e.g., hyperbaric chamber).

Neural control of respiration

The gas exchange takes place in the lungs. Nevertheless, the control of the respiratory system is executed by the central nervous system (CNS) [3]. While we do have some voluntary control of breathing, it is mainly regulated automatically and functions irrespective of our will. Breathing can, however, be suppressed at the neurological level due to narcotic or sedative overdose, as well as brain stem injury [3].

The loci of the CNS that control respiration are located within the brain stem, specifically within the pons and the medulla. These components are responsible for the nervous impulses, which are transmitted via the phrenic and other motor nerves to the diaphragm and intercostal muscles, controlling our basic breathing rhythm. Also located in the brain stem are the central chemoreceptors. These specialized cells signal the body to adjust ventilation based indirectly on the partial pressure of arterial CO_2 ($PaCO_2$) level, thus constituting the primary respiratory drive. Peripheral chemoreceptors, which are located outside of the brain stem in the carotid and aortic arteries, serve as the body's back-up respiratory drive by responding to low levels of O_2. This secondary mechanism is often referred to in COPD patients as a "hypoxic drive" since it takes over as the primary respiratory stimulation after the central chemoreceptors grow numb to chronically elevated $PaCO_2$. A graphic representation of the respiratory centers in the brain stem can be found in Fig. 1.5.

Pulmonary defense mechanisms

During normal breathing most of the particles of more than 10 mm in diameter—such as pollen—are deposited and removed in the nose and nasopharynx. Particles below 1 mm are deposited in the alveoli. Particles between 1 and 10 mm are deposited in the bronchi depending on their diameter. Although sneezing and coughing with expectoration can eliminate many inhaled particles, the mucociliary escalator assisted by bronchus-associated lymphoid tissue (BALT) and alveolar macrophages perform the main clearance of the airways. The airways are protected by humidification all the way to the alveoli with a mucous layer, which prevents dehydration of the epithelium and surrounds the epithelial cilia. This airway mucous consists of polysaccharides from goblet cells and from mucous glands in the bronchial wall and forms a gelatinous blanket on top of the liquid layer. The cilia continuously move the gelatinous blanket with inhaled particles upward toward the pharynx, where they are swallowed [7].

The Respiratory Rhythm Generator

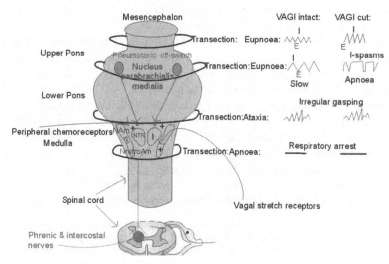

FIGURE 1.5 The respiratory centers responsible for the control of breathing in the brain stem. On the right side of the picture are depicted the consequences of transections in the designated areas on the breathing pattern.

Clearance of the respiratory bronchioles may take days, whereas clearance of the main bronchi is typically accomplished within an hour. Smoking reduces mucociliary transport, and indirectly impairs gas exchange. It also reduces surfactant production and thus increases the work of breathing.

The lung secretions contain bactericidal lysozyme and lactoferrin from granulocytes. The a1-antitrypsin normally neutralizes chymotrypsin, trypsin, elastase, and proteases secreted by granulocytes during inflammation, and thus prevents destruction of lung tissue.

BALT in the walls of the main bronchi is part of the mononuclear phagocytotic system. These tissue aggregates contain macrophages originating from monocytes and lymphocytes. Following sensitization of B-lymphocytes to specific antigens, the cells produce specific antibodies or immunoglobulins (types: IgA, IgG, and IgE) in response to new contact with the antigen. IgA inhibits the attachment of poliovirus, bacteria, and toxins in the respiratory tract. IgE is related to the pathogenesis of allergic disorders.

Lungs additionally have endocrine functions [8]. Alveolar macrophages are amebic cells that swallow particles and bacteria in the alveoli. While they execute microbes in their phagolysosomes, the cells migrate to the mucociliary escalator, or they are removed by the blood or by the lymphatic system. Smoking impairs the normal macrophage activity.

The inactive polypeptide, angiotensin I, is converted into the potent vasoconstrictor, angiotensin II, by the angiotensin-converting enzyme, located on the pulmonary endothelial cells. Angiotensin is important for the regulation of the arterial blood pressure — also during shock.

Adrenergic sympathetic activity (and sympathomimetic drugs) relaxes bronchial smooth muscle via adrenergic b2 receptors, whereas parasympathetic cholinergic activity (and parasympathomimetics) constricts bronchial smooth muscles via muscarinic receptors.

Smoke, dust, and other irritants (perhaps also adenosine, histamine, and substance P) constrict the airway smooth muscles via a reflex triggered by the rapidly adapting irritant receptors. Decreased P_ACO_2, thromboxane, and leukotrienes also act as bronchoconstrictors.

Vasoactive intestinal peptide can dilatate airways and reduce airflow resistance. Substances that dilatate airways include increased P_ACO_2, adrenergic alpha-blockers, catecholamines, and atropine.

Part B: pathology and treatment of lung diseases

Lung diseases are some of the most common medical conditions in the world [9]. They constitute a wide spectrum of pathological conditions, which can be divided into discrete categories, as described in the following sections.

Lung diseases affecting the airways

Asthma

Asthma is a lung disease of the bronchial airways. Inflammation and constriction of the airways (as depicted graphically in Fig. 1.6) result in symptoms. Patients suffer from wheezing, shortness of breath, chest tightness,

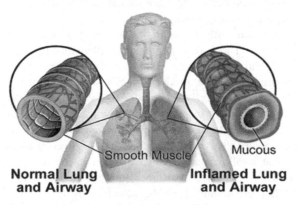

Smooth Muscle   Mucous

Normal Lung and Airway   **Inflamed Lung and Airway**

FIGURE 1.6   Schematic representation of normal (left) and obstructed (right) airways.

and coughing. In general, symptoms are intermittent but when severe, they can be continuous. Allergies, infections, pollution, exercise, and cold temperatures can trigger asthma symptoms.

Diagnosis

Because asthma is a clinical syndrome, there is no gold standard for its diagnosis. Diagnosis is based on physical exam, symptoms, and medical history. More tests are used to examine the pulmonary function:

- *Spirometry.* This simple breathing test measures how much air is blown out and how fast.
- *Peak flow.* This test measures how well the air is pushed out of the lungs. It is less exact than spirometry, but it can be a good way to test the lung function at home. A peak flow meter provides information about treatment effectiveness, factors that worsen symptoms, and when a patient needs emergency care.
- *Methacholine challenge.* During this test, the patient inhales a chemical called methacholine before and after spirometry to see if it makes the airways narrow. If the results fall at least 20%, the diagnosis is asthma. The patient receives medicine at the end of the test to reverse the effects of the methacholine.
- *Exhaled nitric oxide test.* The patient breathes into a tube connected to a machine that measures the amount of nitric oxide in breath. Patient's body makes this gas normally, but levels could be high if the airways are inflamed.

Other tests include:

- *Chest X-ray.* It is not an asthma test, but it can show other diseases.
- *Computed tomography (CT).* A scan of the lungs and sinuses can identify physical problems or diseases (like an infection) that may cause breathing problems or make them worse.
- *Allergy tests.* These can be blood or skin tests. These show allergic reactions to pets, dust, mold, and pollen.
- *Sputum eosinophils.* This test looks for high levels of white blood cells (eosinophils) in the mix of saliva and mucus (sputum) that comes out with cough.

Treatment [10–14]

- *Inhaled corticosteroids.* These medications treat asthma in the long term. They are taken every day to keep the asthma under control.
- *Leukotriene modifiers.* Another long-term asthma treatment. These medications block leukotrienes that trigger an asthma attack. Patient takes them as a pill once a day.

- *Long-acting beta-agonists.* These medications relax the muscle bands that surround the airways (bronchodilators).
- *Combination inhaler.* This device gives an inhaled corticosteroid and a long-acting beta-agonist together to ease the asthma symptoms.
- *Theophylline.* It opens the airways and eases tightness. It is a long-term medication by mouth, either by itself or with an inhaled corticosteroid.
- *Short-acting beta-agonists.* These are known as rescue medicines or rescue inhalers. They loosen the bands of muscle around the airways and ease symptoms.
- *Anticholinergics.* These bronchodilators prevent the muscle bands around the airways from tightening.
- *Oral and intravenous corticosteroids.* These are taken along with a rescue inhaler during an asthma attack. They ease swelling and inflammation in airways. Oral steroids are used for a short time, between 5 days and 2 weeks. Also, the patient is likely to get steroids injected directly into a vein for a severe asthma attack.
- *Biologics.* If there is a severe asthma that does not respond to control medications, biologic factors (omalizumab, benralizumab, mepolizumab, and reslizumab) can be selected to stop the immune cells from causing inflammation.

Chronic obstructive pulmonary disease (COPD)

It is a term that covers two types of chronic (long-term) diseases where the airways (breathing tubes) in the lungs become swollen and partly blocked. COPD gets worse over time. It cannot be cured, but it can be treated and managed.

COPD consists of two major breathing diseases (Fig. 1.7):

- emphysema
- chronic bronchitis

Emphysema damages the tiny alveoli (air sacs) at the tips of the lungs. Normally these air sacs stretch like balloons during breathing in and out. Emphysema makes these air sacs stiff. Because they cannot stretch, air gets trapped inside them. As a result, the inhalation becomes difficult.

Chronic bronchitis makes the airways red, swollen, and irritated. Glands in the airways make extra mucus (phlegm), which blocks some air from passing through. This causes cough, large production of mucus, and shortness of breath. Many people with COPD have both diseases.

Diagnosis [15,16]

A diagnosis of COPD is based on signs and symptoms, history of exposure to lung irritants (such as smoking), and family history.

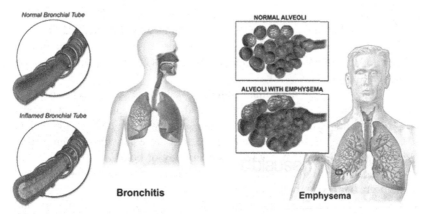

FIGURE 1.7   The two types of COPD.

- *Lung (pulmonary) function tests.* These tests [12] measure the amount of air the patient can inhale and exhale, and whether the lungs deliver enough oxygen to the blood. During the most common test, called spirometry, the patient blows into a large tube connected to a small machine to measure how much air the lungs can hold and how fast he/she can blow the air out of his/her lungs. Other tests include measurement of lung volumes and diffusing capacity, 6-min walk test (to see how far a person can walk in 6 min and how many times he/she needs to stop to catch his/her breath), and pulse oximetry.
- *Chest X-ray.* A chest X-ray can show emphysema, one of the main causes of COPD. An X-ray can also rule out other lung problems or heart failure.
- *CT scan.* A CT scan of lungs can help detect emphysema and help determine if the patient might benefit from surgery for COPD. CT scans can also be used to screen for lung cancer.
- *Arterial blood gas analysis.* This blood test measures how well the lungs are bringing oxygen into blood and removing carbon dioxide.
- *Laboratory tests.* Lab tests are not used to diagnose COPD, but they may be used to determine the cause of symptoms or rule out other conditions. For example, lab tests may be used to determine if the patient has the genetic disorder alpha-1-antitrypsin deficiency, which may be the cause of COPD in some people. This test may be done if a person has a family history of COPD and develops COPD at a young age.

Therapy

- *Quitting smoking*
- *Inhalers:*

(a) Short-acting bronchodilator inhalers: deliver a small dose of medicine directly to the lungs, causing the muscles in airways to relax and open up. They also prevent hyperinflation (over expansion) of the lungs. The inhaler should be used when the patient feels breathless and this should relieve the symptoms.

(b) Long-acting bronchodilator inhalers: each dose lasts for at least 12 h. This can help relieve coughing and shortness of breath and make breathing easier. Depending on the severity of the disease, the patient may need a short-acting bronchodilator before activities, a long-acting bronchodilator that he/she uses every day or both.

- *Inhaled corticosteroid medications* can reduce airway inflammation and help prevent exacerbations.
- *Combination inhalers:* some medications combine bronchodilators and inhaled steroids.
- *Oral corticosteroids* for moderate or severe acute exacerbation.
- *Phosphodiesterase-4 inhibitors* (*roflumilast*): decreases airway inflammation and relaxes the airways.
- *Theophylline* may help improve breathing and prevent episodes of worsening COPD.
- *Antibiotics* help treat respiratory infections, such as acute bronchitis, pneumonia, and influenza, and can aggravate COPD symptoms.
- *Long-term oxygen therapy.* If the oxygen level in the blood is low, the patient may need to take oxygen through nasal tubes, or through a mask. Oxygen is not a treatment for breathlessness, but it is needed for some patients with persistently low oxygen levels in the blood. Oxygen must be taken for at least 16 h a day.
- *Pulmonary rehabilitation program.* These programs generally combine education, exercise training, nutrition advice, and counseling. A variety of specialists can tailor a rehabilitation program to meet a patient's needs. Pulmonary rehabilitation after episodes of worsening COPD may reduce readmission to the hospital, increase the ability to participate in everyday activities, and improve quality of life.
- *In-home noninvasive ventilation therapy* [17,18]: a noninvasive ventilation therapy machine with a mask helps to improve breathing and decrease retention of carbon dioxide (hypercapnia) that may lead to acute respiratory failure and hospitalization.
- *Surgery* is usually only suitable for a small number of people with severe COPD whose symptoms are not controlled with medicine. There are 3 main operations that can be done:

   (a) bullectomy — an operation to remove a pocket of air from one of the lungs, allowing the lungs to work better and make breathing more comfortable.

(b) lung volume reduction surgery — an operation to remove a severely damaged section of lung to allow the healthier parts to work better and make breathing more comfortable.

(c) lung transplant — an operation to remove and replace a damaged lung with a healthy lung from a donor.

Lung diseases affecting the air sacs (alveoli)

Airways branch into tiny tubes (bronchioles) that end in clusters of air sacs called alveoli. These air sacs make up most of the lung tissue. Lung diseases affecting alveoli include:

A. Pneumonia is an infection of alveoli that can cause mild to severe illness in people of all ages. It normally starts with a bacterial, fungal, or viral including the coronavirus that causes COVID-19.

First symptoms of pneumonia are usually similar to those of a cold or flu. The patient develops high fever, chills, cough with sputum, fast breathing and shortness of breath, chest pain that usually worsens when taking a deep breath, fast heartbeat, fatigue and weakness, nausea and vomiting, diarrhea, headache, muscle pain, confusion or delirium, especially in older adults.

Diagnosis is suggested by the medical history and lung examination findings. It should be confirmed by chest radiography or ultrasonography (like the examples in Fig. 1.8). Validated prediction scores for pneumonia severity can guide the decision between outpatient and inpatient therapy.

Pneumonia by a bacterial infection is treated with antibiotics [19].

B Tuberculosis (TB) [20] is a disease caused by bacteria called *Mycobacterium tuberculosis*. The bacteria usually attack the lungs, but they can also damage other parts of the body. Tuberculosis spreads through the air when the patient with TB of the lungs or throat coughs, sneezes, or talks.

TB infection (latent TB): people can have TB bacteria in their body and never develop symptoms. In most people, the immune system can contain the bacteria so that they do not replicate and cause disease. In this case, a person will have TB infection, but not active disease. There is also no risk

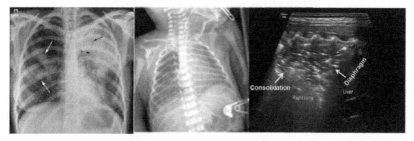

FIGURE 1.8 Radiographic images of pneumonia in an adult female (left), infant (middle), and ultrasound imaging of pneumonia (right).

of passing on a latent infection to another person. However, a person with latent TB still requires treatment.

TB disease: a person with TB disease may experience a persistent cough that produces phlegm with blood, a loss of appetite and weight, fever, a general feeling of fatigue and being unwell, night sweats, chest pain. Symptoms typically worsen over time, but they can also spontaneously go away and return.

Diagnosis [21]: Two tests can show whether TB bacteria are present: the TB skin test and the TB blood test. However, these cannot indicate whether TB is active or latent. To test for active TB disease, the doctor may recommend a sputum test and a chest X-ray (Fig. 1.9).

Therapy: TB is treatable with early detection and appropriate antibiotics. Treatment for active TB may involve taking several drugs for 6—9 months. When a person has a drug resistant strain of TB, the treatment will be more complex.

C. *Pulmonary edema* [22]. Fluid leaks out of the small blood vessels of the lung into the air sacs and the area around them. One form is caused by heart failure and back pressure in the lungs' blood vessels. In another form, injury to lung causes the leak of fluid.

D. *Lung cancer.* It has many forms and may start in any part of the lungs. It most often happens in the main part of the lung, in or near the air sacs.

E. *Acute respiratory distress syndrome (ARDS).* It is a severe, sudden injury to the lungs from a serious illness. COVID-19 is one example. Many people who have ARDS need help breathing from a ventilator until their lungs recover (Fig. 1.10).

F. *Pneumoconiosis.* This is a category of conditions caused by inhaling something that injures the lungs. Examples include black lung disease from coal dust and asbestosis from asbestos dust.

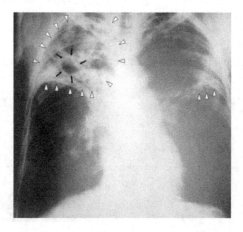

FIGURE 1.9 Chest X-ray showing pulmonary infiltrations due to Tb (*white arrows*) and Tb cavity in the right upper lobe (*black arrows*).

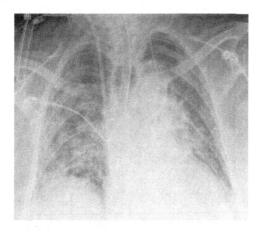

FIGURE 1.10 Chest X-ray showing ARDS due to COVID-19 in an intubated patient.

Lung diseases affecting the interstitium

The interstitium is the thin, delicate lining between the alveoli. Tiny blood vessels run through the interstitium and let gas transfer between the alveoli and blood. Various lung diseases affect the interstitium:

A. **Interstitial lung disease (ILD)**. This is a group of lung conditions that includes idiopathic pulmonary fibrosis [23], sarcoidosis, alveolitis, and autoimmune disease. The most common symptoms of ILD are cough and shortness of breath. This is often accompanied by a dry cough, chest discomfort, fatigue, and occasionally weight loss. Many etiologic factors have been associated with ILD (Fig. 1.11).

   Diagnosis
   - A physical exam with a stethoscope can detect dry, crackling sounds in the chest
   - Blood tests to check for connective tissue diseases
   - Bronchoscopy, with or without biopsy
   - Chest X-ray
   - CT scan of the chest
   - Echocardiogram
   - Open lung biopsy
   - Measurement of the blood oxygen level at rest or when active
   - Pulmonary function tests
   - Six-minute walk test (to see how far the patient can walk in 6 min and how many times he/she needs to stop to catch his/her breath)

B. **Pneumonia** and **pulmonary edema** can also affect the interstitium.

Lung diseases affecting blood vessels

A. **Pulmonary embolism (PE)** [24−26]. A blood clot (usually in a deep leg vein, a condition called deep vein thrombosis) breaks off, travels to the

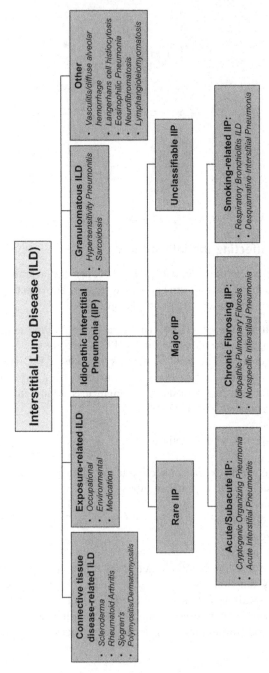

FIGURE 1.11 Etiologic factors related to interstitial lung disease.

heart, and gets pumped into lungs. The clot sticks in a pulmonary artery, often causing shortness of breath and low blood oxygen levels. Pulmonary embolism can be life-threatening.

Diagnosis. Diagnosis of PE is exceedingly difficult because symptoms can vary greatly. Common signs and symptoms include:

- Shortness of breath. This symptom typically appears suddenly and always gets worse with exertion
- Chest pain that may extend into the arm, jaw, neck, and shoulder
- Cough. The cough may produce bloody or blood-streaked sputum
- Rapid or irregular heartbeat
- Lightheadedness or dizziness
- Excessive sweating
- Fever
- Leg pain or swelling, or both, usually in the calf caused by a deep vein thrombosis
- Clammy or discolored skin (cyanosis)
- "Gold standard" test for diagnosis is computed tomography pulmonary angiogram (CTPA − Fig. 1.12)

B. Pulmonary hypertension. Many conditions can cause high blood pressure in pulmonary arteries. This can lead to shortness of breath and chest pain. If a cause cannot be found, it will be called idiopathic pulmonary arterial hypertension.

Lung diseases affecting the pleura

The pleura is the thin lining that surrounds lung and lines the inside of the chest wall. A tiny layer of fluid lets the pleura on lung's surface slide along the chest wall with each breath. Lung diseases of the pleura include:

A. Pleural effusion. Fluid collects in the space between lung and the chest wall. Pneumonia or heart failure usually causes this. Large pleural

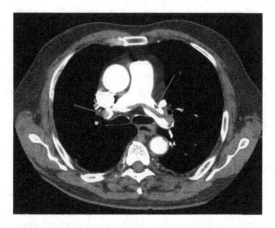

FIGURE 1.12 Computed Tomography Pulmonary Angiography (CTPA) image revealing multiple emboli in the two pulmonary arteries (red arrows (gray in print version)).

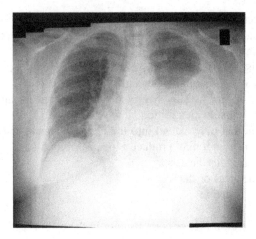

FIGURE 1.13   Chest X-ray showing a large left pleural effusion.

effusions can make it hard to breathe and may need to be drained. The tests most commonly used to diagnose and evaluate pleural effusion include:
- Chest X-ray (Fig. 1.13)
- CT scan of the chest
- Ultrasound of the chest [27]
- Thoracentesis (a needle is inserted between the ribs to remove a biopsy, or sample of fluid)
- Pleural fluid analysis (an examination of the fluid removed from the pleura space)

B. Pneumothorax [28]. Air may get into the space between chest wall and the lung, collapsing the lung. A pneumothorax can be caused by a blunt or penetrating chest injury, certain medical procedures, or damage from underlying lung disease. Or it may occur for no obvious reason. Symptoms usually include sudden chest pain and shortness of breath. On some occasions, a collapsed lung can be a life-threatening event. Diagnosis is based on lung ultrasound or chest X-ray (Fig. 1.14).

Treatment for a pneumothorax usually involves inserting a needle or chest tube between the ribs to remove the excess air. However, a small pneumothorax may heal on its own.

C. Mesothelioma. This is a rare form of cancer that forms on the pleura. Mesothelioma tends to happen several decades after a person comes into contact with asbestos.

Lung diseases affecting the chest wall

Chest wall plays an important role in breathing. Muscles connect the ribs to each other, helping the chest expand. The diaphragm descends with each breath, also causing chest expansion. Diseases that affect chest wall include:

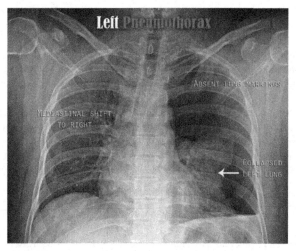

FIGURE 1.14 Chest X-ray showing a total left pneumothorax.

A. **Obesity hypoventilation syndrome**. Extra weight on the chest and belly can make it hard for the chest to expand. This may cause serious breathing problems.

B. **Neuromuscular disorders**. Breathing problems appear when the nerves that control respiratory muscles do not work the way they should. Amyotrophic lateral sclerosis and myasthenia gravis are examples of neuromuscular lung diseases.

References

[1] D. Venes (Ed.), Taber's Cyclopedic Medical Dictionary, twenty-first ed., F.A. Davis Company, Philadelphia, PA, 2009.

[2] T. Des Jardins, Cardiopulmonary Anatomy & Physiology: Essentials of Respiratory Care, fifth ed., Delmar, Clifton Park, NY, 2008.

[3] Martin, Breathe Easy A Guide to Lung and Respiratory Diseases for Patients and Their Families. www.lakesidepress.com/pulmonary/books/breathe/.

[4] R.L. Wilkins, J.K. Stoller, R.M. Kacmarek, Egan's Fundamentals of Respiratory Care, ninth ed., Mosby, St. Louis, MO, 2009.

[5] J.E. Campbell (Ed.), International Trauma Life Support for Prehospital Care Providers, sixth ed., Pearson Education, Upper Saddle River, NJ, 2008.

[6] B.E. Bledsoe, R.S. Porter, R.A. Cherry, Essentials of Paramedic Care, second ed., Prentice Hall, Upper Saddle River, NJ, 2006.

[7] D. Shier, J. Butler, R. Lewis, Hole's Human Anatomy & Physiology, tenth ed., McGraw-Hill, New York, NY, 2009.

[8] P.E. Paulev, Textbook in Medical Physiology and Pathophysiology: Essentials and Clinical Problems, Copenhagen Medical Publishers, 1999.

[9] L. Sihletidis, Pneumonology, University Studio Press, 2009.

[10] GINA, 2020.

[11] J.F. Murray, Ch. 38 asthma, in: R.J. Mason, J.F. Murray, V.C. Broaddus, J.A. Nadel, T.R. Martin, King Jr., E. Talmadge, D.E. Schraufnagel (Eds.), Murray and Nadel's Textbook of Respiratory Medicine, fifth ed., Elsevier, 2010, ISBN 978-1-4160-4710-0.

[12] American Thoracic Society and European Respiratory Society Task Force, R. Pellegrino, G. Viegi, V. Brusasco, R.O. Crapo, F. Burgos, R. Casaburi, A. Coates, C. van der Grinten, P. Gustafsson, et al., Standardization of lung function testing, Eur. Respir. J. 26 (2005) 948–968.

[13] L. Antonicelli, C. Bucca, M. Neri, F. De Benedetto, P. Sabbatani, F. Bonifazi, H.G. Eichler, Q. Zhang, D.D. Yin, Asthma severity and medical resource utilisation, Eur. Respir. J. 23 (5) (2004) 723–729.

[14] R.H. Green, C.E. Brightling, S. McKenna, B. Hargadon, D. Parker, P. Bradding, A.J. Wardlaw, I.D. Pavord, Asthma exacerbations and sputum eosinophil counts: a randomised controlled trial, Lancet 360 (9347) (2002) 1715–1721.

[15] M.F. Murphy, Pulse oximetry, in: R.M. Walls, M.F. Murphy, R.C. Luten, R.E. Schneider (Eds.), Manual of Emergency Airway Management, second ed., Lippincott, Williams, & Wilkins, Philadelphia, PA, 2004.

[16] S.K. Jindal (Ed.), Textbook of Pulmonary and Critical Care Medicine, Jaypee Brothers Medical Publishers, New Delhi, 2011, ISBN 978-93-5025-073-0, p. 242. Archived from the original on 2016-04-24.

[17] B. Rochwerg, L. Brochard, M.W. Elliott, D. Hess, N.S. Hill, S. Nava, P. Navalesi, Members of the Steering Committee, M. Antonelli, J. Brozek, G. Conti, M. Ferrer, K. Guntupalli, S. Jaber, S. Keenan, J. Mancebo, S. Mehta, Raoof S Members of The Task Force, Noninvasive ventilation in adults with acute respiratory failure: benefits and contraindications, Eur. Respir. J. 50 (2) (2017). Epub August 31, 2017.

[18] P.K. Lindenauer, M.S. Stefan, M.S. Shieh, P.S. Pekow, M.B. Rothberg, N.S. Hill, Noninvasive ventilation in adults with acute respiratory failure: benefits and contraindications, JAMA Intern. Med. 174 (12) (December 2014) 1982–1993.

[19] J.P. Metlay, Treatment of community-acquired pneumonia in adults who require hospitalization, Curr. Opin. Infect. Dis. 15 (2) (2002) 163.

[20] M.D. Iseman, A Clinician's Guide to Tuberculosis, Lippincott Williams & Wilkins, 2000, p. 69.

[21] G.B. Migliori, A. Borghesi, P. Rossanigo, et al., Proposal of an improved score method for the diagnosis of pulmonary tuberculosis in childhood in developing countries, Tuber. Lung Dis. 73 (1992) 145–149.

[22] A.J. Singer, C. Emerman, D.M. Char, J.T. Heywood, et al., Bronchodilator therapy in acute decompensated heart failure patients without a history of chronic obstructive pulmonary disease, Ann. Emerg. Med. 51 (1) (2008) 25–34.

[23] C.D. Fell, F.J. Martinez, L.X. Liu, S. Murray, M.K. Han, E.A. Kazerooni, B.H. Gross, J. Myers, W.D. Travis, T.V. Colby, G.B. Toews, K.R. Flaherty, Clinical manifestations and diagnosis of idiopathic pulmonary fibrosis, Am. J. Respir. Crit. Care Med. 181 (8) (2010) 832. Epub January 7, 2010.

[24] A.L. Dalton, D. Limmer, J.J. Mistovich, H.A. Werman, Advanced Medical Life Support, third ed., Pearson Education, Upper Saddle River, NJ, 2007.

[25] Task Force on Pulmonary Embolism, Guidelines on diagnosis and management of acute pulmonary embolism, Eur. Heart J. 21 (2000) 1301.

[26] M.J. Koschel, Pulmonary embolism, Am. J. Nurs. 104 (2004) 46–50.

[27] G. Volpicelli, M. Elbarbary, M. Blaivas, et al., International evidence-based recommendations for point-of-care lung ultrasound, Intensive Care Med. 38 (2012) 577–591, https://doi.org/10.1007/s00134-012-2513-4.

[28] C. Zhan, M. Smith, D. Stryer, Accidental iatrogenic pneumothorax in hospitalized patients, Med. Care 44 (2) (2006) 182–186.

Part II

Wearable sensing

Chapter 2

Respiratory management in daily life: needs and gaps

Alda Marques, Sara Souto-Miranda

Lab3R-Respiratory Research and Rehabilitation Laboratory, School of Health Sciences (ESSUA), Institute of Biomedicine (iBiMED), University of Aveiro, Aveiro, Portugal

Introduction

Chronic respiratory diseases (CRDs), including asthma, chronic obstructive pulmonary disease (COPD), bronchiectasis, cystic fibrosis, interstitial lung disease, lung cancer, and pulmonary hypertension, were highly neglected until 20 years ago, as it was believed that nothing could be done other than persuading the patient to quit smoking and/or use medication [1]. Nowadays, the impact of CRD is so overwhelming on individuals, society, and health systems, that alleviating the burden of CRD is recognized as a leading goal for nations to achieve [2].

Daily life of those living with a CRD goes around juggling between the burden of the disease and the burden of treatment influenced by personal and environmental factors. In the burden of disease, impairments of body functions (e.g., breathing, exercise tolerance, muscle strength) and structures (e.g., respiratory system, trunk, muscles) lead to limitations in activities (e.g., walking, moving around, lifting and carrying objects) and restrictions in participation (e.g., engaging in recreation and leisure activities, carrying out daily routine, employment) of daily life. Burden of treatments comprises learning about the disease and its management, adhering to complex treatment regimens, changing lifestyle behaviors, appointments with multiple health care professionals, and undertaking clinical assessments, among others. Both burdens are highly influenced by diverse contextual, environmental (e.g., air pollution, climate), and personal (e.g., age, psychosocial status) factors.

On a daily basis the pulmonary and extra-pulmonary impairments of body functions and structures of people with CRD imply high symptom burden, including many symptoms beyond dyspnea [3], but also exercise intolerance,

Wearable Sensing and Intelligent Data Analysis for Respiratory Management
https://doi.org/10.1016/B978-0-12-823447-1.00010-5

muscle dysfunction, functional status decline, and physical inactivity, which lead to major challenges on managing work and daily life, stigma, isolation, giving up things they love, and impacts on loved ones [3,4].

Often health-care professionals focus on the pulmonary features, which ignore largely the complexity and heterogeneity of these diseases, and thus improvements of meaningful aspects for the daily life of people with CRDs, such as functioning, independence, self-management, which are highly related with activities, and participation in daily life, are poorly addressed [5,6]. Yet, people with CRDs are being asked to have greater autonomy about their health care as the society evolves [5,7].

This chapter will therefore focus on the unmet needs related with further understanding and addressing the daily needs of people with CRDs.

Specifically, it will focus on:

(1) Daily needs of those living with chronic respiratory disease;
(2) Comprehensive assessments to better tailor therapies for people with CRDs;
(3) Daily management of people with CRDs beyond the use of oximeters and performing bronchial hygiene;
(4) Future avenues for daily assessment and management through wearables and applications.

Daily needs of those living with chronic respiratory diseases

Living with a CRD imposes tremendous daily challenges. Although some needs may be specific according to the CRDs and phases of the disease itself (acute, stable, terminal), there is some common ground across CRDs and between patients and loved ones. Common needs include family-related; health system/patient-family communication with health-care professionals; information/education; interpersonal/intimacy; physical/cognitive; practical/daily living; psychosocial/emotional; social/societal; and spiritual needs (Table 2.1) [8–15].

Nevertheless, there is often a mismatch between what health-care professionals believe to be needed and what people with CRDs and their loved ones feel they need for their daily management [16–19]. Bridging this gap, which implies giving voice to patients and families, is fundamental for meaningful care [20]. It involves the development of patient-centric health-care systems with strong partnerships (collaborative relationship between two or more parties based on trust, equality, and mutual understanding for achievement of a specified goal [8]) between health-care professionals, people with CRDs, and their loved ones [5].

TABLE 2.1 Summary of the needs of people with chronic respiratory diseases and their loved ones [8–15].

Needs	Description
1. Family-related	Communication with friends and within the family, concerns for family's future, family coping, (dys)functional relationships
2. Health system/patient-family–health professionals communication	Access and use to services and professionals, communication between patient-family and health professionals, participation of the patient-family in decision-making and preferences in communication, patient-family preferences of health care, satisfaction with care
3. Information/education	Adherence to therapy and healthy lifestyles, knowledge about the disease (etiology, prognosis, treatments) and how to manage it (self-care), end-of-life and palliative care
4. Interpersonal/intimacy	Body image, connectedness, intimacy with the partner, sexuality
5. Physical/cognitive	Symptom management, cognitive (dys) function
6. Practical/daily living	Advanced directives, disease management (symptoms, medication), equipment handling, exercise, finance, funeral preparation, housekeeping, maintaining independence, out-of-hours accessibility, physical activity, transportation, work
7. Psychosocial/emotional	Anxiety, burden, burnout, depression, despair, end-of-life issues, mood, fear/worry, optimism, satisfaction, stress, thinking about the future
8. Social/societal	Recreational, socialization, social support
9. Spiritual	Religious

It basically means going beyond the pathophysiological aspects of the disease and consider the whole health experience of each person. Therefore, assessments need to be patient-centered, comprehensive and include multiple areas of life, to reveal the unique needs of people with CRDs and their loved ones and guide decision-making within multidisciplinary teams and personalize interventions. This falls into the International Classification of Functioning, Disability, and Health (ICF) developed by the World Health Organization [21–23].

Comprehensive assessment for meaningful daily management

The ICF covers the whole life span and consists of three key components: body functions and structures, activities, and participation [23]. Body function and structures refer to physiologic functions and anatomic parts. Loss or deviations from normal are referred to as "impairments." In a person with CRD typical examples would be airflow limitation, exercise intolerance, and muscle weakness. Activities refer to the ability of a person to perform specific and isolated tasks, and difficulties experienced performing a specific task in a controlled environment are referred to "limitations." In a person with CRD typical examples would be limitation in climbing stairs, putting down objects, and using oxygen. Participation refers to involvement in daily life activities. Problems experienced by each person in this domain are referred as "restrictions." Examples of a person with CRD would be restrictions in doing housework, shopping, dressing, talking with friends, playing with children, and taking care of the garden. Body function and structures, activities, and participation are summarized under functioning and disability but they are related to and interact with the health condition and contextual, personal (e.g., age, emotional status), and environmental (e.g., indoor and outdoor air quality, climate) factors [22,23].

Health-care professionals have been focused on assessing and consequently, managing impairments of body functions and structures of people with CRDs, specifically, dyspnea—defined as a subjective experience of breathing discomfort that consists of qualitatively distinct sensations that vary in intensity [24], exercise intolerance, and muscle dysfunction but also some activity limitations, e.g., physical inactivity. This focus makes sense as these features have shown to be present even at early stages of CRDs [25,26], influence each other (disease spiral or vicious circle of dyspnea-inactivity) [27—31], and have been associated with mortality [32—37]. Their management is therefore fundamental.

However, many other symptoms and functional status, much less addressed in research and clinical practice, are fundamental for the daily life of people with CRDs.

Symptoms beyond dyspnea

People with CRDs often report many other symptoms beyond dyspnea such as fatigue, cough, sputum, cognitive decline, pain, depression, anxiety, sleep disturbances, and loneliness [3,38—71]. Fatigue is defined as the subjective feeling of tiredness, exhaustion, or lack of energy, that occurs on a daily basis, and it can be physical or emotional [72—74]. Cough is an inspiratory effort followed by a forced expiration against a transiently closed glottis, which results in a rapid expulsion of air [75,76]. It is a defense mechanism that protects the airways from aspiration and clears secretions. Sputums are the expectorated secretions. Mucus is normal; however, when increased due to injury of the bronchial tree or lung parenchyma, caused by inhaled noxious substances or by inflammation (asthma, bronchitis, pneumonia), it has higher viscosity and elasticity and is less easily cleared by cilia secretions [77] and needs cough to be expectorated. Cognition is commonly defined as any brain function that enables an individual to perceive, register, store, retrieve, and use information in order to adapt behavior to new situations and function in our environment [78]. Cognitive impairment, i.e., memory loss, frequent word-finding pauses or substitutions, frequently asking the same question or repeating the same story, inability to recognize familiar people and places and lack of orientation, trouble in making actions, judging distances not related with eyesight and in completing step-by-step tasks, and loss of executive function, among others, is inevitable with advance aging, but acute and CRD can accelerate it [79,80]. Pain is an unpleasant sensory and emotional experience associated with, or resembling, actual or potential tissue damage [81]. Depressive symptoms are more than sadness and commonly include the experience of lack of interest and pleasure in daily activities, significant weight loss or gain, insomnia or excessive sleeping, lack of energy, inability to concentrate, feelings of worthlessness or excessive guilt, and recurrent thoughts of death or suicide [82,83]. Anxiety symptoms commonly include feelings of tension, worried thoughts, and physical changes like increased blood pressure [83,84]. Sleep disturbances are commonly described as difficulties in initiating or maintaining sleep, insomnias, increased light sleep and reduced rapid eye movement sleep, nonrestorative sleep, frequent sleep stage shifts, arousals, and daytime sleepiness [61,85]. Loneliness is not necessarily the same as being socially isolated. It is a subjective perception of how one feels while isolation is the objective measure of how large the social network is. Loneliness can be social (e.g., absence of a social network or friends) and/or emotional (e.g., absence of a close, intimate attachment to another person) [70,71,86]. People can have many friends and be lonely, or no friends and feeling fulfilled [86].

All these highly prevalent symptoms limit the daily life of people with CRDs and their loved ones [4,73,87] and their productivity [88], but have also been associated with increased health-care costs and burden to society [89−91]. They also interact with each other and impact and are influenced by the debilitating physical, psychological, and behavioral consequences of CRDs, often aggravated by the presence of multiple comorbidities [92], restricting activities, and affecting severely participation in daily life [71−75,79,80,84,93−96]. Although frequently found and reported by people with CRDs, the underlying causes of most of these symptoms and how they behave and can be managed during the trajectory of CRDs are much less understood than dyspnea [3,38−41]. Further research to improve our understanding of these symptoms is essential as they are in the center of the burden of the disease perceived by people with CRDs and perpetuate it [3,42].

Functional status

Other aspects critical to activity and participation in daily life are rarely assessed and/or addressed such as dressing and housework, which fall into the concept of functional status [6,97]. Indeed, one of the most frequent impacts on daily life reported by people with CRD is decreased functional status, which includes struggling to perform basic (e.g., climbing stairs, getting dressed, housework), work, and leisure activities [4,98−100]. Functional status is an individual's ability to perform normal daily activities required to meet basic needs, fulfill usual roles, and maintain health and well-being [101]. Decline on functional status has been associated with poor respiratory disease control [102], increase dependency on others [4], productivity loss [88], high disease-related costs [103], and premature death [104]. In fact, decreased functional status is a predictor of exacerbations, hospital admissions (mainly responsible for disease-related costs) and mortality, contributing to high individual and societal burden [105]. Despite being highly meaningful to individuals and society, functional status has been overlooked from assessments and treatment options.

Functional status includes functional capacity—individual's maximum capacity to perform daily life activities; functional performance—activities people actually do during the course of their daily lives; functional reserve—difference between capacity and performance; and capacity utilization—extent to which functional potential is used.

Commonly, health-care professionals assess functional capacity (field walk tests, e.g., 6 minute walking test or incremental shuttle walk test) and exercise capacity such as physiological maximal response to exercise (e.g., maximal oxygen consumption or heart rate) or the body structure's maximal ability to fulfill its own function (e.g., maximal voluntary contraction of a skeletal muscle) [106]. However, functional performance reflects the participation in daily life activities and is usually performed at a level that does not require nor meet maximal exercise capacity [106]. Although it is likely that improvements in physiological exercise capacity may result into some improvements in the performance of daily activities (as ventilation decreases during submaximal activity, symptoms such as dyspnea are reduced), the extension of this translation is still to be determined. Therefore, functional status assessment needs to include functional capacity and performance [101] as described in the ICF "Mobility" chapter, e.g., maintaining a standing position, changing basic body position, walking and moving, as well as carrying, moving, and handling objects, among others [106], to identify the impairments (in body function and structures) but also the limitations of activities and restrictions in participation that are relevant for the daily living of people with CRDs.

Person-centered comprehensive assessments are needed to identify the daily needs of people with CRDs and their loved ones and optimize their well-being. These assessments need to include physiological/objective but also patient-reported and patient-experienced outcome measures. Only then meaningful interventions can be developed and implemented with direct benefit translated to the daily management of people with CRDs and their loved ones, ultimately resulting in reduced burden of disease and treatment, and better health status and quality of life.

Patient-centric assessment for daily management

People with CRDs are normally assessed with physiological/objective measures (e.g., lung function—Forced expiratory volume in one second, forced vital capacity; exercise capacity—cardiopulmonary exercise test; muscle dysfunction—one repetition maximum or multiple repetitions). However, patient-reported outcome measures (PROMs) and patient-reported experience measures (PREMs) are essential measures to provide a patient-centric view of the health care needed [107].

PROMs are self-reported questionnaires, completed by people with CRDs that can be generic or disease- specific. These measures seek to gather the perceptions of health status and health-related quality of life of people with CRDs. Their content tends to focus on one or more of the following: physical functioning, symptoms, social well-being, psychological well-being, cognitive function, and role activities [107]. These measures have the advantage of being developed with the purpose of reflecting meaningful outcomes for people with CRDs and can be useful for assessing the quality of interventions and guiding necessary adjustments, taking into consideration the views of people with CRDs [108].

However, PROMs are commonly administered only to people with CRDs, and similar but tailored ones should be developed and used in loved ones (c.g., health-related quality of life, burden), since these stakeholders have a direct influence on the self-management of people with CRDs and they often, directly or indirectly, participate in interventions [109,110].

PREMs assess the perception of the personal experience of the health care that people with CRDs have received. PREMs should focus on aspects of the care that matter to the people with CRDs. They can be relational (patients experience of their relationships during treatment), or functional (practical issues such as availability of facilities) [107]. PREMs are crucial to better personalize interventions, as they can indicate what could be changed for a specific person, be used as quality control indicators to guarantee consistency of the quality of care and/or guide improvements to interventions if necessary. Similarly to PROMs, these measures should also be explored with loved ones, since they are key to keep people with CRD adherent to interventions and their experience might enlighten barriers/facilitators to the interventions. However, PREMs are not commonly used in respiratory conditions and few have been developed [107,111].

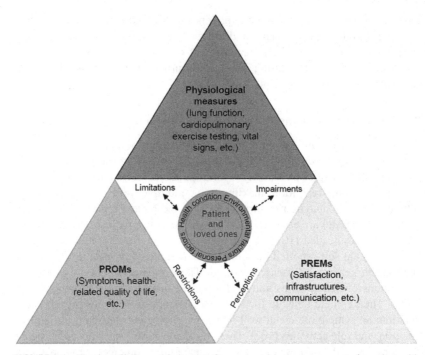

FIGURE 2.1 The three fundamental elements for a comprehensive assessment of a patient with chronic respiratory disease. *PREMs*, patient-reported experience measures; *PROMs*, patient-reported outcome measures.

PROMs and PREMs can alter the dynamic of health-care communication based upon what is important to the person with CRD and thus are essential to inform clinical guidelines [107].

PROMs provide insight into the impact of the disease or an intervention on the person with CRD, while PREMs provide insight into the quality of care during the intervention. Hence, the two should be used in parallel to present the perceptions of both the process and outcome of the care according to the views of people with CRDs and their loved ones.

Fig. 2.1 shows the relationship between the different types of measures to include in a comprehensive assessment.

Meaningful daily management

People with CRDs and their loved ones have multiple physical (pulmonary and extra-pulmonary), emotional, and/or social limitations; thus, patient-centric comprehensive assessments followed by personalized interventions, tackling the identified treatable traits, are needed for improvements to be translated to meaningful daily management, independently of the disease severity [112].

Meaningful daily management of people with CRDs needs to go far beyond pharmacology as this will have no or limited effect on the physical, emotional, and social conditions of people with CRDs [112]. The degree of disease complexity should determine the type of intervention (monodisciplinary or interdisciplinary) as well as its setting (inpatient, outpatient, home) [112]. Self-management, healthy lifestyle recommendations (e.g., physical activity), target therapies (e.g., physiotherapy), or pulmonary rehabilitation might be applied to people without clear symptom burden and daily life limitations, to those with a single physical, emotional, or social limitation or to those with multiple limitations, according to their identified daily needs [112].

Self-management

Daily management of people with CRDs has a good dose of self-management. Self-management has been defined as the ability of a patient to deal with all that a chronic disease entails, including symptoms, treatment, physical and social consequences, and lifestyle changes [113]. On a daily basis this is supported by loved ones [114]. People with CRDs and their loved ones have to understand and adhere to treatment regimens (e.g., medication, physiotherapy, pulmonary rehabilitation), handle equipment, recognize early changes in daily symptoms and an exacerbation and instigate treatment, organize appointments with health-care professionals and annual vaccinations, manage symptoms to perform activities of daily living, ensure performance of airway clearance techniques (if needed), keep physically active, quit smoking, and maintain a healthy diet [115].

A behavioral model that describes self-management is patient activation [116] which emphasizes the need for knowledge, skills, and confidence to manage health and health care [117]. Education alone is therefore not sufficient; monitoring and assessment of progress is also essential. A consensus for COPD has been reached describing self-management interventions as structured but personalized and often multicomponent, with goals of motivating, engaging, and supporting the patients to positively adapt their health behavior(s) and develop skills to better manage their disease [118]. Collaboration among health-care professionals and people with CRDs in self-management interventions is vital. These patient-centered interactions focus on: (1) identifying needs, health beliefs, and enhancing intrinsic motivations; (2) eliciting personalized goals; (3) formulating appropriate strategies (e.g., exacerbation management) to achieve these goals; and if required (4) evaluating and readjusting strategies. Behavior change techniques are used to elicit patient motivation, confidence, and competence. Literacy-sensitive approaches are used to enhance comprehensibility [118]. Strategies to promote self-efficacy include personal experience and practice, feedback and reinforcement, analysis of causes of failure, and shared experience with successful peers [119].

Self-management interventions in people with CRDs should not be an optional extra [120] and have been associated with improvements in health-related quality of life, reduced health-care utilization, and lower probability of respiratory-related hospital admissions [117,121,122].

Healthy lifestyles and target therapies

A healthy lifestyle is a way of living that lowers the risk of being seriously ill or dying prematurely [123,124]. Unhealthy lifestyles are commonly associated with smoking, excessive alcohol consumption, poor diet, and physical inactivity. People with CRDs should therefore avoid tobacco smoking with smoking cessation therapy being often essential [125,126]; avoid excessive alcohol consumption and follow dietary recommendations including nutritional supplements and/or appetite stimulants, thus nutritional therapy is often needed; keep physically active, adhere to exercise regimens [127]; and avoid prolonged time in sedentary behavior, in which physiotherapy is often required.

Smoking cessation reduces the risk of developing many diseases and increases life expectancy [125]. Effective treatments exist and have been highly successful in helping many smokers to quit, especially if combining psychosocial counseling and pharmacotherapy [126].

Malnourishment or undernourishment in people with CRDs are common [127] and important risk factors for poor quality of life and exercise intolerance, with increased risk of exacerbations [128]. Nutritional interventions commonly target fat loss, muscle loss, bone mineral density loss, adiposity, and acute exacerbations [127,129].

Physical activity is defined as any bodily movement produced by skeletal muscles that requires energy expenditure and can be performed at a variety of intensities, as part of recreation and leisure (play, games, sports, or planned exercise), transportation (wheeling, walking, and cycling), work or household chores, in the context of daily occupational, educational, home, and community settings [130]. Different recommendations exist according to age. Focusing on adults, people can be considered physically active if they undertake at least 150–300 min of moderate-intensity aerobic physical activity per week; or 75–150 min of vigorous-intensity aerobic physical activity per week; or an equivalent combination of moderate and vigorous-intensity activity throughout the week, for substantial health benefits [130]. All minutes performed count and more time spent seems to translate into additional health benefits. Adults should also do resistance training at moderate or greater intensity that involve all major muscle groups on 2 or more days a week, as these provide additional health benefits [130].

Sedentary behavior is defined as any waking behavior characterized by an energy expenditure of 1.5 METS or lower while sitting, reclining, or lying [130].

Physical inactivity has been associated with all-cause mortality and multiple chronic diseases (cardiovascular diseases, CRDs, cancer, and diabetes) [131,132]. Prolonged sedentary behavior has also been acknowledged as an independent risk factor for an increased incidence rate of many chronic diseases, hospitalizations, and mortality [133].

People with CRDs are highly inactive [134–138] and sedentary [139,140]. These behaviors are especially aggravated during hospitalizations [141] and their deleterious effects are widely acknowledged [142,143] in people with CRDs.

Engaging in regular physical activity participation requires motivation, capability, and opportunity [144]. Simply advising people to be more physically active is ineffective [145]. Living with others and having an active caregiver [146,147] and/or having a dog and grandchildren [148] have been associated with higher levels of physical activity in people with COPD. Therefore, physical activity advice to people with CRDs needs to be specific; individualized; supported by a behavior change framework; taking into consideration the specific respiratory disease, each person body function and structures impairments, activity limitations, and participation restrictions within his/her personal and environmental factors; the person goals, his/her preferences, and priorities. Therefore, involvement of a health professional such as a physiotherapist is commonly observed [149] and recommended.

Promoting physical activity with a health coaching intervention might be fundamental in people with CRDs. Physical activity coaching can be defined as the use of evidence-based skillful conversation, clinical strategies, and interventions to actively and safely engage (people with CRDs) in (physical activity) behavior change to better self-manage their health, health risk(s), and acute or chronic health conditions resulting in optimal wellness, improved health outcomes, lowered health risk, and decreased health-care costs [150]. Although it is well documented that physical activity improves health outcomes with no increased risk of harm in people with CRDs the optimal way to encourage, improve, and/or maintain physical activity and break sedentary time in this population is still unknown [138,140,151,152].

Comprehensive nonpharmacological intervention—pulmonary rehabilitation

Pulmonary rehabilitation is a nonpharmacological, patient-centered intervention with the participation of multiple professionals according to the identified needs of people with CRDs. It is defined as "a comprehensive intervention based on a thorough patient assessment followed by patient-tailored therapies, which include, but are not limited to, exercise training, education, and behavior change, designed to improve the physical and psychological condition of people with CRDs and to promote the long-term adherence of health-enhancing behaviors" [119].

The exercise training component normally comprises endurance (in a treadmill, step, or cycloergometer), resistance/strength training (against gravity, body weight, fixed or free weights), and flexibility training but several other additional training strategies such as respiratory muscle training, balance training, functional training, breathing, and airway clearance techniques can be implemented if needed [153–155]. The nonexercise components commonly include education and behavior change, i.e., knowledge about the disease, symptom control, healthy lifestyles, nutrition, and psychological support [119,154].

Pulmonary rehabilitation has been compared to a Swiss army knife, activating each specific tool according to the needs of each person, i.e., based on a comprehensive assessment, physical, emotional, and social treatable traits are identified, and can then be addressed by a dedicated, interdisciplinary team using targeted therapies [112], ultimately, improving the physical and psychological condition of people with CRDs and promoting their long-term adherence to health-enhancing behaviors [119].

Despite being implemented in various settings (inpatient, outpatient, home, community) with a high heterogeneity in terms of the frequency, duration, and content of the sessions as well as total length of the program and type of assessments, the evidence of benefit and thus, for providing pulmonary rehabilitation to people with CRDs is overwhelming [156–159]. In COPD, a Cochrane review has stated that there is no need to conduct more randomized controlled trials as pulmonary rehabilitation is so effective that having a group without access to it would be unethical [157]. Physical, psychological, and social improvements have been largely reported in people with CRDs [3,119,156,158], and benefits for the whole family to cope and psychosocially adjust to illness have also been demonstrated [160]. Pulmonary rehabilitation has been found to be more cost-effective than pharmacological treatments [161] across different settings [162]. Recently, evidence also supports that it reduces the risk of mortality 1 year after initiating pulmonary rehabilitation within 90 days post COVID-19 discharge [163]. Given the benefits of pulmonary rehabilitation for individuals, families, and society, this intervention has been proposed to be part of the standard care offered to people with CRDs [119,156], and all efforts around the world are of paramount importance to increase access to it [164,165].

In sum, living with a CRD involves a wide variety of needs and meaningful daily management is complex and needs to be tailored. A one-size-fits-all approach will not meet the demands of people with CRDs and their loved ones. Often, self-monitoring and different treatment regimens will need to be involved. Patient-managed devices, remote monitoring, and telecare [7] implying not just how the person takes care of himself/herself but also the way he/she makes demands on services will constitute future challenges for the near future.

Future avenues for daily assessment and management

Wearable technology and smart applications are currently well accepted by people with CRDs as a way of monitoring and daily manage their condition and thus, hold future promise [166,167]. Wearables and applications may integrate several patient-held devices, such as smartphones or tablets and be useful to assess and monitor physical (e.g., vital signs, respiratory sounds, oxygen levels, functional status, physical activity), psychosocial (e.g., anxiety, depression, mood), and behavioral (nocturnal awakening, coping with the CRDs, consumption of caffeine, tobacco, alcohol) features as well as symptoms of people with CRDs and/or their loved ones during daily life over time [168,169]. Specifically, wearables might be an important technology in the future, since their sensors could be used not only to monitor patients, but also to provide alerts, detect emergency events, and personalize interventions. Nevertheless, the reliability of such wearables is not yet optimal for some features (e.g., energy expenditure, sleep pattern) [172,173] and their role in delivering patient health-care benefits needs further investigation [174].

Regarding daily assessments, this technology can integrate the triad of physiological, PROMs, and PREMs (Fig. 2.1) and provide direct feedback to different stakeholders, thus constituting an important opportunity to consider in patient-centered comprehensive assessments. There are some specific features of wearables that are of most importance to the daily assessment of CRDs which will be briefly discussed below.

Wearables allow a comprehensive assessment of the physical features of people with CRDs. Despite being less reliable than conventional electrocardiography, heart rate monitoring through photoplethysmography might be useful to provide health-care professionals with long data on heart rate patterns, especially in people with cardiac comorbidities, to detect events of tachycardia or bradycardia and ascertain the need of interventions [175].

Although there is still no commercially available wearable for collecting respiratory sounds, this is an important feature for CRDs. Respiratory sounds' characteristics change according to the person's characteristics (e.g., age, gender, height, position, and airflow), local of sound heard, and position in the respiratory phase [182]. They are particularly different between children and adults, in the absence/presence of a respiratory condition [182], between stable and acute exacerbations periods of the respiratory condition [183] and are highly sensitive to very small airway changes (changes as little as 10% in forced expiratory volume in one second—FEV_1) [184] and responsive to respiratory treatments [185,186]. Respiratory sounds therefore remain one of the most valuable information for diagnosing and monitoring respiratory diseases over time [186,187]. Their acquisition can still be challenging (e.g., due to presence of surrounding noise) and the amount of data generated can be overwhelming and difficult to interpret in a timely manner.

Nevertheless, given their importance to provide noninvasive information about the regional ventilation and airflow obstruction within the lung and trachea; several wearables are currently under development to collect, store, automatically analyze, and display respiratory sounds in a friendly and real-time manner [187].

Measurement of peripheral oxygen saturation is of most importance in CRD; however, few market devices are coupled with such technology. This feature might be helpful in the future to detect abnormal values (i.e., hypoxia) on a daily basis, which can prompt individuals to seek health-care services [177].

Since physical inactivity is one of the features of people with CRDs, wearables might be an important vehicle to assess the levels of physical activity, patterns that influence such levels (e.g., weather, time of the day) and provide feedback to people with CRDs and health-care professionals, while being less expensive than established accelerometers [176]. Nonetheless, only step counts seem to be reliable with other important parameters such as energy expenditure being less accurate [172,173].

In terms of behavioral features, sleep is one of the few elements explored in wearables. Sleep disorders are very common in people with CRDs, namely reduced quality of sleep, nocturnal hypoxia, and obstructive sleep apnea and their monitoring is essential to prevent serious events [61,178]. However, this is usually accomplished through polysomnography with complex and expensive equipment and difficult interpretation which could be overcome with more comfortable, simpler devices such as wrist bands or smartwatches [179,180]. Although wearables might be an easy way to assess sleep, their sensors need to be optimized as they overestimate the sleep stages and few studies have been conducted to explore their ability to detect sleep apnea [180,181].

Assessment of psychosocial features and symptoms (e.g., dyspnea, fatigue, anxiety, depression) is usually not performed through wearables. However, this would be a valuable asset to provide real-time details of symptoms while the individual is performing an activity (e.g., fatigue while walking or running) without the need of a smartphone or a paper-based scale.

Wearables can also be used to improve self-management, promote healthy lifestyles, and complement comprehensive nonpharmacological therapies, such as pulmonary rehabilitation.

The measurement of heart rate may allow patients to detect the need to visit their doctors, and help physicians to adjust pharmacological treatment, and it can be used as an inexpensive alternative to monitor exercise intensity [175,188] which might be particularly useful in telerehabilitation settings.

Measuring respiratory sounds can also be of most value for daily management of the disease, as their ability to detect acute exacerbations at their initial stage could prompt patients to seek its early management with respiratory interventions, such as physiotherapy and pulmonary rehabilitation [183,185].

In parallel, monitoring SpO_2 during physical activity may help patients to understand the need to increase or decrease its intensity, and health-care professionals to ensure safety in exercise-based interventions [177].

Wearables are also a promising technology to support physical activity interventions and the adoption of healthier lifestyles, though some technical problems as memory storage exist [176]. These devices commonly called "activity trackers" are usually coupled with smartphone applications, which can serve as motivators through sharing achievements with peers and getting positive reinforcement [189], although these have been little explored in people with CRDs.

Sleep tracking might be particularly useful for the management of sleep apnea, through the setting of alarm systems on patient devices and allowing remote control of health-care professionals in continuous positive airway pressure devices to adjust titration [190]. Additionally, some devices coupled with applications have claimed to improve the quality of sleep, although no clinical validation has been performed [191]. Fig. 2.2 illustrates the different clinical applications of a wearable device (i.e., smartwatch).

Beyond wearable technologies, there are a large number of smartphone/tablet applications that can be used by people with CRDs to improve their knowledge about the disease, track symptoms and respiratory sounds, manage medication, support smoking cessation and nutrition, and promote physical activity [187,192,193]. However, most of these apps have not been clinically validated [193] and although there seems to be benefits with their use to improve self-management, more studies are needed [194,195].

For CRDs, there exists a large amount of mobile/tablet applications, with about 91 apps available from both the Android and Apple stores mainly for the monitoring and management of asthma, COPD, and rhinitis [196]. However, only 14% of the available apps integrate validated PROMs such as the Asthma Control Test or the COPD Assessment Test, although 38% have some type of symptom evaluation [196]. Moreover, most of the apps are focused on respiratory symptoms, such as dyspnea or cough [196]. In order to make comprehensive assessments, future apps should contain the assessment of other frequent and limiting symptoms such as fatigue, pain, anxiety, depression, poor sleep quality, among others, as well as daily functional status using validated and disease-specific PROMs.

Furthermore, the use of a single app for the management of each specific CRD integrating physiological outcomes, PROMs, and PREMs is currently unavailable and could be of most value for health-care providers and patients since they could access all the clinical information and transport it through different health-care services.

Wearable technologies accompanied by a dedicated support system with smart applications are promising tools for the daily assessment and management of CRDs. While wearables seem to be ready to use to assess and monitor

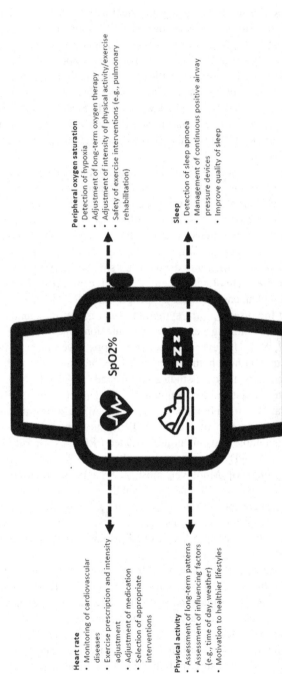

Heart rate
- Monitoring of cardiovascular diseases
- Exercise prescription and intensity adjustment
- Adjustment of medication
- Selection of appropriate interventions

Physical activity
- Assessment of long-term patterns
- Assessment of influencing factors (e.g., time of day, weather)
- Motivation to healthier lifestyles

Peripheral oxygen saturation
- Detection of hypoxia
- Adjustment of long-term oxygen therapy
- Adjustment of intensity of physical activity/exercise
- Safety of exercise interventions (e.g., pulmonary rehabilitation)

Sleep
- Detection of sleep apnoea
- Management of continuous positive airway pressure devices
- Improve quality of sleep

FIGURE 2.2 Example of a wearable device (smartwatch) and its future utilities in clinical practice with chronic respiratory disease patients.

heart rate and number of steps, other clinical features require further technological developments. Improving the quality of sensors and validating smartphone/tablet apps are crucial before their routine use in clinical practice, and challenges with managing big data, data protection, and costs still need to be overcome [168,170,171].

References

[1] W.W. Labaki, M.K. Han, Chronic respiratory diseases: a global view, Lancet Respir. Med 8 (6) (2020) 531−533.

[2] Forum of International Respiratory Societies, The Global Impact of Respiratory Disease, second ed., European Respiratory Society, Sheffield, 2017.

[3] A. Machado, A. Marques, C. Burtin, Extra-pulmonary manifestations of COPD and the role of pulmonary rehabilitation: a symptom-centered approach, Expert Rev. Respir. Med. 15 (1) (2021) 131−142.

[4] Food and Drug Administration, The Voice of the Patient: Idiopathic Pulmonary Fibrosis, ForIndustry/UserFees/PrescriptionDrugUserFee/UCM440829pdf 17, 2015.

[5] E.F.M. Wouters, B. Wouters, I.M.L. Augustin, S. Houben-Wilke, L. Vanfleteren, F.M.E. Franssen, Personalised pulmonary rehabilitation in COPD, Eur. Respir. Rev. 27 (147) (2018).

[6] A.W. Vaes, J.M.L. Delbressine, R. Mesquita, et al., Impact of pulmonary rehabilitation on activities of daily living in patients with chronic obstructive pulmonary disease, J. Appl. Physiol. 126 (3) (2019) 607−615.

[7] C.R. May, D.T. Eton, K. Boehmer, et al., Rethinking the patient: using Burden of Treatment Theory to understand the changing dynamics of illness, BMC Health Serv. Res. 14 (2014) 281.

[8] M. Clari, D. Ivziku, R. Casciaro, M. Matarese, The unmet needs of people with chronic obstructive pulmonary disease: a systematic review of qualitative findings, COPD 15 (1) (2018) 79−88.

[9] A.C. Gardener, G. Ewing, I. Kuhn, M. Farquhar, Support needs of patients with COPD: a systematic literature search and narrative review, Int. J. Chronic Obstr. Pulm. Dis. 13 (2018) 1021−1035.

[10] K.L.M. Hester, J. Newton, T. Rapley, A. De Soyza, Patient information, education and self-management in bronchiectasis: facilitating improvements to optimise health outcomes, BMC Pulm. Med. 18 (1) (2018) 80.

[11] L.Y. Hsieh, F.J. Chou, S.E. Guo, Information needs of patients with lung cancer from diagnosis until first treatment follow-up, PLoS One 13 (6) (2018) e0199515.

[12] M.L. Kong, C. Armour, K. LeMay, L. Smith, Information needs of people with asthma, Int. J. Pharm. Pract. 22 (3) (2014) 178−185.

[13] J. Kuon, J. Vogt, A. Mehnert, et al., Symptoms and needs of patients with advanced lung cancer: early prevalence assessment, Oncol. Res. Treat. 42 (12) (2019) 650−659.

[14] J.Y.T. Lee, G. Tikellis, T.J. Corte, et al., The supportive care needs of people living with pulmonary fibrosis and their caregivers: a systematic review, Eur. Respir. Rev. 29 (156) (2020).

[15] A. Marques, R.S. Goldstein, Living with chronic lung disease: the experiences and needs of patients and caregivers. Pulmonary Rehabilitation, 2020, p. 281.

[16] D. Ramadurai, S. Corder, T. Churney, et al., Idiopathic pulmonary fibrosis: educational needs of health-care providers, patients, and caregivers, Chron. Respir. Dis. 16 (2019), 1479973119858961.

[17] S. Souto-Miranda, A. Marques, Triangulated perspectives on outcomes of pulmonary rehabilitation in patients with COPD: a qualitative study to inform a core outcome set, Clin. Rehabil. 33 (4) (2019) 805–814.

[18] V.D. Dinglas, C.M. Chessare, W.E. Davis, et al., Perspectives of survivors, families and researchers on key outcomes for research in acute respiratory failure, Thorax 73 (1) (2018) 7–12.

[19] F. Early, M. Lettis, S.J. Winders, J. Fuld, What matters to people with COPD: outputs from Working Together for Change, NPJ Prim. Care Respir. Med. 29 (1) (2019) 11.

[20] A. Agusti, E. Bel, M. Thomas, et al., Treatable traits: toward precision medicine of chronic airway diseases, Eur. Respir. J. 47 (2) (2016) 410–419.

[21] T.H. Dahl, International classification of functioning, disability and health: an introduction and discussion of its potential impact on rehabilitation services and research, J. Rehabil. Med. 34 (5) (2002) 201–204.

[22] G. Stucki, International Classification of Functioning, Disability, and Health (ICF): a promising framework and classification for rehabilitation medicine, Am. J. Phys. Med. Rehabil. 84 (10) (2005) 733–740.

[23] World Health Organization, International Classification of Functioning, Disability and Health, World Health Organization, Geneva, 2001.

[24] M.B. Parshall, R.M. Schwartzstein, L. Adams, et al., An official American Thoracic Society statement: update on the mechanisms, assessment, and management of dyspnea, Am. J. Respir. Crit. Care Med. 185 (4) (2012) 435–452.

[25] D.E. O'Donnell, A.F. Elbehairy, D.C. Berton, N.J. Domnik, J.A. Neder, Advances in the evaluation of respiratory pathophysiology during exercise in chronic lung diseases, Front. Physiol. 8 (2017), 82-82.

[26] D. Shrikrishna, M. Patel, R.J. Tanner, et al., Quadriceps wasting and physical inactivity in patients with COPD, Eur. Respir. J. 40 (5) (2012) 1115–1122.

[27] C.B. Cooper, Exercise in chronic pulmonary disease: limitations and rehabilitation, Med. Sci. Sports Exerc. 33 (7 Suppl. 1) (2001) S643–S646.

[28] M.A. Ramon, G. Ter Riet, A.E. Carsin, et al., The dyspnoea-inactivity vicious circle in COPD: development and external validation of a conceptual model, Eur. Respir. J. 52 (3) (2018).

[29] J.Z. Reardon, S.C. Lareau, R. ZuWallack, Functional status and quality of life in chronic obstructive pulmonary disease, Am. J. Med. 119 (10 Suppl. 1) (2006) 32–37.

[30] M.I. Polkey, J. Moxham, Attacking the disease spiral in chronic obstructive pulmonary disease, Clin. Med. 6 (2) (2006) 190–196.

[31] M. Decramer, S. Rennard, T. Troosters, et al., COPD as a lung disease with systemic consequences–clinical impact, mechanisms, and potential for early intervention, COPD 5 (4) (2008) 235–256.

[32] B. Vainshelboim, J. Oliveira, S. Izhakian, A. Unterman, M.R. Kramer, Lifestyle behaviors and clinical outcomes in idiopathic pulmonary fibrosis, Respiration 95 (1) (2018) 27–34.

[33] G. Raghu, H.R. Collard, J.J. Egan, et al., An official ATS/ERS/JRS/ALAT statement: idiopathic pulmonary fibrosis: evidence-based guidelines for diagnosis and management, Am. J. Respir. Crit. Care Med. 183 (6) (2011) 788–824.

[34] A.W. Vaes, J. Garcia-Aymerich, J.L. Marott, et al., Changes in physical activity and all-cause mortality in COPD, Eur. Respir. J. 44 (5) (2014) 1199–1209.

[35] R.-J. Shei, K.A. Mackintosh, J.E. Peabody Lever, M.A. McNarry, S. Krick, Exercise physiology across the lifespan in cystic fibrosis, Front. Physiol. 10 (2019), 1382–1382.

[36] B. Waschki, A. Kirsten, O. Holz, et al., Physical activity is the strongest predictor of all-cause mortality in patients with COPD: a prospective cohort study, Chest 140 (2) (2011) 331–342.

[37] F. Maltais, M. Decramer, R. Casaburi, et al., An official American Thoracic Society/European Respiratory Society statement: update on limb muscle dysfunction in chronic obstructive pulmonary disease, Am. J. Respir. Crit. Care Med. 189 (9) (2014) e15–62.

[38] J. Garner, P.M. George, E. Renzoni, Cough in interstitial lung disease, Pulm. Pharmacol. Therapeut. 35 (2015) 122–128.

[39] S. Carvajalino, C. Reigada, M.J. Johnson, M. Dzingina, S. Bajwah, Symptom prevalence of patients with fibrotic interstitial lung disease: a systematic literature review, BMC Pulm. Med. 18 (1) (2018) 78.

[40] S. Iyer, A. Roughley, A. Rider, G. Taylor-Stokes, The symptom burden of non-small cell lung cancer in the USA: a real-world cross-sectional study, Support. Care Cancer 22 (1) (2014) 181–187.

[41] A. Harle, A. Molassiotis, O. Buffin, et al., A cross sectional study to determine the prevalence of cough and its impact in patients with lung cancer: a patient unmet need, BMC Cancer 20 (1) (2020) 9.

[42] N.S. Cook, K. Kostikas, J.-B. Gruenberger, et al., Patients' perspectives on COPD: findings from a social media listening study, ERJ Open Res. 5 (1) (2019) 00128–02018.

[43] M. Van Herck, M.A. Spruit, C. Burtin, et al., Fatigue is highly prevalent in patients with asthma and contributes to the burden of disease, J. Clin. Med. 7 (12) (2018) 471.

[44] S. Carnio, R.F. Di Stefano, S. Novello, Fatigue in lung cancer patients: symptom burden and management of challenges, Lung Cancer 7 (2016) 73–82.

[45] A.E.M. Bloem, R.L.M. Mostard, N. Stoot, et al., Severe fatigue is highly prevalent in patients with IPF or sarcoidosis, J. Clin. Med. 9 (4) (2020) 1178.

[46] K.L. Hester, J.G. Macfarlane, H. Tedd, et al., Fatigue in bronchiectasis, QJM 105 (3) (2012) 235–240.

[47] D. Puebla-Neira, W.J. Calhoun, Why do asthma patients cough? New insights into cough in allergic asthma, J. Allergy Clin. Immunol. 144 (3) (2019) 656–657.

[48] M. Mac Aogáin, S.H. Chotirmall, Bronchiectasis and cough: an old relationship in need of renewed attention, Pulm. Pharmacol. Therapeut. 57 (2019) 101812.

[49] A.S.M. Harle, F.H. Blackhall, A. Molassiotis, et al., Cough in patients with lung cancer: a longitudinal observational study of characterization and clinical associations, Chest 155 (1) (2019) 103–113.

[50] F. Irani, J.M. Barbone, J. Beausoleil, L. Gerald, Is asthma associated with cognitive impairments? A meta-analytic review, J. Clin. Exp. Neuropsychol. 39 (10) (2017) 965–978.

[51] E.F. van Dam van Isselt, K.H. Groenewegen-Sipkema, M. Spruit-van Eijk, et al., Pain in patients with COPD: a systematic review and meta-analysis, BMJ Open 4 (9) (2014) e005898.

[52] S. Mercadante, V. Vitrano, Pain in patients with lung cancer: pathophysiology and treatment, Lung Cancer 68 (1) (2010) 10–15.

[53] P.T. King, S.R. Holdsworth, M. Farmer, N.J. Freezer, P.W. Holmes, Chest pain and exacerbations of bronchiectasis, Int. J. Gen. Med. 5 (2012) 1019–1024.

[54] S.C. Wynne, S. Patel, R.E. Barker, et al., Anxiety and depression in bronchiectasis: response to pulmonary rehabilitation and minimal clinically important difference of the Hospital Anxiety and Depression Scale, Chron. Respir. Dis. 17 (2020), 1479973120933292-1479973120933292.

[55] M. Bedolla-Barajas, J. Morales-Romero, J.C. Fonseca-López, N.A. Pulido-Guillén, D. Larenas-Linnemann, D.D. Hernández-Colín, Anxiety and depression in adult patients with asthma: the role of asthma control, obesity and allergic sensitization, J. Asthma (2020) 1−9.

[56] A.M. Yohannes, G.S. Alexopoulos, Depression and anxiety in patients with COPD, Eur. Respir. Rev. 23 (133) (2014) 345−349.

[57] K. Carlsen, A.B. Jensen, E. Jacobsen, M. Krasnik, C. Johansen, Psychosocial aspects of lung cancer, Lung Cancer 47 (3) (2005) 293−300.

[58] C.G. Brown Johnson, J.L. Brodsky, J.K. Cataldo, Lung cancer stigma, anxiety, depression, and quality of life, J. Psychosoc. Oncol. 32 (1) (2014) 59−73.

[59] M. Nishiura, A. Tamura, H. Nagai, E. Matsushima, Assessment of sleep disturbance in lung cancer patients: relationship between sleep disturbance and pain, fatigue, quality of life, and psychological distress, Palliat. Support Care 13 (3) (2015) 575−581.

[60] T.M. Alanazi, H.S. Alghamdi, M.S. Alberreet, et al., The prevalence of sleep disturbance among asthmatic patients in a tertiary care center, Sci. Rep. 11 (1) (2021) 2457.

[61] W.T. McNicholas, J. Verbraecken, J.M. Marin, Sleep disorders in COPD: the forgotten dimension, Eur. Respir. Rev. 22 (129) (2013) 365.

[62] C.S. Phua, T. Wijeratne, C. Wong, L. Jayaram, Neurological and sleep disturbances in bronchiectasis, J. Clin. Med. 6 (12) (2017) 114.

[63] J.-G. Cho, A. Teoh, M. Roberts, J. Wheatley, The prevalence of poor sleep quality and its associated factors in patients with interstitial lung disease: a cross-sectional analysis, ERJ Open Res. 5 (3) (2019) 00062−02019.

[64] Y.J. Lee, S.M. Choi, Y.J. Lee, et al., Clinical impact of depression and anxiety in patients with idiopathic pulmonary fibrosis, PLoS One 12 (9) (2017) e0184300-e0184300.

[65] Q. Shen, T. Guo, M. Song, et al., Pain is a common problem in patients with ILD, Respir. Res. 21 (1) (2020) 297.

[66] M.J.G. van Manen, M.S. Wijsenbeek, Cough, an unresolved problem in interstitial lung diseases, Curr. Opin. Support. Palliat. Care 13 (3) (2019) 143−151.

[67] E. Tudorache, D. Traila, M. Marc, et al., Cognitive impairment in patients with idiopathic pulmonary fibrosis - obstructive sleep apnea overlap, Eur. Respir. J. 54 (Suppl. 63) (2019) PA1305.

[68] U.C. Ojha, D.P. Singh, O.K. Choudhari, D. Gothi, S. Singh, Correlation of severity of functional gastrointestinal disease symptoms with that of asthma and chronic obstructive pulmonary disease: a multicenter study, Int. J. Appl. Basic Med. Res. 8 (2) (2018) 83−88.

[69] Z.-W. Hu, Z.-G. Wang, Y. Zhang, et al., Gastroesophageal reflux in bronchiectasis and the effect of anti-reflux treatment, BMC Pulm. Med. 13 (2013) 1−9.

[70] T. Reijnders, M. Schuler, D. Jelusic, et al., The impact of loneliness on outcomes of pulmonary rehabilitation in patients with COPD, COPD 15 (5) (2018) 446−453.

[71] K.A. Hyland, B.J. Small, J.E. Gray, et al., Loneliness as a mediator of the relationship of social cognitive variables with depressive symptoms and quality of life in lung cancer patients beginning treatment, Psycho Oncol. 28 (6) (2019) 1234−1242.

[72] S. Small, M. Lamb, Fatigue in chronic illness: the experience of individuals with chronic obstructive pulmonary disease and with asthma, J. Adv. Nurs. 30 (2) (1999) 469−478.

[73] M.A. Spruit, J.H. Vercoulen, M.A.G. Sprangers, E.F.M. Wouters, Fatigue in COPD: an important yet ignored symptom, Lancet Respir. Med. 5 (7) (2017) 542–544.

[74] E. Ream, A. Richardson, Fatigue in patients with cancer and chronic obstructive airways disease: a phenomenological enquiry, Int. J. Nurs. Stud. 34 (1) (1997) 44–53.

[75] A.K. Pandya, K.K. Lee, S.S. Birring, Cough, Medicine 44 (4) (2016) 213–216.

[76] R.S. Irwin, M.H. Baumann, D.C. Bolser, et al., Diagnosis and management of cough executive summary: ACCP evidence-based clinical practice guidelines, Chest 129 (1 Suppl. l) (2006) 1s–23s.

[77] J.V. Fahy, B.F. Dickey, Airway mucus function and dysfunction, N. Engl. J. Med. 363 (23) (2010) 2233–2247.

[78] F.A.H.M. Cleutjens, D.J.A. Janssen, R.W.H.M. Ponds, J.B. Dijkstra, E.F.M. Wouters, Ognitive-pulmonary disease, BioMed Res. Int. 2014 (2014), 697825-697825.

[79] M. van Beers, D.J.A. Janssen, H.R. Gosker, A.M.W.J. Schols, Cognitive impairment in chronic obstructive pulmonary disease: disease burden, determinants and possible future interventions, Expert Rev. Respir. Med. 12 (12) (2018) 1061–1074.

[80] V. Andrianopoulos, R. Gloeckl, I. Vogiatzis, K. Kenn, Cognitive impairment in COPD: should cognitive evaluation be part of respiratory assessment? Breathe 13 (1) (2017) e1–e9.

[81] S.N. Raja, D.B. Carr, M. Cohen, et al., The revised International Association for the Study of Pain definition of pain: concepts, challenges, and compromises, Pain 161 (9) (2020) 1976–1982.

[82] J.C. Tolentino, S.L. Schmidt, DSM-5 criteria and depression severity: implications for clinical practice, Front. Psychiatr. 9 (2018) 450.

[83] A.P. Association, Diagnostic and Statistical Manual of Mental Disorders (DSM–5), 2020.

[84] E. Atlantis, P. Fahey, B. Cochrane, S. Smith, Bidirectional associations between clinically relevant depression or anxiety and COPD: a systematic review and meta-analysis, Chest 144 (3) (2013) 766–777.

[85] R. Jen, Y. Li, R.L. Owens, A. Malhotra, Sleep in chronic obstructive pulmonary disease: evidence gaps and challenges, Cancer Res. J. 2016 (2016) V.

[86] R. Benzo, Satori: awakening to outcomes that matter: the impact of social support in chronic obstructive pulmonary disease, Ann. Am. Thorac. Soc. 14 (9) (2017) 1385–1386.

[87] M. Kentson, K. Tödt, E. Skargren, et al., Factors associated with experience of fatigue, and functional limitations due to fatigue in patients with stable COPD, Ther. Adv. Respir. Dis. 10 (5) (2016) 410–424.

[88] M. Algamdi, M. Sadatsafavi, J.H. Fisher, et al., Costs of workplace productivity loss in patients with fibrotic interstitial lung disease, Chest 156 (5) (2019) 887–895.

[89] L.K. Lee, E. Obi, B. Paknis, A. Kavati, B. Chipps, Asthma control and disease burden in patients with asthma and allergic comorbidities, J. Asthma 55 (2) (2018) 208–219.

[90] M. Axelsson, A. Lindberg, A. Kainu, E. Rönmark, S.-A. Jansson, Respiratory symptoms increase health care consumption and affect everyday life — a cross-sectional population-based study from Finland, Estonia, and Sweden, Eur. Clin. Respir. J. 3 (1) (2016) 31024.

[91] L.K. Lee, K. Ramakrishnan, G. Safioti, R. Ariely, M. Schatz, Asthma control is associated with economic outcomes, work productivity and health-related quality of life in patients with asthma, BMJ Open Respir. Res. 7 (1) (2020).

[92] M.V. López Varela, M. Montes de Oca, R. Halbert, et al., Comorbidities and health status in individuals with and without COPD in five Latin American cities: the PLATINO study, Arch. Bronconeumol. 49 (11) (2013) 468–474.

[93] E. Landt, Y. Çolak, P. Lange, L.C. Laursen, B.G. Nordestgaard, M. Dahl, Chronic cough in individuals with COPD: a population-based cohort study, Chest 157 (6) (2020) 1446–1454.

[94] M. Gruet, Fatigue in chronic respiratory diseases: theoretical framework and implications for real-life performance and rehabilitation, Front. Physiol. 9 (2018), 1285–1285.

[95] M. Shorofsky, J. Bourbeau, J. Kimoff, et al., Impaired sleep quality in COPD is associated with exacerbations: the CanCOLD cohort study, Chest 156 (5) (2019) 852–863.

[96] E.P.A. Rutten, K. Lenaerts, W.A. Buurman, E.F.M. Wouters, Disturbed intestinal integrity in patients with COPD: effects of activities of daily living, Chest 145 (2) (2014) 245–252.

[97] R.J. Kaptain, T. Helle, A. Kottorp, A.H. Patomella, Juggling the management of everyday life activities in persons living with chronic obstructive pulmonary disease, Disabil. Rehabil. (2021) 1–12.

[98] J. Annegarn, K. Meijer, V.L. Passos, et al., Problematic activities of daily life are weakly associated with clinical characteristics in COPD, J. Am. Med. Dir. Assoc. 13 (3) (2012) 284–290.

[99] H.J. Bendixen, E.E. Wæhrens, J.T. Wilcke, L.V. Sørensen, Self-reported quality of ADL task performance among patients with COPD exacerbations, Scand. J. Occup. Ther. 21 (4) (2014) 313–320.

[100] N. Nakken, D.J.A. Janssen, E.F.M. Wouters, et al., Changes in problematic activities of daily living in persons with COPD during 1 year of usual care, Aust. Occup. Ther. J. 67 (5) (2020) 447–457.

[101] N.K. Leidy, Functional status and the forward progress of merry-go-rounds: toward a coherent analytical framework, Nurs. Res. 43 (4) (1994) 196–202.

[102] E.C. Woods, R. O'Conor, M. Martynenko, M.S. Wolf, J.P. Wisnivesky, A.D. Federman, Associations between asthma control and airway obstruction and performance of activities of daily living in older adults with asthma, J. Am. Geriatr. Soc. 64 (5) (2016) 1046–1053.

[103] W.A. Wuyts, S. Papiris, E. Manali, et al., The burden of progressive fibrosing interstitial lung disease: a DELPHI approach, Adv. Ther. 37 (7) (2020) 3246–3264.

[104] M.J. Thomeer, J. Vansteenkiste, E.K. Verbeken, M. Demedts, Interstitial lung diseases: characteristics at diagnosis and mortality risk assessment, Respir. Med. 98 (6) (2004) 567–573.

[105] V.S. Fan, S.D. Ramsey, B.J. Make, F.J. Martinez, Physiologic variables and functional status independently predict COPD hospitalizations and emergency department visits in patients with severe COPD, Chronic Obstr. Pulm. Dis. 4 (1) (2007) 29–39.

[106] K.L. Bui, A. Nyberg, F. Maltais, D. Saey, Functional tests in chronic obstructive pulmonary disease, Part 1: clinical relevance and links to the international classification of functioning, disability, and health, Ann. Am. Thorac. Soc. 14 (5) (2017) 778–784.

[107] M. Hodson, S. Andrew, C.M. Roberts, Towards an understanding of PREMS and PROMS in COPD, Breathe 9 (5) (2013) 358–364.

[108] D. Jahagirdar, T. Kroll, K. Ritchie, S. Wyke, Patient-reported outcome measures for chronic obstructive pulmonary disease : the exclusion of people with low literacy skills and learning disabilities, Patient 6 (1) (2013) 11–21.

[109] K.A. Lippiett, A. Richardson, M. Myall, A. Cummings, C.R. May, Patients and informal caregivers' experiences of burden of treatment in lung cancer and chronic obstructive pulmonary disease (COPD): a systematic review and synthesis of qualitative research, BMJ Open 9 (2) (2019) e020515.

[110] N. Nakken, D.J.A. Janssen, E.H.A. van den Bogaart, et al., Informal caregivers of patients with COPD: home sweet home? Eur. Respir. Rev. 24 (137) (2015) 498.

[111] M. Hodson, C.M. Roberts, S. Andrew, L. Graham, P.W. Jones, J. Yorke, Development and first validation of a patient-reported experience measure in chronic obstructive pulmonary disease (PREM-C9), Thorax 74 (6) (2019) 600−603.

[112] M.A. Spruit, E.F.M. Wouters, Organizational aspects of pulmonary rehabilitation in chronic respiratory diseases, Respirology 24 (9) (2019) 838−843.

[113] J. Barlow, C. Wright, J. Sheasby, A. Turner, J. Hainsworth, Self-management approaches for people with chronic conditions: a review, Patient Educ. Counsel. 48 (2) (2002) 177−187.

[114] M.T. Macdonald, A. Lang, J. Storch, et al., Examining markers of safety in homecare using the international classification for patient safety, BMC Health Serv. Res. 13 (2013) 191.

[115] J. Bourbeau, J. van der Palen, Promoting effective self-management programmes to improve COPD, Eur. Respir. J. 33 (3) (2009) 461−463.

[116] J.H. Hibbard, J. Stockard, E.R. Mahoney, M. Tusler, Development of the Patient Activation Measure (PAM): conceptualizing and measuring activation in patients and consumers, Health Serv. Res. 39 (4 Pt 1) (2004) 1005−1026.

[117] R.E. Jordan, S. Majothi, N.R. Heneghan, et al., Supported self-management for patients with moderate to severe chronic obstructive pulmonary disease (COPD): an evidence synthesis and economic analysis, Health Technol. Assess. 19 (36) (2015) 1−516.

[118] T.W. Effing, J.H. Vercoulen, J. Bourbeau, et al., Definition of a COPD self-management intervention: international expert group consensus, Eur. Respir. J. 48 (1) (2016) 46−54.

[119] M.A. Spruit, S.J. Singh, C. Garvey, et al., An official American Thoracic Society/European Respiratory Society statement: key concepts and advances in pulmonary rehabilitation, Am. J. Respir. Crit. Care Med. 188 (8) (2013) e13−64.

[120] H. Pinnock, Supported self-management for asthma, Breathe 11 (2) (2015) 98−109.

[121] A. Lenferink, M. Brusse-Keizer, P.D. van der Valk, et al., Self-management interventions including action plans for exacerbations versus usual care in patients with chronic obstructive pulmonary disease, Cochrane Database Syst. Rev. 8 (8) (2017) Cd011682.

[122] A. Hodkinson, P. Bower, C. Grigoroglou, et al., Self-management interventions to reduce healthcare use and improve quality of life among patients with asthma: systematic review and network meta-analysis, BMJ 370 (2020) m2521.

[123] World Health Organization, Regional Office for Europe. Healthy living: what is a healthy lifestyle?, in: Copenhagen: WHO Regional Office for Europe, 1999.

[124] Y.V. Chudasama, K. Khunti, C.L. Gillies, et al., Healthy lifestyle and life expectancy in people with multimorbidity in the UK Biobank: a longitudinal cohort study, PLoS Med. 17 (9) (2020) e1003332.

[125] N.A. Rigotti, Smoking cessation in patients with respiratory disease: existing treatments and future directions, Lancet Respir. Med. 1 (3) (2013) 241−250.

[126] L.F. Stead, P. Koilpillai, T.R. Fanshawe, T. Lancaster, Combined pharmacotherapy and behavioural interventions for smoking cessation, Cochrane Database Syst. Rev. 3 (2016) Cd008286.

[127] J. Gea, A. Sancho-Muñoz, R. Chalela, Nutritional status and muscle dysfunction in chronic respiratory diseases: stable phase versus acute exacerbations, J. Thorac. Dis. 10 (Suppl. 12) (2018) S1332−s1354.

[128] G. Rawal, S. Yadav, Nutrition in chronic obstructive pulmonary disease: a review, J. Transl. Int. Med. 3 (4) (2015) 151−154.

[129] A.M. Schols, I.M. Ferreira, F.M. Franssen, et al., Nutritional assessment and therapy in COPD: a European Respiratory Society statement, Eur. Respir. J. 44 (6) (2014) 1504−1520.

[130] World Health Organization, WHO Guidelines on Physical Activity and Sedentary Behaviour, World Health Organization, Geneva, 2020.

[131] S.A. Lear, W. Hu, S. Rangarajan, et al., The effect of physical activity on mortality and cardiovascular disease in 130000 people from 17 high -income,middle -income,and low-income countries :the PURE study, Lancet 390 (10113) (2017) 2643−2654.

[132] I.M. Lee, E.J. Shiroma, F. Lobelo, P. Puska, S.N. Blair, P.T. Katzmarzyk, Effect of physical inactivity on major non-communicable diseases worldwide: an analysis of burden of disease and life expectancy, Lancet 380 (9838) (2012) 219−229.

[133] A. Biswas, P.I. Oh, G.E. Faulkner, et al., Sedentary time and its association with risk for disease incidence, mortality, and hospitalization in adults: a systematic review and meta-analysis, Ann. Intern. Med. 162 (2) (2015) 123−132.

[134] F. Pitta, T. Troosters, M.A. Spruit, V.S. Probst, M. Decramer, R. Gosselink, Characteristics of physical activities in daily life in chronic obstructive pulmonary disease, Am. J. Respir. Crit. Care Med. 171 (9) (2005) 972−977.

[135] S.N. Vorrink, H.S. Kort, T. Troosters, J.W. Lammers, Level of daily physical activity in individuals with COPD compared with healthy controls, Respir. Res. 12 (1) (2011) 33.

[136] P.D. Freitas, R.F. Xavier, V.M. McDonald, et al., Identification of asthma phenotypes based on extrapulmonary treatable traits, Eur. Respir. J. 57 (1) (2021).

[137] P.S.P. Cho, S. Vasudevan, M. Maddocks, et al., Physical inactivity in pulmonary sarcoidosis, Lung 197 (3) (2019) 285−293.

[138] A.T. Burge, N.S. Cox, M.J. Abramson, A.E. Holland, Interventions for promoting physical activity in people with chronic obstructive pulmonary disease (COPD), Cochrane Database Syst. Rev. 4 (4) (2020) Cd012626.

[139] L. Cordova-Rivera, P.G. Gibson, P.A. Gardiner, V.M. McDonald, A systematic review of associations of physical activity and sedentary time with asthma outcomes, J. Allergy Clin. Immunol. Pract. 6 (6) (2018) 1968−1981, e1962.

[140] W. Geidl, J. Carl, S. Cassar, et al., Physical activity and sedentary behaviour patterns in 326 persons with COPD before starting a pulmonary rehabilitation: a cluster analysis, J. Clin. Med. 8 (9) (2019).

[141] M.W. Orme, T.C. Harvey-Dunstan, I. Boral, et al., Changes in physical activity during hospital admission for chronic respiratory disease, Respirology 24 (7) (2019) 652−657.

[142] L. Cordova-Rivera, P.G. Gibson, P.A. Gardiner, S.A. Hiles, V.M. McDonald, Extrapulmonary associations of health status in severe asthma and bronchiectasis: comorbidities and functional outcomes, Respir. Med. 154 (2019) 93−101.

[143] H. Watz, F. Pitta, C.L. Rochester, et al., An official European Respiratory Society statement on physical activity in COPD, Eur. Respir. J. 44 (6) (2014) 1521−1537.

[144] S. Michie, M.M. van Stralen, R. West, The behaviour change wheel: a new method for characterising and designing behaviour change interventions, Implement. Sci. 6 (2011) 1−12.

[145] M. Hillsdon, M. Thorogood, I. White, C. Foster, Advising people to take more exercise is ineffective: a randomized controlled trial of physical activity promotion in primary care, Int. J. Epidemiol. 31 (4) (2002) 808−815.

[146] Z. Chen, V.S. Fan, B. Belza, K. Pike, H.Q. Nguyen, Association between social support and self-care behaviors in adults with chronic obstructive pulmonary disease, Ann. Am. Thorac. Soc. 14 (9) (2017) 1419−1427.

[147] R. Mesquita, N. Nakken, D.J.A. Janssen, et al., Activity levels and exercise motivation in patients with COPD and their resident loved ones, Chest 151 (5) (2017) 1028−1038.

[148] A. Arbillaga-Etxarri, E. Gimeno-Santos, A. Barberan-Garcia, et al., Socio-environmental correlates of physical activity in patients with chronic obstructive pulmonary disease (COPD), Thorax 72 (9) (2017) 796−802.

[149] M.A. Spruit, A. Van't Hul, H.L. Vreeken, et al., Profiling of patients with COPD for adequate referral to exercise-based care: the Dutch model, Sports Med. 50 (8) (2020) 1421−1429.

[150] M.H. Huffman, Advancing the practice of health coaching: differentiation from wellness coaching, Workplace Health Saf. 64 (9) (2016) 400−403.

[151] E. Gimeno-Santos, A. Frei, C. Steurer-Stey, et al., Determinants and outcomes of physical activity in patients with COPD: a systematic review, Thorax 69 (8) (2014) 731−739.

[152] D.B. Coultas, B.E. Jackson, R. Russo, et al., Home-based physical activity coaching, physical activity, and health care utilization in chronic obstructive pulmonary disease. Chronic obstructive pulmonary disease self-management activation research trial secondary outcomes, Ann. Am. Thorac. Soc. 15 (4) (2018) 470−478.

[153] S. Jenkins, K. Hill, N.M. Cecins, State of the art: how to set up a pulmonary rehabilitation program, Respirology 15 (8) (2010) 1157−1173.

[154] A. Nakazawa, N.S. Cox, A.E. Holland, Current best practice in rehabilitation in interstitial lung disease, Ther. Adv. Respir. Dis. 11 (2) (2017) 115−128.

[155] M. Armstrong, I. Vogiatzis, Personalized exercise training in chronic lung diseases, Respirology 24 (9) (2019) 854−862.

[156] C.L. Rochester, C. Fairburn, R.H. Crouch, Pulmonary rehabilitation for respiratory disorders other than chronic obstructive pulmonary disease, Clin. Chest Med. 35 (2) (2014) 369−389.

[157] B. McCarthy, D. Casey, D. Devane, K. Murphy, E. Murphy, Y. Lacasse, Pulmonary rehabilitation for chronic obstructive pulmonary disease, Cochrane Database Syst. Rev. 2 (2015) Cd003793.

[158] L. Dowman, C.J. Hill, A. May, A.E. Holland, Pulmonary rehabilitation for interstitial lung disease, Cochrane Database Syst. Rev. (2) (2021).

[159] C.L. Granger, N.R. Morris, A.E. Holland, Practical approach to establishing pulmonary rehabilitation for people with non-COPD diagnoses, Respirology 24 (9) (2019) 879−888.

[160] A. Marques, C. Jácome, J. Cruz, R. Gabriel, D. Brooks, D. Figueiredo, Family-based psychosocial support and education as part of pulmonary rehabilitation in COPD: a randomized controlled trial, Chest 147 (3) (2015) 662−672.

[161] IE ATSOC, IMARY, IMPRESS Guide to the Relative Value of COPD Interventions, 2012.

[162] S. Liu, Q. Zhao, W. Li, X. Zhao, K. Li, The Cost-Effectiveness of Pulmonary Rehabilitation for COPD in Different Settings: A Systematic Review, Applied Health Economics and Health Policy, 2020.

[163] P.K. Lindenauer, M.S. Stefan, P.S. Pekow, et al., Association between initiation of pulmonary rehabilitation after hospitalization for COPD and 1-year survival among medicare beneficiaries, Jama 323 (18) (2020) 1813−1823.

[164] I. Vogiatzis, C.L. Rochester, M.A. Spruit, T. Troosters, E.M. Clini, Increasing implementation and delivery of pulmonary rehabilitation: key messages from the new ATS/ERS policy statement, Eur. Respir. J. 47 (5) (2016) 1336−1341.

[165] R. Casaburi, Pulmonary rehabilitation: where we've succeeded and where we've failed, COPD 15 (3) (2018) 219−222.

[166] R. Kayyali, V. Savickas, M.A. Spruit, et al., Qualitative investigation into a wearable system for chronic obstructive pulmonary disease: the stakeholders' perspective, BMJ Open 6 (8) (2016) e011657.

[167] R.C. Wu, S. Ginsburg, T. Son, A.S. Gershon, Using wearables and self-management apps in patients with COPD: a qualitative study, ERJ Open Res. 5 (3) (2019).

[168] A. Angelucci, A. Aliverti, Telemonitoring systems for respiratory patients: technological aspects, Pulmonology 26 (4) (2020) 221–232.

[169] J. Buekers, P. De Boever, A.W. Vaes, et al., Oxygen saturation measurements in telemonitoring of patients with COPD: a systematic review, Expert Rev. Respir. Med. 12 (2) (2018) 113–123.

[170] H. Pinnock, B. McKinstry, Telehealth for chronic obstructive pulmonary disease: promises, populations, and personalized care, Am. J. Respir. Crit. Care Med. 198 (5) (2018) 552–554.

[171] P.P. Walker, P.P. Pompilio, P. Zanaboni, et al., Telemonitoring in chronic obstructive pulmonary disease (CHROMED). A randomized clinical trial, Am. J. Respir. Crit. Care Med. 198 (5) (2018) 620–628.

[172] D. Fuller, E. Colwell, J. Low, et al., Reliability and validity of commercially available wearable devices for measuring steps, energy expenditure, and heart rate: systematic review, JMIR Mhealth Uhealth 8 (9) (2020) e18694.

[173] K.R. Evenson, M.M. Goto, R.D. Furberg, Systematic review of the validity and reliability of consumer-wearable activity trackers, Int. J. Behav. Nutr. Phys. Activ. 12 (2015) 159.

[174] A. Vegesna, M. Tran, M. Angelaccio, S. Arcona, Remote patient monitoring via non-invasive digital technologies: a systematic review, Telemed. J. E Health 23 (1) (2017) 3–17.

[175] C.C. Cheung, A.D. Krahn, J.G. Andrade, The emerging role of wearable technologies in detection of arrhythmia, Can. J. Cardiol. 34 (8) (2018) 1083–1087.

[176] P. Pericleous, T.P. van Staa, The use of wearable technology to monitor physical activity in patients with COPD: a literature review, Int. J. Chronic Obstr. Pulm. Dis. 14 (2019) 1317–1322.

[177] D. Dias, J. Paulo Silva Cunha, Wearable health devices—vital sign monitoring, systems and technologies, Sensors 18 (8) (2018) 2414.

[178] K.J. Myall, A. West, B.D. Kent, Sleep and interstitial lung disease, Curr. Opin. Pulm. Med. 25 (6) (2019) 623–628.

[179] D. Withrow, T. Roth, G. Koshorek, T. Roehrs, Relation between ambulatory actigraphy and laboratory polysomnography in insomnia practice and research, J. Sleep Res. 28 (4) (2019) e12854-e12854.

[180] M. De Zambotti, N. Cellini, A. Goldstone, I.M. Colrain, F.C. Baker, Wearable sleep technology in clinical and research settings, Med. Sci. Sports Exerc. 51 (7) (2019).

[181] K.G. Baron, J. Duffecy, M.A. Berendsen, I. Cheung Mason, E.G. Lattie, N.C. Manalo, Feeling validated yet? A scoping review of the use of consumer-targeted wearable and mobile technology to measure and improve sleep, Sleep Med. Rev. 40 (2018) 151–159.

[182] A. Marques, A. Oliveira, Normal versus adventitious respiratory sounds, in: Breath Sounds, Springer, 2018, pp. 181–206.

[183] C. Jácome, A. Oliveira, A. Marques, Computerized respiratory sounds: a comparison between patients with stable and exacerbated COPD, Clin. Res. J 11 (5) (2017) 612–620.

[184] L.P. Malmberg, A.R. Sovijärvi, E. Paajanen, P. Piirilä, T. Haahtela, T. Katila, Changes in frequency spectra of breath sounds during histamine challenge test in adult asthmatics and healthy control subjects, Chest 105 (1) (1994) 122–131.

[185] A. Marques, A. Oliveira, C. Jácome, Computerized adventitious respiratory sounds as outcome measures for respiratory therapy: a systematic review, Respir. Care 59 (5) (2014) 765–776.

[186] A. Oliveira, J. Rodrigues, A. Marques, Enhancing our understanding of computerised adventitious respiratory sounds in different COPD phases and healthy people, Respir. Med. 138 (2018) 57−63.

[187] A. Marques, C. Jácome, Future prospects for respiratory sound research, in: Breath Sounds, Springer, 2018, pp. 291−304.

[188] D.L. Weeks, G.L. Sprint, V. Stilwill, A.L. Meisen-Vehrs, D.J. Cook, Implementing wearable sensors for continuous assessment of daytime heart rate response in inpatient rehabilitation, Telemed. J. E Health 24 (12) (2018) 1014−1020.

[189] J.M. Petersen, I. Prichard, E. Kemps, A comparison of physical activity mobile apps with and without existing web-based social networking platforms: systematic review, J. Med. Internet Res. 21 (8) (2019) e12687.

[190] J.A. Villanueva, M.C. Suarez, O. Garmendia, V. Lugo, C. Ruiz, J.M. Montserrat, The role of telemedicine and mobile health in the monitoring of sleep-breathing disorders: improving patient outcomes, Smart Homecare Technol. TeleHealth 4 (2017) 1−11.

[191] M.T. Bianchi, Sleep devices: wearables and nearables, informational and interventional, consumer and clinical, Metabolism 84 (2018) 99−108.

[192] G.E. MacKinnon, E.L. Brittain, Mobile health technologies in cardiopulmonary disease, Chest 157 (3) (2020) 654−664.

[193] U. Katwa, E. Rivera, Asthma management in the era of smart-medicine: devices, gadgets, apps and telemedicine, Indian J. Pediatr. 85 (9) (2018) 757−762.

[194] J.S. Marcano Belisario, K. Huckvale, G. Greenfield, J. Car, L.H. Gunn, Smartphone and tablet self management apps for asthma, Cochrane Database Syst. Rev. 2013 (11) (2013). CD010013-CD010013.

[195] C. McCabe, M. McCann, A.M. Brady, Computer and mobile technology interventions for self-management in chronic obstructive pulmonary disease, Cochrane Database Syst. Rev. 5 (5) (2017) Cd011425.

[196] K. Sleurs, S.F. Seys, J. Bousquet, et al., Mobile health tools for the management of chronic respiratory diseases, Allergy 74 (7) (2019) 1292−1306.

Chapter 3

Sensor technologies for mobile and wearable applications in mobile respiratory management

Josias Wacker, Benjamin Bonnal, Fabian Braun, Olivier Chételat, Damien Ferrario, Mathieu Lemay, Michaël Rapin, Philippe Renevey and Gürkan Yilmaz
Swiss Center for Electronics and Microtechnology (CSEM, Centre Suisse d'Electronique et de Microtechnique), Neuchâtel, Switzerland

Introduction

Instruments for the monitoring of respiratory functions play an important role in various settings, ranging from intensive care units to sport laboratories and even track fields. By measuring and computing key parameters like the breathing rate (BR), the forced expiratory volume (FEV), or the expired volume per minute, they support health-care professionals in taking the right therapeutic decisions or athletes in properly adapting their training. With the advent of ever smaller and ever lighter electronics it became possible to integrate the functions of formerly massive, stationary devices in garments which are comfortable to wear and easy to use and thus allow the recording of breathing-related signals in daily life and for a prolonged period. Such ambulatory monitoring improves the assessment of a person's health state — notably, it allows to react quickly in the case of exacerbations of chronically ill patients [1] and can even play a role in preventing the outbreak of infectious diseases [2].

This chapter presents the most important wearable and portable devices for monitoring respiratory functions. The coarse framework of this text follows the monitored organs; its finer structure is guided by the physics employed to record a physiological signal. For further reading, we refer to the following literature: Refs. [1,3,4].

Wearable Sensing and Intelligent Data Analysis for Respiratory Management
https://doi.org/10.1016/B978-0-12-823447-1.00006-3
59

Assessment of respiratory functions through monitoring of the lungs

The lungs are the central organ in breathing. Driven by the movement of the diaphragm and the intercostal muscles, they inflate when we breathe in (inspiration) and deflate when we breathe out (expiration). During each respiration cycle, the chest and the abdominal wall move in a characteristic pattern. As will be shown in the following paragraphs, these patterns can be recorded to monitor breathing with various noninvasive methods.

Optical methods

A very unobtrusive way of monitoring respiration is optoelectronic plethysmography (OEP) which is based on the optical observation of respiration-induced movements, e.g., through the use of cameras and reflective markers on the thorax [5] or laser distance sensors [6]. Typical features extracted from these optical signals include the BR, chest wall asynchrony [7], and other respiratory movement patterns [8]. Protocols and algorithms that only require cameras (and no markers on the body) reduce the needed preparatory efforts [9]. The presented studies were, however, done using stationary installations. The use of smartphone cameras [10] makes OEP portable, but requires monitored subjects to stay essentially motionless. In the following paragraphs, we present technologies through which respiration-linked body signals can be measured when subjects move and even during high physical activity.

Methods that measure changes of the circumference of the chest and the abdomen

Respiration-induced movements of the chest and of the abdominal wall have also been measured through the strain that these movements induce on tightly fitting chest and belly belts, T-shirts, and other suited garments for the upper body. These strain variations can be measured with various mechanoelectrical transducers:

(1) Transducers which change **resistivity** when stretched (piezoresistive strain sensors): These sensors exploit the fact that the initial electrical resistance R_0 of a conductive element changes (by ΔR) when its initial length L_0 is stretched by ΔL: $\frac{\Delta R}{R_0} = k_G \frac{\Delta L}{L_0}$, where k_G is the gauge factor. Pacelli and colleagues have proposed a tightly fitting T-shirt which contains flat knitted piezoresistive patches made of a conductive and an elastic yarn [11]. The resistivity of these patches changes when they are stretched due to the movements of the chest during a respiration cycle (Fig. 3.1). A similar concept has been proposed, where single stretchable

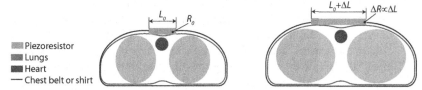

FIGURE 3.1 Working principle of resistive measurement of chest movements, shown in a schematic transverse cross section through the thorax: The resistance of a piezoresistive strain sensor which is integrated in a chest belt or a tightly fitting shirt varies during a breathing cycle with the circumference of the chest.

conductive yarns (in contrast to knitted patches) are used to follow respiratory movements [12]. In addition to these textile solutions, strain sensors have also been realized as metal thin films [13] which can be fixed with adhesive tape to precisely defined spots on the abdomen and thus allow measuring not only the BR, but also the respiration volume [14]. Like other techniques which follow the movements of the chest walls to measure the breathing, piezoresistive strain sensors are sensitive to movements which are not related to respiration [15]. In addition, their performance deteriorates over time, e.g., due to folding of the fabric and washing [16].

(2) Transducers which change **capacity** when being compressed: The core pieces of a capacitive sensor are two layers of a conductive material separated by an isolating material. The capacitance C of this construction is proportional to the inverse of the thickness d of the isolating material: $C \propto \frac{A}{d}$ (with A being the surface area of the conductive layers). When the isolating layer is compressible, a compression of the capacitive sensor therefore appears as a change in capacitance (Fig. 3.2). In a fully textile implementation [17], the conductive layers are made of a tissue containing 75% silver. For the isolating layer, foams and spacer fabrics have been tested. Spacer fabrics have been chosen in a final implementation since they have shown less hysteresis in capacitance when compressed

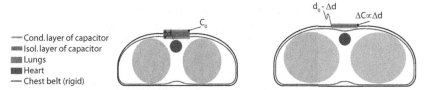

FIGURE 3.2 Working principle of capacitive measurement of chest movements, shown in a schematic transverse cross section through the thorax: The thickness d of a capacitive strain sensor which is integrated in a rigid chest belt varies during a breathing cycle with the circumference of the chest which leads to a measurable change in capacitance ΔC.

and released. In contrast to the resistive sensors, capacitive sensors require a relatively rigid chest belt, which holds the sensor against the chest, so that chest movements are directly converted into a compression of the sensor. The setup has been shown capable of monitoring BR. Even though it is theoretically possible to follow the volumetric expansion of the chest during breathing with a capacitive sensor, attempts to deduce the respiration volume from these measurements failed [17]. On the one hand, it is not sufficient to monitor only movements of the chest to estimate the volume change of the lungs, since breathing movements are not limited to the chest, but involve the full thorax (including the abdomen). On the other hand, measurements require a constant counterforce of the chest belt which is difficult to implement in practice.

(3) Measurements of changes in **inductance**: In so-called respiratory inductive plethysmography (RIP), sinusoid coils are placed around the chest and the abdomen. The voltage V_L induced by a current flowing through the coils is proportional to the change of the magnetic flux Φ caused by this current: $V_L \propto \frac{d\Phi}{dt}$. The magnetic flux varies with the changes of the cross section A of the chest and the abdomen during breathing (Fig. 3.3). RIP is frequently used to measure BR and volumetric parameters, for example, in sleep laboratories performing polysomnography recordings [18]. An example of a wearable with RIP-based breathing monitoring is the LifeShirt (Vivometrics, Ventura, US). A small cohort study showed that wearable RIP accurately measures the BR and can be used to monitor average tendencies in changes of the tidal volume and the minute ventilation volume [19]. For the measurement of absolute values, calibration of the wearable RIP device is necessary.

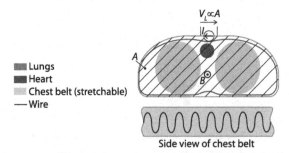

FIGURE 3.3 Working principle of inductive measurement of chest movements, shown in a schematic transverse cross section through the thorax: The voltage V_L, induced by the current I flowing through wire coils in a stretchable chest belt, depends on the cross-sectional surface A which changes during breathing.

Impedance-based methods

This section addresses the measurement of respiratory function using bio-impedance signals with a particular focus on the hardware aspects of this sensing technology. The challenges and potentials of wearable bioimpedance measurements are explained by the example of electrical impedance tomography (EIT), a complex, multichannel bioimpedance modality for the monitoring of regional lung function.

Measurement principle and safety considerations

The physiological process of breathing not only changes the volume of the lungs, but also their electrical conductivity and thus their electrical impedance [20,21], also denoted as bioimpedance [22]. The increase of air volume in the lungs during inspiration leads to a higher impedance which decreases again during expiration. These variations can be measured to assess the BR, to estimate the lung air volume, or even to monitor regional lung function, depending on the complexity of the measurement modality used.

Since the impedance Z is the only part of Ohm's law that cannot be measured directly, it must be derived from an excitation and a measurement. This excitation is usually a current I of known amplitude and the measured quantity the resulting voltage V. The impedance Z is related to V and I through Ohm's law: $V = Z \cdot I$. Analyzing the phase offset between the current and the voltage allows to differentiate between the resistive and capacitive part of the impedance, denoted as resistance R and reactance X, with $Z = R + jX$ [22,23], j being the imaginary unit.

To ensure a safe measurement for short- and long-term use, the amplitude of the injected excitation current is limited according to the international standards for medical devices [24]. A second aspect to consider for the selection of the excitation current properties (frequency and amplitude) is the electrical properties of the tissue. The capacitive nature of the cells [22,25] makes the 50–200 kHz frequency range most appropriate for bioimpedance measurements. In this frequency range the current will follow a path through the body that is partially intra- (at higher frequencies) and extracellular (at lower frequencies) [23].

For reliable bioimpedance measurements a tetrapolar setup is required. This is a four electrodes setup, where one electrode pair is used for stimulation (i.e., to inject the excitation current I) and the other pair for acquisition (i.e., to measure the resulting voltage V) [22,23]. The use of this setup ensures that there is no current flowing through the two electrodes measuring the voltage V, avoiding undesired voltage drops across the skin. This further requires these two electrodes to have a very high input impedance to limit potential leakage currents. With the tetrapolar setup a single bioimpedance channel can be acquired. Adding more electrodes allows for multichannel measurements as explained in the following section.

Single versus multichannel bioimpedance measurement modalities

Different bioimpedance measurement modalities exist ranging from simple single-channel systems (to estimate the BR) to complex multichannel systems (to make images of regional lung function).

Single-channel: impedance pneumography

One of the simplest applications of bioimpedance-based respiratory monitoring is the assessment of the BR using a single measurement channel. In this procedure — known as impedance pneumography [23] — a small current is injected through electrodes on the skin, e.g., electrodes used to record electrocardiograms (ECGs), to then derive the respiratory-induced impedance changes. When the electrodes are placed at dedicated body locations, not only the BR but also the respiratory volume can be assessed using bioimpedance measurements [23,26]. However, this requires prior calibration which is known to be affected by body posture [27].

Further applications of single-channel measurements — which are out of scope of the current chapter —include impedance cardiography for assessing cardiac function [23,28] and bioelectrical impedance analysis for estimating body composition [23].

Multichannel: electrical impedance tomography

By placing multiple electrodes around the body, multichannel bioimpedance measurements can be performed. The most common application of multichannel bioimpedance measurements is EIT [20,21,29] which is often used for the measurement of respiration-linked activities, since it allows for reconstruction of tomographic image sequences representing the temporal changes of the intrathoracic impedance. Notably, EIT allows recording without ionizing radiation CT images with high temporal resolution, thus providing figurative insights in the respiratory system's performance. As a downside, EIT has a lower spatial resolution than other CT modalities (e.g., X-ray, MRI) and the hardware is relatively complex (see illustration in Fig. 3.4 and explanations below).

In the remainder of this chapter, we mainly address these hardware challenges and the potential of EIT as sensing technology for the use in mobile and wearable applications. The interested reader is also referred to Chapter 6, which is entirely dedicated to EIT addressing aspects such as image reconstruction or EIT-based cardiovascular monitoring.

The currently commercially available EIT systems for clinical use (including Dräger's Pulmovista 500, Timpel's Enlight 1800, Sentec's LuMon, and Elisa 800 VIT from Löwenstein Medical) mainly target the monitoring and optimization of mechanical ventilation. Thus, these are stationary devices to be used at bedside. Most of these clinical EIT systems perform the EIT

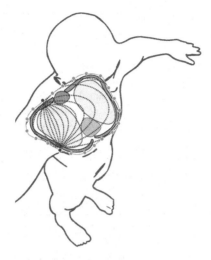

FIGURE 3.4 EIT working principle. Black *dots* represent electrodes which inject a current (*dotted lines*); *grey dots* represent electrodes which measure the resulting voltages. The classical EIT measuring method is a simple scaling of the basic tetrapolar impedance acquisition. The electrodes are multiplexed to perform current injections and voltage measurements according to predetermined patterns [29].

acquisition via a "classical" electrode arrangement, where all electrodes are wired individually to a centralized electronic system (see Fig. 3.5, left). This topology is derived from the classical 12-lead ECG. In the context of EIT, the higher number of electrodes and the need for extensive shielding (see next section) make the wiring challenging. While in clinical EIT systems this aspect has been solved by bundling wires in multicore cables or in flexible circuit board interfaces, this solution is not practical for wearable EIT systems

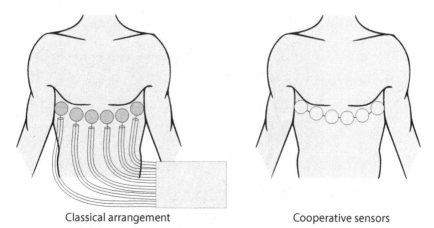

Classical arrangement   Cooperative sensors

FIGURE 3.5 Comparison between electrode arrangements of: (Left) a classical (centralized) and (right) a cooperative (distributed) EIT system.

targeted for mobile use since a multitude of wires may interfere with natural body movements and make a wearable system more rigid as well as more difficult to produce.

In the following, we will first review in detail the challenges of EIT hardware, then give an overview of existing wearable EIT systems, and finally introduce the concept of cooperative sensors (CSs), an architecture which allows connecting a multitude of sensor electrodes with minimal wiring effort without any loss in signal quality (see Fig. 3.5, right).

Technological challenges of bioimpedance measurements for electrical impedance tomography

In EIT, impedances and impedance variations of organs (located inside the body) are measured from the body surface which is challenging for various reasons. The impedances of organs and their amplitude variations are relatively small compared to the impedance of the skin and compared to the variation of the impedance due to changes in electrode contact pressure (i.e., the pressure at which the electrode is pressed against the skin). To compensate for that, the measuring electrodes must have a very high input impedance so that no disturbing current flows between the electrodes and the skin, potentially inducing disturbances in the voltage measurement.

Fig. 3.6 illustrates a tetrapolar impedance measurement setup with no measures taken for noise mitigation. There are three main problems [30] related to this setup:

(1) The voltage measuring electrodes couple capacitively to external disturbances and induce a voltage across the electrode–skin interface. Note that the frequencies of the disturbances (e.g., mains hum at 50 Hz)

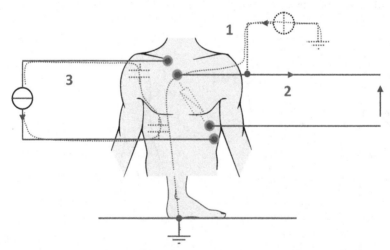

FIGURE 3.6 Basic tetrapolar body impedance measurement setup. The numbers in the figure correspond to the three main metrological problems of the setup listed in the text.

are typically well separated from the measurement frequencies (which are made, for example, at 50 kHz). Yet, external disturbances are problematic since they easily saturate amplifiers of the measuring circuit.

(2) The input impedance of the measuring electrodes is finite and thus the electrode draws a small current from the body resulting in an induced voltage across the electrode—skin interface.

(3) The current injecting electrodes couple capacitively to the body, offering the excitation current an alternative path, resulting in a current discrepancy through the impedance under test.

Problems (1) and (3) are usually solved by shielding wires and problem (2) by increasing the input impedance through an active supply of the acquisition input bias current [31]. An improved setup solving the abovementioned problems (1) to (3) is shown in Fig. 3.4 which also reveals the high importance of shielding of the cables.

Driven cable shielding is necessary at various locations. First, for the voltage acquisition (right side of body in Fig. 3.7), to prevent any capacitively induced current from flowing through the electrodes, creating a voltage disturbance. Second, for the current injection (left side of body in Fig. 3.7), to guarantee that all current is forced through the electrode and no capacitive loss occurs. As also illustrated in Fig. 3.7, proper shielding of a current does not only require single- but double-shielded cables [32]. This is because a single shield to ground would allow capacitively coupled losses through the grounded shield, while a single driven shield would create unwanted capacitively induced current through the body. To solve this issue, an inner driven shield and an external grounded shield must be used simultaneously to perform the

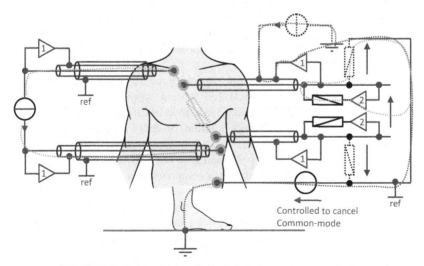

FIGURE 3.7 Improved tetrapolar body impedance measurement setup.

task of forcing an exact amount of current through the injection electrodes only [31]. The depicted solution for increasing the input impedance consists in applying the measured voltage through an impedance of the same value to that of the acquisition input impedance. Thanks to this method (also known as "bootstrapping") [33], none of the acquisition bias current is provided through the electrode—skin interface.

When using a classical electrode arrangement (Fig. 3.5, left) each cable connecting the electrodes to the central electronic system requires double shielding which results in a bulky system that is inappropriate for wearable EIT systems targeted for mobile use.

Wearable electrical impedance tomography systems

This section treats only the impedance part of multisensor systems. For more information on other measuring modalities of multisensor systems, please refer to the section Multimodal systems for parallel monitoring of several organs.

Different approaches to overcome the aforementioned issues with classical, stationary EIT devices were undertaken in the past years. A variety of wearable EIT systems were developed which are listed in Table 3.1. All these systems use an active electrode circuit which — like shielded cables — minimizes the noise induced by capacitive coupling with the external environment. Since the analogue signals are amplified and digitized close to each individual EIT electrode, no driven shielding of cables is needed and thus the wiring of the sensors can be simplified when compared to traditional EIT systems.

These wearable systems differ from the clinical EIT systems in that they target another use case: the ambulatory and long-term monitoring of cardio-pulmonary diseases. This results in new requirements for this type of EIT systems. First, the ergonomics of the systems need to address the long-term comfort and ease of use for the patient. Second, the systems need to allow for battery operation over a long period of time. Finally, wireless and autonomous data communication shall enable medical staff the remote monitoring and diagnosis of ambulatory patients.

Based on these new requirements mainly three strategies emerged. First, the miniaturization of the classical EIT system topology with the starlike arrangement connecting the electrodes via individual cables to a centralized EIT system [34,35]. Second, a hybrid solution of daisy chaining the sensors [36—38]. And finally, a bus — or parallel — interconnection of the sensors [39—41]. The CS approach [42] — detailed in the next section — is a refined version of the third strategy where the bus comprises only two unshielded wires. Leveraging only two wires for a bus opens a world of possibilities in terms of the ergonomics of a wearable system. For instance, in a scenario where both faces of a belt or shirt are coated with a conducting layer (see Fig. 3.8), it enables great freedom in size, shape, sensor count, and sensor placement.

TABLE 3.1 State of the art of wearable EIT-based respiratory monitoring systems.

Wearable EIT for lung function monitoring	Swisstom 2012 [36]	KAIST 2014 [34]	KAIST 2 2015 [35]	KAIST 3 2017 [39,40]	CRADL v1.0 2018 [38]	WELCOME 2018 [41]	CRADL v2.0 2019 [37]
Number of electrodes	32	32	32	3 × 16	32	8 current +8 voltage	16
Max frame rate (fps)	30	5	30	70	107	80	122
Number of wires	16	32	32	(4 + 3) per row	4 + 9 + 2 (estimated)	2	6
Interconnecting topology	Daisy chained	Star	Star	Parallel	Daisy chained	Parallel	Daisy chained
Active acquisition (High impedance input path < 5 cm)	Yes	No	No	Yes	Yes	Yes	Yes
Local digitalization	No	N/A	N/A	Yes	No	Yes	No
Chip technology	Discrete	180 nm	180 nm	65 nm HV	350 nm	Discrete	350 nm HV
Local current excitation	No	N/A	N/A	Yes	Yes	Yes	Yes
Multi-injection stimulation	No	No	No	No	No	Yes	No
Complex I&Q impedance acquisition	Yes	Yes	Yes	Yes	Yes	No	Yes
3D acquisition	No	No	No	Yes	No	No	No
Power supply	Remote DC	N/A	N/A	Remote DC 3.7 V	Remote DC ±9 V	Local DC 3.7 V	Remote DC ±9 V

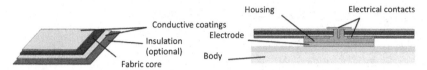

FIGURE 3.8 Example of a dual-sided conductive fabric and its interface to a cooperative sensor.

The wearability and proper fit of an EIT system is of paramount importance for its performance since the signal quality relies on the electrode preload and alignment against the skin.

Cooperative sensors for electrical impedance tomography

CSs leverage electronics that are local to each individual EIT electrode (see illustration in Fig. 3.2, right) without the necessity of having a centralized electronic system which interconnects each electrode with double-shielded cables [30]. Instead, CSs work in concert to acquire EIT data [31,42,43]. To do so, their synchronization and powering as well as concentration of acquired data are achieved by an interconnecting bus composed of two unshielded wires. The CS technology — as developed by Centre Suisse d'Electronique et de Microtechnique (CSEM) — offers the possibility for an ambulatory EIT system with high signal quality. By limiting the bus to solely two unshielded wires, the integration of a CS-based EIT system into textiles is highly facilitated [42,43]. This strategy not only improves the wiring but also makes the high impedance input immune to capacitive coupling, thanks to its "zero" length input path. Finally, the maximum achievable input impedance can be high enough so that CSs are able to acquire EIT signals using dry electrodes (usually made of a simple stainless-steel disk). In contrast, classical EIT systems use gel electrodes or conductive liquids to minimize the artifacts induced by the skin—electrode interface.

A CS-based EIT system has the advantage that it can easily be extended to a multimodal sensor system for the simultaneous acquisition of additional signals including ECG, lung sounds, etc. [42,43] (see Section Multimodal systems for parallel monitoring of several organs).

Assessment of respiratory functions through monitoring of the airways

This section provides an overview of the sensor technologies employed to assess the health of the respiratory system via monitoring of the airways and their relevant use cases. The sensor technologies can be categorized into two groups: acoustic and airflow based. Acoustic methods mainly refer to the systems based on a microphone or stethoscope, thus converting pressure waves into electrical signals. Airflow-based methods typically employ spirometers.

Acoustic methods

The turbulent flow of air entering and leaving the lungs creates sounds which fall in the human audible range. In the presence of certain respiratory diseases, these sounds tend to have specific patterns that can be used as biomarkers for diagnosing and monitoring the diseases. For instance, *wheezes* are characterized by high-frequency components and associated with obstructions in the respiratory tract, whereas *crackles* are characterized as discontinuous popping sounds associated with explosive opening of airways [44]. Since the invention of the stethoscope about 200 years ago, it is a common clinical practice to listen to these abnormal sounds, also known as adventitious sounds, for diagnostic purposes, a procedure known as auscultation.

Conventional stethoscopes (or acoustic stethoscopes) are designed to acquire sounds via chestpieces, which have different shapes optimized for cardiac and respiratory auscultation purposes, and transmit the sounds via tubes to the practitioner's ears. This method, however, remains subjective since it relies on the auscultation skills which depend on training, the number of different cases encountered, and eventually the health of the auditory system of the practitioner. Additionally, conventional acoustic stethoscopes do not allow for the recording or streaming of these audio signals, thus demanding a second opinion on the sounds acquired during a visit is not possible. With emerging sensor technologies and the trend toward digitalization in health care, electronic stethoscopes have been introduced to address this need [45]. Electronic stethoscopes have not only enabled recording of the auscultation data but also challenged the form factor of acoustic stethoscopes by means of wearable stethoscope heads, and created an opportunity to process the acquired sound signals for supporting clinicians' diagnosis. In line with the scope of this chapter, we limit the examples to mobile and wearable applications and focus on the sensing technologies behind these examples.

Air microphones

Due to their intuitive nature, air microphones are common practice for wearable stethoscope developments [46–48]. As the name indicates, sound waves propagate in the air, and instead of directly reaching the listener's ear as in the case of conventional stethoscopes, they are captured by an air-coupled microphone. While certain applications employ micromachined capacitive or piezoelectric microphones to lower the cost [46], electret condenser microphones provide a higher SNR (up to 80 dB) [47]. An imminent drawback of air microphones is their inherent susceptibility to capture ambient noise as well. Unless a very rigid wall or multilayer sound absorption solutions are incorporated, ambient noise penetrates through the sidewalls into the air chamber and significantly decreases the signal quality. On the other hand, air microphones inserted into the stethoscope heads benefit from the optimized mechanical design to guide the pressure waves. The work by Kraman and colleagues [49] provides a detailed analysis on the effect of microphone air cavity width, shape, and venting to guide this design effort.

Contact microphones

The central element of the sensor module, the acoustic transducer, can also be realized by means of a contact microphone [50–53] (Fig. 3.9). While air microphones necessitate a diaphragm [54], contact microphones can directly be interfaced with the skin. More precisely, sound waves can be coupled to the sensor via a solid or liquid medium rather than air. Contact microphones, if designed properly, can provide a better isolation from ambient noise as their operation principle requires matching specific acoustic impedances of the transducer element and the source [55].

In summary, incorporating air microphones into electronic stethoscopes benefits from the legacy of the conventional stethoscopes such as established mechanical and acoustic design and reduces the design and development time. Nevertheless, the low-frequency (<100 Hz) performance of the microphone could be a limiting factor as most low-cost microphones on the market are developed for speech applications, commonly starting from 100 to 150 Hz rather than 20 Hz which is known as the limit of the human audible range. Contact microphones, on the other hand, allow much higher imped-ance matching, which can only be achieved with unpractical dimensions for air microphones. Consequently, the sound waves reach the sensors with minimum attenuation improving the signal-to-noise ratio. This high

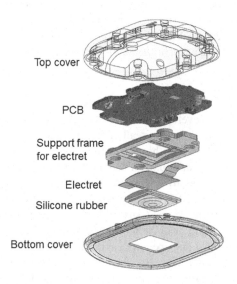

Top cover

PCB

Support frame
for electret

Electret

Silicone rubber

Bottom cover

FIGURE 3.9 Exploded view of a multimodal sensor including a contact microphone realized with a silicone rubber and an electret.

performance, however, requires a custom multiphysics design, i.e., prolonged design and development time.

Although sensor technologies are evolving very rapidly and various products are readily available on the market for the electronic stethoscopes, in the context of wearable solutions, two issues remain unsolved: positioning and stabilization of the sensors in the absence of a clinician. This is obviously not a primary concern for electronic stethoscopes as they are used by the clinicians themselves. Positioning of the transducers on the chest is therefore of vital importance, as it defines which respiratory sounds are captured. Stabilization of the sensors is particularly essential for applications where the patients are monitored for prolonged durations and when they are mobile as it reduces the low-frequency noise stemming from movement artifacts. Addressing these two questions will lead to reliable respiratory health monitoring solutions which are easily deployable and useable by the patients while they continue their life as usual.

Spirometers

Spirometers, the second type of airway-monitoring devices, measure volumes and flow rates of breathed air. Literature distinguishes between closed and open spirometers. While the former require a closed volume into which a subject is breathing, the latter do not have such volume and allow users to freely exchange air with the surroundings. Open spirometers are therefore better suited for integration in portable devices. For the sake of completeness, we are also introducing closed spirometers. For further reading, including minimum recommendations for the construction of spirometers and test procedures, we refer to Refs. [3,56−59].

Closed-circuit spirometers

Early spirometers were based on the principle of displacement of water by inhaled and exhaled air. In modern versions of these so-called wet spirometers, subjects breathe through a hose which is connected to an air chamber that is partially submerged in water (Fig. 3.10). The vertical movements of the air chamber are then a direct measure of the air moved by the lungs. Hutchinson's version of this apparatus is worth mentioning because it allowed him to define and study the vital capacity, a physiological measurement which is still used today for the diagnosis of respiratory diseases. Today, wet spirometers are mostly used for teaching purposes (e.g., Student Wet Spirometer from Phipps&Bird). Dry closed-circuit spirometers consist of a piston which moves inside a cylinder. The seal between the piston and the cylinder is made by a flexible plastic sheet which rolls on itself. The third kind of volume measuring mechanical spirometers are bellows-type spirometers. Here, subjects breathe into the chamber of a wedge-bellow. One side of the bellow is kept fix, while

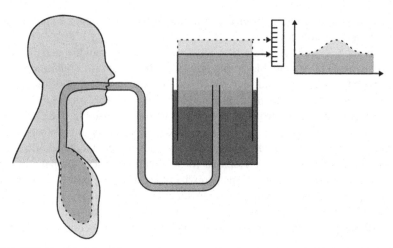

FIGURE 3.10 Principle of the wet spirometer: a subject breathes in an air-filled chamber which is partially submerged in a water tank and moves with the displaced air volume.

the other side can move freely and thus indicates the displacement of breathed air. Wet, dry, and bellow-type spirometers are typically bulky, not wearable devices and have been replaced in clinical practice by smaller open-circuit flow-measurement spirometers.

Open-circuit spirometers

Open-circuit flow-measuring spirometers (pneumotachographs) do not require as much space as volume-measuring spirometers and therefore lend themselves to the integration in portable, handheld devices. Because they are open, air volumes such as (forced) vital capacity and FEV need to be deduced by integration of the measured air flows. The structure of the following paragraphs which present different types of open-circuit spirometers is given by the measuring principle employed in the devices.

The most straightforward realization of a flow-measuring spirometer measures the rotation of a turbine which is driven by inhaled or exhaled air. The speed of the turbine can then be read out by optical means. The inertia and the deformation of the moving parts limit, however, the temporal precision of turbine-based spirometers. Advantages of turbines are, on the other hand, their insensitivity to the composition and humidity of the exhaled gas [57]. Due to their small size and light weight, turbine systems are often integrated in handheld devices which are easy to use at primary care points or even at home, e.g., to train lung functions or to prevent exacerbations (e.g., MIR's Smart One). Turbines are also integrated in portable systems for assessing athletes' energy expenditure like Cosmed's K5 Metabolic system and Cortex's spiroergometers (MetaMax). Note that these calorimeters contain additional sensors, e.g., for measuring inhaled oxygen and exhaled carbon dioxide [60].

Differential pressure flowmeters are an interesting alternative to turbine-based sensors, as they do not contain moving parts. In the context of spirometers, devices based on the measurement of differential pressures are also called pneumotachometers.[1] They exploit the observation that the static pressure of a fluid current drops when the speed of the fluid increases. In pneumotachographs, an increase in fluid velocity is achieved by integrating a flow resistor (e.g., a porous membrane or parallel capillaries) in a short tube through which subjects breathe. Note that a direct (linear) relation between the flow volume and the pressure drop is only valid for laminar flows. Therefore, pneumotachographs are formed in a way to reduce turbulences as much as possible. The accuracy of pneumotachographs is mainly limited by obstructions of the flow resistors (e.g., by spittle or condensing air). In some constructions, the resistors are therefore heated (which in turn changes the density of the air and other variables and overall does not increase the accuracy) [61]. An example of a handheld pneumotachograph is the In2itive Spirometer from Vitalograph.

Hot wire anemometers have the advantage of being purely electrical. In this type of spirometer, a wire which is heated by an electrical current flowing through it is placed in the flow of breathed air (see Fig. 3.11). The airflow cools the wire by convection and thus changes its electrical resistance. This change in resistance is then a direct measure for the speed of the airflow. The main benefit of hot wire anemometers is the simple readout and automation. The disadvantages are the nonlinear dependency between the airflow velocity

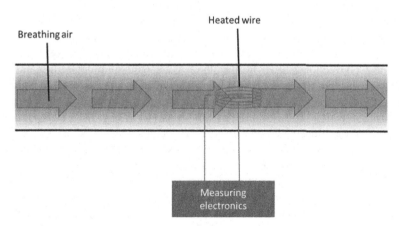

FIGURE 3.11 Working principle of the hot wire anemometer. The flow of breathing air cools a heated wire whose resistance depends on its temperature. The measuring electronics records the changes in wire resistance and thus in the speed of airflow.

---

1. Sometimes, the term "pneumotachograph" is used for any type of open-circuit spirometer.

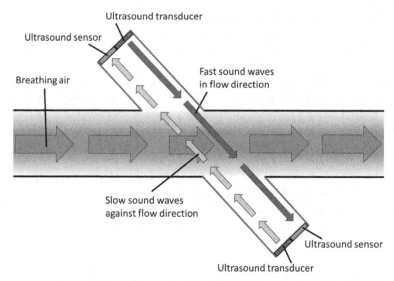

Ultrasound transducer

Ultrasound sensor

Breathing air

Fast sound waves
in flow direction

Slow sound waves
against flow direction

Ultrasound sensor

Ultrasound transducer

FIGURE 3.12 Working principle of the ultrasonic flow meter. Ultrasonic sound waves emitted and received by transducers and sensors, respectively, propagate faster in the direction of the flow of breathing air than in the opposite direction. The difference in travel time is directly linked to the speed of the breathing airflow.

and the resistance of the heated wire, the sensitivity of the setup to electrical resistances in the overall circuit, as well as their insensitivity to the flow direction (inspiration vs. expiration). Sendsor's Lufttacho is based on a hot wire anemometer.

The spirometers presented so far measure a substitute signal for the flow (the rotational speed of a turbine; the pressure drop caused by a constriction; the heat convection caused by the flow). Ultrasonic spirometers in contrast measure directly the speed of air in a tube. In short, they are based on the principle that sound waves travel faster in the direction of flow than in the opposite direction. The working principle is shown in Fig. 3.12: Two pairs of ultrasound transducers and sensors are arranged at an angle with respect to the flow direction. For each pair of transducers/sensors, the time between sound emission and sound reception is measured. The faster the air flows in the spirometer tube, the bigger the difference is in sound speed. Ultrasonic spirometers have none of the other spirometers' drawbacks. However, they are relatively expensive. As an example, the calorimeter MedGem from microlife uses ultrasound to measure the speed of inspired and expired air [62].

In conclusion, open-circuit spirometers allow compact integration in handheld devices. Since they assess breathed air volumes through the integration of the measured flow rates, even small errors quickly sum up, and the accuracy of the final result depends critically on the solid calibration of the device.

Assessment of respiratory functions through monitoring of the cardiovascular system

Due to the tight physiological links between the respiratory and the cardiovascular system, breathing functions can be monitored based on cardiovascular signals. In the following sections, we present how ECGs are measured in wearable applications and how they contain information on breathing. We then show how similar information can be extracted from photoplethysmography (PPG) signals. Finally, we present pulse oximeters, which use PPG to assess the oxygen saturation in the blood.

Electrocardiogram

ECGs are the classical method to assess the health state of the heart. They contain a wealth of information, not only pertaining to the cardiovascular system, but also to respiratory functions, as we show below.

Electrocardiogram measurement basics

An ECG is a graphical recording of the electrical signal originating from the depolarization and repolarization of the cardiac muscle. The size and the shape of the ECG signal, as well as the time intervals between the various peaks, contain useful information about the health and the function of the heart [63]. In addition, the baseline wander, amplitude modulation, and frequency modulation of the ECG signal can be used to assess respiration-related information, as will be explained below [64,65].

ECGs are classically measured on the body surface. The amplitude of the measured voltage is in the order of 1 mV [66]. Given the small amplitude of the signal, accurate amplification is necessary. Therefore, the amplifier must have good performances in terms of linearity, absence of drift, noise, and interference rejection [67]. In practice, wearable ECG monitoring systems — also called Holter monitors — require a low bias current of the analog frontend to guarantee a good measurement quality. Indeed, the bias current flowing through the skin impedance generates a voltage drop. Consequently, any changes in the skin impedance values, for instance, due to motion in the electrode—body interface, may generate artifacts on the measured voltages [68]. It is therefore important to minimize both the skin impedance and the bias current of the analog frontends. Another issue related to the sensing circuits is the voltage divider embodied by the skin impedance and the input impedance of the voltage-sensing analog frontend [69] which attenuates the measured ECG signal. The mismatch between the skin impedances and the input impedance of the different sensing channels also affects the common-mode rejection of the system. For this reason, it is important to maximize the input impedance of the analog frontends, and to minimize the skin impedance.

Processing of electrocardiogram signals to assess respiratory function

ECG measurements can be used to indirectly monitor the respiratory function. This indirect assessment is possible due to the following phenomena. The respiratory process induces a modulation of the relative intrathoracic pressure through the interaction of the intercostal muscles and the diaphragm. During inspiration, the volume of the thorax is increased resulting in a decrease of the internal pressure allowing air to fill the lungs. During expiration, the opposite mechanism takes place. These changes in the shape of the thorax also affect the electrical conduction from the heart to the surface electrodes and result in respiration-induced modulations which can be observed in the ECG signal [64]. The modulations include changes in the baseline of the ECG signal (baseline modulation or baseline wander) and in the amplitude (amplitude modulation). A third type of observable respiration-induced modulation is caused by the regulation of the autonomous nervous system (ANS). During inspiration, the relative drop in intrathoracic pressure also increases the venous return (deoxygenated blood returning from the body to the heart through the venous system). As more blood is available in the venous system the volume of blood ejected by the heart increases, producing an increase of blood pressure. As a result, the baroreceptors located around the aorta send a nervous signal to the brain where the ANS increases the heart rate to counteract the increase of blood pressure (the reduced time between consecutive cardiac contractions reduces the ejected blood volume allowing the blood pressure to return to homeostasis values). During expiration the opposite occurs: increase of intrathoracic pressure, reduction of venous return, reduction of ejected blood volume, drop in blood pressure, and decrease in heart rate. The time series representing the duration between consecutive contractions of the heart (also known as RR intervals) thus contains − among other components − a modulation which is related to the respiratory function (frequency modulation). In contrast to the first two types of modulation (baseline and amplitude) which are related to electromechanical changes, the frequency modulation − also known as heart rate variability (HRV) − is limited by the implication of the ANS control and not necessarily always observable. If the BR is higher than the ANS bandwidth (typically above 30 breaths per minute) the frequency modulation of the RR intervals will not be observable. Similarly, if the subject suffers from cardiac arrhythmia, abnormal blood pressure, or other dysfunctions of the cardiovascular system the indirect observation of the BR through frequency modulation (of the RR intervals) can be significantly impaired.

The estimation of the BR from these three types of respiration-induced changes (baseline, amplitude, and frequency modulations) implies a two-step processing. First, features of interest must be extracted from the ECG signal which are specific to the type of modulation to be assessed:

(1) The modulation of the baseline is obtained by either low-pass filtering the ECG signal or by detecting fiducial points in the signal such as the onset of the ST segment.

(2) The modulation of the amplitude requires the extraction of two fiducial points from the signal (generally the R peak and the ST segment onset) and to subtract their amplitudes.

(3) The modulation of the frequency implies the detection of the R peaks and the calculation of the time interval between consecutive heartbeats.

The second processing step is identical for the three different types of modulation. It includes the rejection of nonphysiological values in the time series (outlier detection and rejection). The time series are then uniformly resampled and filtered to restrict the frequency band to possible respiratory frequencies and reject nonrespiratory-related phenomena. Finally, the frequency spectrum of the resulting time series is analyzed to determine the spectral peak that corresponds to the BR. The review by Charlton and colleagues [64] provides an excellent overview of these methods together with a list of available algorithms which were also assessed in their previous work [70].

Photoplethysmography

PPG devices present a promising opportunity for unobtrusive and long-term cardiopulmonary monitoring [71] since they can be worn over several days. In the context of wearable monitoring, PPG has the additional advantage over methods based on ECG and impedance measurements that no cabling of sensors is needed since measurements can be done on one single spot. The output of a PPG device is the photoplethysmogram, a signal that originates from the circulation of blood and the change of its volume in tissues [72]. This signal is obtained noninvasively with a system composed of a light emitter (LED) and a light receiver (photodiode). The light penetrating tissues are partially absorbed by the blood and the resulting signal is a function of blood volume and composition. The exact origin of PPG signals (changes of light propagation in tissue) is still subject of investigation. Yet it is known that the PPG signal has a strong morphological similarity with the blood pressure waveform [73,74]. Besides, several factors may affect or deteriorate PPG waveforms. These include skin properties, device-skin pressure [72,75], and postural changes [76].

The PPG technology is best known from fingertip pulse oximeters where it is used to estimate blood oxygen saturation (see next Section: Pulse oximetry). These pulse oximeters traditionally work in transmission mode (applicable at the finger, toe, or ear) which results in a higher signal quality when compared to reflectance mode sensors (applicable at various locations including the wrists or the forehead) [71]. However, the latter have the big advantage that they can be placed at virtually any body location rendering the measurement even more unobtrusive. With the advent of smartwatches and fitness trackers the PPG technology has experienced a strong boost. Besides simply measuring heart rate these devices allow for the assessment of other cardiovascular parameters including HRV [71], blood pressure [77,78], or respiration [64,79,80], as detailed in the following paragraph.

The PPG-based assessment of respiratory activity is based on the same three types of modulation as observed for ECG [64]: (a) baseline modulation; (b) amplitude modulation; (c) frequency modulation. However, in particular for (a) and (b), the resulting respiration-induced modulations are not necessarily comparable between PPG and ECG. This is because the ECG signal reflects electromechanical changes while the PPG signal originates from cardiovascular changes in the peripheral blood vessels [71]. The interested reader is referred to the review paper by Charlton and colleagues [64] and their performance assessment of different algorithms [70]. More recently, PPG-derived respiration was tested on wearable PPG devices — such as smartwatches — which showed the feasibility for nocturnal monitoring of BR [79,80] which could pave the way toward an ambulatory and thus early screening of respiratory sleep disorders or other pulmonary diseases.

Pulse oximetry

Monitoring of peripheral oxygen saturation (SpO_2) by pulse oximetry is used to estimate arterial blood oxygen saturation (SaO_2) and provides vital information about a patient's respiratory function [81]. The commonly used "unit" for oxygen saturation is percent, indicating the amount of oxygen-saturated hemoglobin relative to the total amount of hemoglobin in the blood. The noninvasive nature of pulse oximetry and its advantage of providing continuous monitoring has led to its widespread and growing popularity since its commercialization in the early 1980s. It is now universally used for monitoring patients in a wide range of clinical applications, such as emergency medicine, intensive care, anesthesia, postoperative recovery, and sleep apnea monitoring [82] (see Chapter 9).

Pulse oximetry uses two PPG wavelengths — red (660 nm) and infrared (940 nm) — to differentiate oxyhemoglobin (HbO_2) and deoxyhemoglobin (Hb) concentration of the illuminated tissues. The PPG signal pulsation (AC) is used to distinguish arterial blood absorbance from other absorbers: skin, tissues, bones, and other fluids reflected in the static part of the signal (DC).

For both wavelengths, the pulsatile amplitude is normalized to its value at end-diastole, and the ratio of both wavelengths is computed. This ratio of perfusion indices (AC/DC) is then calibrated against an invasive measurement of SaO_2 to obtain the SpO_2 [83].

The accuracy of pulse oximetry should be evaluated as part of a clinical evaluation involving a controlled desaturation study taken over the full range over which the accuracy is claimed [84]. According to the ISO 80601-2-61 standard, a root-mean-square difference (A_{rms}) between the estimated SpO_2 and a reference SaO_2 obtained via a CO-oximeter shall be less than or equal to 4%: $A_{rms} \leq 4\%$, and the FDA recommends typical A_{rms} values of 3% for transmission pulse oximeters, and 3.5% for ear clip and reflectance sensors [85]. The accuracy claimed by conventional pulse oximeter devices involving transmission PPG is, however, smaller and typically below 2% ($A_{rms} \leq 2\%$) over the range of 70%–100%. Pulse oximeters based on a two-wavelength approach can distinguish oxyhemoglobin and deoxyhemoglobin but cannot identify carboxyhemoglobin (COHb) or methemoglobin (MetHb). Another limitation was revealed in a recent study where pulse oximeters showed a biased result for people with darker skin color [86].

Classical pulse oximeters are based on transmission PPG requiring the light to go through the tissue. As discussed above, reflectance pulse oximetry can be placed on virtually any cutaneous surface of the body. However, compared to transmission pulse oximetry, reflectance-based SpO_2 is challenging due to the following reasons [71]:

(1) Lower level of the pulsatile component, which depends on the perfusion of the underlying tissue [87],
(2) Difference in the optical path between red and infrared wavelengths [88],
(3) In some circumstances, a venous pulsation may occur and disrupt the SpO_2 estimation [89].

The majority of the companies providing medical pulse oximeters now also offer a reflectance version allowing the monitoring of SpO_2 at the forehead with accuracies similar to those of conventional fingertip sensors [90,91]. More recently, bracelets, watches, and most of the smartwatches on the market incorporate a pulse oximeter sensor and provide an estimate of SpO_2, with the apparent benefit of being worn on the wrist (e.g., Garmin, Apple Fitbit, Oxitone, and Biostrap).

Guber et al. evaluated the first FDA-cleared wrist-located pulse oximeter (Oxitone-1000, Oxitone Medical) against a traditional fingertip pulse oximeter on 15 healthy volunteers and 23 patients, and reported an $A_{rms} < 3\%$ [92]. Proença et al. have developed and validated a reflectance pulse oximeter sensor both at the chest with an A_{rms} of 3.1% and at the forehead with an A_{rms} of 1.9% (in 8 healthy adults) and at the forehead of 25 neonatal patients with an A_{rms} of 3.9% [90]. A similar approach was used in 57 subjects to assess SpO_2 at the wrist during sleep with $A_{rms} = 2.7\%$ [93] or at the ear of 20 hypoxemic patients with chronic respiratory diseases showing an $A_{rms} = 2.5\%$ [94].

Except for Oxitone, most of these devices do not have a medical certification and provide an SpO_2 estimate with "Fitness Tracker Accuracy" not intended for medical purposes. This highlights that smartwatch-derived SpO_2 estimates should be interpreted with caution and rigorous clinical and ambulatory validation studies are required in the future to avoid unnecessary doctor visits and false alarms.

Multimodal systems for parallel monitoring of several organs

As shown in the previous sections, respiration includes a multitude of organs of which all can be affected by respiratory diseases. An ideal disease management therefore monitors the health state of several organs in parallel. Traditional apparatuses which measure several body functions concurrently are relatively bulky (e.g., the polysomnograph used in sleep laboratories). In recent years, various groups have undertaken efforts to integrate sensors which measure a variety of physiological signals in simple garments, patches, and other wearables. We are going to present some of these systems, focusing on the connection methods between the sensors.

The most straightforward method to connect distributed sensors is to cable them to a central box (see illustration in Fig. 3.5 left and Section Multichannel: electrical impedance tomography). Wearables which implement this principle with a multitude of sensors for monitoring various physiological parameters have been developed for more than a decade. A few examples of commercial multiparameter systems following this approach are listed in Table 3.2.

TABLE 3.2 Commercial multimodal systems.

Name	Measured signals	Producer
MicroPaq	Up to 2 ECG leads, SpO_2	Welch Allyn (withdrawn)
ApexPro	Multi-lead ECG, SpO_2, blood pressure	GE Healthcare
LifeShirt	ECG, RIP, possibility to extend with additional devices	Vivometrics
EX eqO2+ LifeMonitor	ECG, BR, body activity, skin temperature	Equivital
BioHarness	ECG, BR, body activity	Zephyr activity

Another approach to measure signals from distant spots of the body is to distribute individual devices over the body and to assure wireless communication (e.g., via Bluetooth) to a central computer. Note that this approach is not possible if a galvanic connection between the distant spots is needed (e.g., ECG, impedance measurements). An example for a wireless system is presented in Ref. [95]. This research demonstrator collects environmental and physiological data which is relevant to asthma patients. It consists of three devices: (1) a wrist band which measures air temperature and humidity, the level of ozone and volatile organic compounds in the air, as well as the heart rate through PPG; (2) a chest patch which is equipped with local skin impedance, local ECG, strain, and PPG sensors as well as an accelerometer and a microphone; (3) a spirometer which measures the flow of breathed air. The combination of physiological and environmental data is promising for future epidemiological studies and to warn persons at risk from dangerous situations. The three devices act, however, as standalone instruments, without the possibility for communication between them.

The patch concept allows integrating many sensors in a single device. While patches have been mostly used to measure ECG only, Orglmeister's group presented a patch which records a 1-lead ECG, thoracic impedance, and chest sounds from a stethoscope [46]. By combining the signals, other parameters were estimated including BR, additional ECG leads, and indicators pertaining to the functioning of the heart.

To achieve simple sensor-to-sensor communication, we have developed a new architecture of wearable sensor network, the CS (see Section Cooperative sensors for electrical impedance tomography). In contrast to the systems described above, CS can measure signals from distant parts of the body even for measurement modalities which require a galvanic connection (e.g., ECG, EIT). In addition, the architecture lends itself to the extension to a wide variety of physiological signals, without adding any complication in the connection between the sensors (see illustration in Fig. 3.5, right). The first system based on CS was the Sense vest (see Fig. 3.13) which contains two sensors that are equipped with an accelerometer and measure a 1-lead ECG, the BR through transthoracic impedance, and the body surface temperature [96]. Note that the sensors are connected through a single, unshielded wire only, which makes the production of the Sense garment extremely easy.

The system was later extended with an additional sensor, which adds another ECG lead as well as PPG to the wearable (called LTMS-S) [97] (Fig. 3.14).

An even more advanced implementation of CS was WELCOME, a vest developed for COPD patients that has 20 sensors and 1 master and measures in parallel 12-lead ECG, EIT on 16 × 16 channels, PPG, and physical activity of the wearer [31,42,43] (Fig. 3.15). All sensors are connected through a two-wire bus only. This makes the integration in a vest much simpler than in the traditional approach with a central box. Thanks to the minimal wiring, the vest

FIGURE 3.13 Sense system and an ECG as well as an impedance signal recorded with the dry sense electrodes.

is comfortable to wear (e.g., in daily life, during telemonitoring sessions, or a visit to a medical doctor). While CS can in principle be used with gel electrodes, their electronics have been developed for the use with dry electrodes,

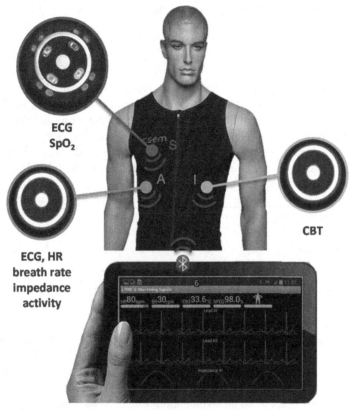

FIGURE 3.14 LTMS-S and recorded signals.

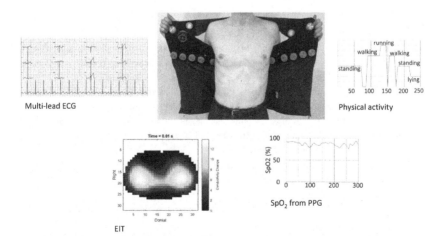

FIGURE 3.15 The WELCOME vest and measured signals.

which are much more comfortable to wear. The quality of ECG signals recorded with dry CS electrodes is as good as the quality of traditional systems using gel electrodes (see Figs. 3.13, 3.14 and 3.15).

This section presented systems which contain several sensors that are distributed over the body for measuring a variety of physiological signals. Classical systems rely on an architecture which includes a central box to which all sensors are connected (with cables or wireless). Such implementation is limited in the maximal number of sensors, as the wiring effort and the size of the central box increase with each added sensor. Wireless implementations are limited to signals which do not require a galvanic line (i.e., ECG and impedance-based measurements are precluded). CSs on the other hand have been integrated in big numbers using one or two wires only to connect all sensors. CSs have been notably used to monitor COPD patients with multi-modal system embedded in a T-shirt recording multichannel ECG and EIT, PPG, and physical activity.

Conclusion

This chapter summarized the state of the art in mobile and wearable monitoring of respiratory function. The presented technologies cover hardware and algorithms. Since breathing involves many organs, a plethora of signals is available which carry information on the health state and functioning of the respiratory tract. Due to this indirect nature of monitoring, a careful assessment of the validity of the signals is even more important than usual. While the past has brought important progress in the miniaturization of sensors, future developments will focus on the exploitation of even more signals and integrate electronics even more seamlessly, so that users will find sensor-equipped wearables as comfortable as traditional garments.

References

[1] K.G. Fan, J. Mandel, P. Agnihotri, M. Tai-Seale, Remote patient monitoring technologies for predicting chronic obstructive pulmonary disease exacerbations: review and comparison, JMIR Mhealth Uhealth 8 (2020) e16147, https://doi.org/10.2196/16147.

[2] D.K. Ming, S. Sangkaew, H.Q. Chanh, P.T.H. Nhat, S. Yacoub, P. Georgiou, A.H. Holmes, Continuous physiological monitoring using wearable technology to inform individual management of infectious diseases, public health and outbreak responses, Int. J. Infect. Dis. 96 (2020) 648–654, https://doi.org/10.1016/j.ijid.2020.05.086.

[3] F. de Jongh, Spirometers, Breathe 4 (2008) 251–254.

[4] C. Massaroni, A. Nicolò, D. Lo Presti, M. Sacchetti, S. Silvestri, E. Schena, Contact-based methods for measuring respiratory rate, Sensors 19 (2019), https://doi.org/10.3390/s19040908.

[5] A. Aliverti, A. Pedotti, Opto-electronic plethysmography, Monaldi Arch. Chest Dis. 59 (2003) 12–16.

[6] F.O. Monika, R. Karin, Reliability of the respiratory movement measuring instrument, RMMI, Clin. Physiol. Funct. Imag. 30 (2010) 349–353.

[7] R. Priori, A. Aliverti, A.L. Albuquerque, M. Quaranta, P. Albert, M. Peter, A. Calverley, The effect of posture on asynchronous chest wall movement in COPD, J. Appl. Physiol. 114 (2013) 1066–1075, https://doi.org/10.1152/japplphysiol.00414.2012.

[8] C. Hagman, C. Janson, A. Alinovschi, H. Hedenström, M. Emtner, Measuring breathing patterns and respiratory movements with the respiratory movement measuring instrument, Clin. Physiol. Funct. Imag. 36 (2015) 1–7, https://doi.org/10.1111/cpf.12302.

[9] F. Braun, A. Lemkaddem, V. Moser, S. Dasen, O. Grossenbacher, M. Bertschi, Contactless Respiration Monitoring in Real-Time via a Video Camera, 2017, https://doi.org/10.1007/978-981-10-5122-7_142.

[10] Y. Nam, Y. Kong, B. Reyes, N. Reljin, K.H. Chon, Monitoring of heart and breathing rates using dual cameras on a smartphone, PLoS One 11 (2016) e0151013, https://doi.org/10.1371/journal.pone.0151013.

[11] M. Pacelli, G. Loriga, R. Paradiso, Flat knitted sensors for respiration monitoring, in: IEEE International Symposium on Industrial Electronics, IEEE, Vigo, Spain, 2007, pp. 2838–2841, https://doi.org/10.1109/ISIE.2007.4375062.

[12] C.-T. Huang, C.-F. Tang, C.-L. Shen, A wearable textile for monitoring respiration, using a yarn-based sensor, in: Presented at the 10th IEEE International Symposium on Wearable Computers, IEEE, Montreux, 2006, pp. 141–142, https://doi.org/10.1109/ISWC.2006.286366.

[13] J.D. Pegan, J. Zhang, M. Chu, T. Nguyen, S.-J. Park, A. Paul, J. Kim, M. Bachman, M. Khine, Skin-mountable stretch sensor for wearable health monitoring, Nanoscale 8 (2016) 17295–17303, https://doi.org/10.1039/C6NR04467K.

[14] M. Chu, T. Nguyen, V. Pandey, Y. Zhou, H.N. Pham, R. Bar-Yoseph, S. Radom-Aizik, R. Jain, D.M. Cooper, M. Khine, Respiration rate and volume measurements using wearable strain sensors, Npj Digit. Med. 2 (2019) 8, https://doi.org/10.1038/s41746-019-0083-3.

[15] O. Atalay, W. Kennon, E. Demirok, Weft-knitted strain sensor for monitoring respiratory rate and its electro-mechanical modeling, Sens. J. IEEE 15 (2015) 110–122, https://doi.org/10.1109/JSEN.2014.2339739.

[16] C.-T. Huang, C.-L. Shen, C.-F. Tang, S.-H. Chang, A wearable yarn-based piezo-resistive sensor, Sens. Actuators Phys. 141 (2008) 396–403, https://doi.org/10.1016/j.sna.2007.10.069.

[17] T. Hoffmann, B. Eilebrecht, S. Leonhardt, Respiratory monitoring system on the basis of capacitive textile force sensors, IEEE Sensor. J. 11 (2011) 1112–1119.

[18] R.B. Berry, S.F. Quan, A.R. Abreu, M.L. Bibbs, L. DelRosso, S.M. Harding, M.-M. Mao, D.T. Plante, M.R. Pressman, M.M. Troester, B.V. Vaughn, The AASM Manual for the Scoring of Sleep and Associated Events: Rules, Terminology and Technical Specifications, Version 2.6, American Academy of Sleep Medicine, Darien, IL, USA, 2020.

[19] P. Grossmann, F.H. Wilhelm, M. Brutsche, Accuracy of ventilatory measurement employing ambulatory inductive plethysmography during tasks of everyday life, Biol. Psychol. 84 (2010) 121–128.

[20] I. Frerichs, M.B.P. Amato, A.H. van Kaam, D.G. Tingay, Z. Zhao, B. Grychtol, M. Bodenstein, H. Gagnon, S.H. Böhm, E. Teschner, O. Stenqvist, T. Mauri, V. Torsani, L. Camporota, A. Schibler, G.K. Wolf, D. Gommers, S. Leonhardt, A. Adler, TREND study group, Chest electrical impedance tomography examination, data analysis, terminology, clinical use and recommendations: consensus statement of the TRanslational EIT develompeNt stuDy group, Thorax 72 (2017) 83–93, https://doi.org/10.1136/thoraxjnl-2016-208357.

[21]   Electrical impedance tomography: methods, history, and applications, in: D. Holder (Ed.), Medical Physics and Biomedical Engineering, Institute of Physics Publishing, Bristol, UK, 2005.

[22]   S. Grimnes, Ø.G. Martinsen, Bioimpedance and Bioelectricity Basics, third ed., Academic Press, 2015 https://doi.org/10.1016/C2012-0-06951-7.

[23]   D. Naranjo-Hernández, J. Reina-Tosina, M. Min, Fundamentals, recent advances, and future challenges in bioimpedance devices for healthcare applications, J. Sens. 2019 (2019) 1–42, https://doi.org/10.1155/2019/9210258.

[24]   IEC 60601-1, Medical Electrical Equipment - Part 1 : General Requirements for Basic Safety and Essential Performance, 3.0. Ed. Geneva, Switzerland, 2005.

[25]   K.S. Cole, Permeability and impermeability of cell membranes for ions, Cold Spring Harbor Symp. Quant. Biol. 8 (1940) 110–122, https://doi.org/10.1101/SQB.1940.008.01.013.

[26]   V.-P. Seppä, J. Hyttinen, M. Uitto, W. Chrapek, J. Viik, Novel electrode configuration for highly linear impedance pneumography, Biomed. Tech. Eng. 58 (2013), https://doi.org/10.1515/bmt-2012-0068.

[27]   M. Młyńczak, W. Niewiadomski, M. Żyliński, G. Cybulski, Assessment of calibration methods on impedance pneumography accuracy, Biomed. Eng. Biomed. Tech. 61 (2016), https://doi.org/10.1515/bmt-2015-0125.

[28]   L.A. Critchley, Minimally invasive cardiac output monitoring in the year 2012, in: W.S. Aronow (Ed.), Artery Bypass, InTech, Rijeka, Croatia, 2013, https://doi.org/10.5772/54413.

[29]   A. Adler, A. Boyle, Electrical impedance tomography: tissue properties to image measures, IEEE Trans. Biomed. Eng. (2017), https://doi.org/10.1109/TBME.2017.2728323 (in press).

[30]   O. Chételat, R. Gentsch, J. Krauss, J. Luprano, Getting rid of the wires and connectors in physiological monitoring, in: Presented at the 2008 30th Annual International Conference of the IEEE Engineering in Medicine and Biology Society, IEEE, Vancouver, BC, 2008, pp. 1278–1282, https://doi.org/10.1109/IEMBS.2008.4649397.

[31]   M. Rapin, M. Proença, F. Braun, C. Meier, J. Solà, D. Ferrario, O. Grossenbacher, J.-A. Porchet, O. Chételat, Cooperative dry-electrode sensors for multi-lead biopotential and bioimpedance monitoring, Physiol. Meas. 36 (2015) 767–783, https://doi.org/10.1088/0967-3334/36/4/767.

[32]   R.D. Cook, G.J. Saulnier, D.G. Gisser, J.C. Goble, J.C. Newell, D. Isaacson, ACT3: a high-speed, high-precision electrical impedance tomograph, IEEE Trans. Biomed. Eng. 41 (1994) 713–722, https://doi.org/10.1109/10.310086.

[33]   O. Chételat, B. Bonnal, A. Fivaz, EP19213839, Remotely Powered Cooperative Sensor Device, vol. 4, 2019.

[34]   S. Hong, J. Lee, J. Bae, H.-J. Yoo, A 10.4 mW Electrical Impedance Tomography SoC for Portable Real-Time Lung Ventilation Monitoring System 4, 2014.

[35]   Y. Lee, K. Song, H.-J. Yoo, A 4.84mW 30fps dual frequency division multiplexing electrical impedance tomography SoC for lung ventilation monitoring system, in: Presented at the 2015 Symposium on VLSI Circuits, IEEE, Kyoto, Japan, 2015, pp. C204–C205, https://doi.org/10.1109/VLSIC.2015.7231259.

[36]   P.O. Gaggero, A. Adler, J. Brunner, P. Seitz, Electrical impedance tomography system based on active electrodes, Physiol. Meas. 33 (2012) 831–847, https://doi.org/10.1088/0967-3334/33/5/831.

[37]   Y. Wu, D. Jiang, A. Bardill, R. Bayford, A. Demosthenous, A 122 fps, 1 MHz bandwidth multi-frequency wearable EIT belt featuring novel active electrode architecture for neonatal thorax vital sign monitoring, IEEE Trans. Biomed. Circuits Syst. 13 (2019) 927–937, https://doi.org/10.1109/TBCAS.2019.2925713.

[38] Y. Wu, D. Jiang, A. Bardill, S. de Gelidi, R. Bayford, A. Demosthenous, A high frame rate wearable EIT system using active electrode ASICs for lung respiration and heart rate monitoring, IEEE Trans. Circuits Syst. Regul. Pap. 65 (2018) 3810−3820, https://doi.org/10.1109/TCSI.2018.2858148.

[39] M. Kim, J. Bae, H.-J. Yoo, Wearable 3D lung ventilation monitoring system with multi frequency electrical impedance tomography, in: Presented at the 2017 IEEE Biomedical Circuits and Systems Conference (BioCAS), IEEE, Torino, 2017, pp. 1−4, https://doi.org/10.1109/BIOCAS.2017.8325163.

[40] M. Kim, H. Kim, J. Jang, J. Lee, J. Lee, J. Lee, K. Lee, K. Kim, Y. Lee, H. Yoo, 21.2 A 1.4mΩ-sensitivity 94dB-dynamic-range electrical impedance tomography SoC and 48-channel Hub SoC for 3D lung ventilation monitoring system, in: Presented at the 2017 IEEE International Solid- State Circuits Conference - (ISSCC), IEEE, San Francisco, CA, USA, 2017, pp. 354−355, https://doi.org/10.1109/ISSCC.2017.7870407.

[41] M. Rapin, A Wearable Sensor Architecture for High-Quality Measurement of Multilead ECG and Frequency-Multiplexed EIT (Doctoral thesis), ETH Zurich, 2018, https://doi.org/10.3929/ethz-b-000302712.

[42] M. Rapin, F. Braun, A. Adler, J. Wacker, I. Frerichs, B. Vogt, O. Chételat, Wearable sensors for frequency-multiplexed EIT and multilead ECG data acquisition, IEEE Trans. Biomed. Eng. 1−1 (2018), https://doi.org/10.1109/TBME.2018.2857199.

[43] I. Frerichs, B. Vogt, J. Wacker, R. Paradiso, F. Braun, M. Rapin, L. Caldani, O. Chételat, N. Weiler, Multimodal remote chest monitoring system with wearable sensors: a validation study in healthy subjects, Physiol. Meas. 41 (2020) 015006, https://doi.org/10.1088/1361-6579/ab668f.

[44] P. Forgacs, Crackles and wheezes, Lancet 290 (1967) 203−205, https://doi.org/10.1016/S0140-6736(67)90024-4.

[45] S. Leng, R.S. Tan, K.T.C. Chai, C. Wang, D. Ghista, L. Zhong, The electronic stethoscope, Biomed. Eng. Online 14 (2015) 66, https://doi.org/10.1186/s12938-015-0056-y.

[46] M. Klum, M. Urban, T. Tigges, A.-G. Pielmus, A. Feldheiser, T. Schmitt, R. Orglmeister, Wearable cardiorespiratory monitoring employing a multimodal digital patch stethoscope: estimation of ECG, PEP, LVET and respiration using a 55 mm single-lead ECG and phonocardiogram, Sensors 20 (2020) 2033, https://doi.org/10.3390/s20072033.

[47] E. Messner, M. Hagmüller, P. Swatek, F. Pernkopf, A robust multichannel lung sound recording device, in: Proceedings of the 9th International Joint Conference on Biomedical Engineering Systems and Technologies. Presented at the 9th International Conference on Biomedical Electronics and Devices, SCITEPRESS - Science and and Technology Publications, Rome, Italy, 2016, pp. 34−39, https://doi.org/10.5220/0005660200340039.

[48] I. Sen, Y.P. Kahya, A multi-channel device for respiratory sound data acquisition and transient detection, in: Presented at the 2005 IEEE Engineering in Medicine and Biology 27th Annual Conference, IEEE, Shanghai, China, 2005, pp. 6658−6661, https://doi.org/10.1109/IEMBS.2005.1616029.

[49] S.S. Kraman, G.R. Wodicka, Y. Oh, H. Pasterkamp, Measurement of respiratory acoustic signals, Chest 108 (1995) 1004−1008, https://doi.org/10.1378/chest.108.4.1004.

[50] Y. Cotur, M. Kasimatis, M. Kaisti, S. Olenik, C. Georgiou, F. Güder, Stretchable composite acoustic transducer for wearable monitoring of vital signs, Adv. Funct. Mater. 30 (2020) 1910288, https://doi.org/10.1002/adfm.201910288.

[51] S.S. Kraman, G.R. Wodicka, G.A. Pressler, H. Pasterkamp, Comparison of lung sound transducers using a bioacoustic transducer testing system, J. Appl. Physiol. 101 (2006) 469−476, https://doi.org/10.1152/japplphysiol.00273.2006.

[52] B. Panda, S. Mandal, S.J.A. Majerus, Flexible, skin coupled microphone array for point of care vascular access monitoring, IEEE Trans. Biomed. Circuits Syst. 13 (2019) 1494−1505, https://doi.org/10.1109/TBCAS.2019.2948303.

[53] G. Yilmaz, P. Starkov, M. Crettaz, J. Wacker, O. Chételat, A low-cost USB-compatible electronic stethoscope unit for multi-channel lung sound acquisition, in: J. Henriques, N. Neves, P. de Carvalho (Eds.), IFMBE Proceedings Presented at the XV Mediterranean Conference on Medical and Biological Engineering and Computing − MEDICON 2019, Springer International Publishing, Coimbra, Portugal, 2020, pp. 1299−1303, https://doi.org/ 10.1007/978-3-030-31635-8_159.

[54] H. Eshach, A. Volfson, Explanatory model for sound amplification in a stethoscope, Phys. Educ. 50 (2015) 75−80, https://doi.org/10.1088/0031-9120/50/1/75.

[55] M. Toda, M.L. Thompson, Contact-type vibration sensors using curved clamped PVDF film, IEEE Sensor. J. 6 (2006) 1170−1177, https://doi.org/10.1109/JSEN.2006.881407.

[56] F. García-Río, M. Calle Rubio, F. Burgos, P. Casan, F. del Campo, J. Gáldiz, J. Giner, N. González-Mangado, F. Ortega, L. Puente-Maestu, Spirometry, Arch. Bronconeumol. 49 (2013), https://doi.org/10.1016/j.arbres.2013.04.001.

[57] D.J. Macfarlane, Open-circuit respirometry: a historical review of portable gas analysis systems, Eur. J. Appl. Physiol. 117 (2017) 2369−2386, https://doi.org/10.1007/s00421-017-3716-8.

[58] M.R. Miller, J. Hankinson, V. Brusasco, F. Burgos, R. Casaburi, A. Coates, R. Crapo, P. Enright, C.P.M. van der Grinten, P. Gustafsson, R. Jensen, D.C. Johnson, N. MacIntyre, R. McKay, D. Navajas, O.F. Pedersen, R. Pellegrino, G. Viegi, J. Wanger, Standardisation of spirometry, Eur. Respir. J. 26 (2005) 319−338, https://doi.org/10.1183/ 09031936.05.00034805.

[59] B.S. Overstreet, D.R.J. Bassett, S.E. Crouter, B.C. Rider, B.B. Parr, Portable open-circuit spirometry systems, J. Sports Med. Phys. Fit. 57 (2017) 227−237, https://doi.org/ 10.23736/S0022-4707.16.06049-7.

[60] D.J. Macfarlane, P. Wong, Validity, reliability and stability of the portable Cortex Metamax 3B gas analysis system, Eur. J. Appl. Physiol. 112 (2012) 2539−2547, https://doi.org/ 10.1007/s00421-011-2230-7.

[61] Vitalograph, The Fleisch Pneumotachograph [WWW Document], 2021, https://vitalograph. com/education/fleisch (Accessed 30 January 21).

[62] S.O. McDoniel, A systematic review on use of a handheld indirect calorimeter to assess energy needs in adults and children, Int. J. Sport Nutr. Exerc. Metabol. 17 (2007) 491−500, https://doi.org/10.1123/ijsnem.17.5.491.

[63] U.R. Acharya, Advances in Cardiac Signal Processing, Springer-Verlag, Berlin Heidelberg, 2007.

[64] P.H. Charlton, D.A. Birrenkott, T. Bonnici, M.A.F. Pimentel, A.E.W. Johnson, J. Alastruey, L. Tarassenko, P.J. Watkinson, R. Beale, D.A. Clifton, Breathing rate estimation from the electrocardiogram and photoplethysmogram: a review, IEEE Rev. Biomed. Eng. 11 (2018) 2−20, https://doi.org/10.1109/RBME.2017.2763681.

[65] H. Liu, J. Allen, D. Zheng, F. Chen, Recent development of respiratory rate measurement technologies, Physiol. Meas. 40 (2019) 07TR01, https://doi.org/10.1088/1361-6579/ab299e.

[66] J.R. Levick, An Introduction to Cardiovascular Physiology, fourth ed., Hodder Arnold, London, UK, 2003.

[67] B. Hayes, Non-invasive Cardiovascular Monitoring, Principles and Practice Series, BMJ Publishing, London, 1997.

[68] P. Zipp, H. Ahrens, A model of bioelectrode motion artefact and reduction of artefact by amplifier input stage design, J. Biomed. Eng. 1 (1979) 273−276, https://doi.org/10.1016/0141-5425(79)90165-1.

[69] T. Degen, H. Jäckel, Continuous monitoring of electrode−skin impedance mismatch during bioelectric recordings, IEEE Trans. Biomed. Eng. 55 (2008) 1711−1715, https://doi.org/10.1109/TBME.2008.919118.

[70] P.H. Charlton, T. Bonnici, L. Tarassenko, D.A. Clifton, R. Beale, P.J. Watkinson, An assessment of algorithms to estimate respiratory rate from the electrocardiogram and photoplethysmogram, Physiol. Meas. 37 (2016) 610−626, https://doi.org/10.1088/0967-3334/37/4/610.

[71] M. Lemay, M. Bertschi, J. Sola, P. Renevey, E. Genzoni, M. Proença, D. Ferrario, F. Braun, J. Parak, I. Korhonen, Applications of optical cardiovascular monitoring, in: E. Sazonov (Ed.), Wearable Sensors, Elsevier, 2020.

[72] A. Reisner, P.A. Shaltis, D. McCombie, H.H. Asada, Utility of the photoplethysmogram in circulatory monitoring, Anesthesiology 108 (2008) 950−958, https://doi.org/10.1097/ALN.0b013e31816c89e1.

[73] G. Martínez, N. Howard, D. Abbott, K. Lim, R. Ward, M. Elgendi, Can photoplethysmography replace arterial blood pressure in the assessment of blood pressure? J. Clin. Med. 7 (2018) https://doi.org/10.3390/jcm7100316.

[74] S.C. Millasseau, F.G. Guigui, R.P. Kelly, K. Prasad, J.R. Cockcroft, J.M. Ritter, P.J. Chowienczyk, Noninvasive assessment of the digital volume pulse, Hypertension 36 (2000) 952−956, https://doi.org/10.1161/01.HYP.36.6.952.

[75] T. Tamura, Y. Maeda, M. Sekine, M. Yoshida, Wearable photoplethysmographic sensors—past and present, Electronics 3 (2014) 282−302, https://doi.org/10.3390/electronics3020282.

[76] S.P. Linder, S.M. Wendelken, E. Wei, S.P. McGrath, Using the morphology of photoplethysmogram peaks to detect changes in posture, J. Clin. Monit. Comput. 20 (2006) 151−158, https://doi.org/10.1007/s10877-006-9015-2.

[77] Y. Ghamri, M. Proença, G. Hofmann, P. Renevey, G. Bonnier, F. Braun, A. Axis, M. Lemay, P. Schoettker, Automated pulse oximeter waveform analysis to track changes in blood pressure during anesthesia induction: a proof-of-concept study, Anesth. Analg. 130 (2020) 1222−1233, https://doi.org/10.1213/ANE.0000000000004678.

[78] M. Proença, P. Renevey, F. Braun, G. Bonnier, R. Delgado-Gonzalo, A. Lemkaddem, C. Verjus, D. Ferrario, M. Lemay, Pulse wave analysis techniques, in: J. Solà, R. Delgado-Gonzalo (Eds.), The Handbook of Cuffless Blood Pressure Monitoring: A Practical Guide for Clinicians, Researchers, and Engineers, Springer International Publishing, Cham, 2019, pp. 107−137, https://doi.org/10.1007/978-3-030-24701-0_8.

[79] G.B. Papini, P. Fonseca, M.M. van Gilst, J.W. Bergmans, R. Vullings, S. Overeem, Respiratory activity extracted from wrist-worn reflective photoplethysmography in a sleep-disordered population, Physiol. Meas. 41 (2020) 065010, https://doi.org/10.1088/1361-6579/ab9481.

[80] P. Renevey, R. Delgado-Gonzalo, A. Lemkaddem, C. Verjus, S. Combertaldi, B. Rasch, B. Leeners, F. Dammeier, F. Kuubler, Respiratory and cardiac monitoring at night using a wrist wearable optical system, in: Presented at the 2018 40th Annual International Conference of the IEEE Engineering in Medicine and Biology Society (EMBC), IEEE, Honolulu, HI, 2018, pp. 2861−2864, https://doi.org/10.1109/EMBC.2018.8512881.

[81] J.G. Webster (Ed.), Design of Pulse Oximeters, IOP Publishing Ltd, 1997, https://doi.org/10.1887/0750304677.

[82] J. Moyle, Pulse Oximetry, Blackwell Pub., Oxford, 2002.

[83] A. Jubran, Pulse oximetry, Crit. Care 19 (2015) 272, https://doi.org/10.1186/s13054-015-0984-8.

[84] ISO 80601-2-61:2017, Medical Electrical Equipment - Part 2-61: Particular Requirements for Basic Safety and Essential Performance of Pulse Oximeter Equipment, International Organization for Standardization (ISO), Geneva, Switzerland, 2017.

[85] Pulse Oximeters - Premarket Notification Submissions [510(k)s]: Guidance for Industry and Food and Drug Administration Staff, U.S. Department of Health and Human Services, Food and Drug Administration, Silver Spring, USA, 2013.

[86] M.W. Sjoding, R.P. Dickson, T.J. Iwashyna, S.E. Gay, T.S. Valley, Racial bias in pulse oximetry measurement, N. Engl. J. Med. 383 (2020) 2477−2478, https://doi.org/10.1056/NEJMc2029240.

[87] Y. Mendelson, B.D. Ochs, Noninvasive pulse oximetry utilizing skin reflectance photoplethysmography, IEEE Trans. Biomed. Eng. 35 (1988) 798−805, https://doi.org/10.1109/10.7286.

[88] S. Chatterjee, P. Kyriacou, Monte Carlo analysis of optical interactions in reflectance and transmittance finger photoplethysmography, Sensors 19 (2019) 789, https://doi.org/10.3390/s19040789.

[89] K.H. Shelley, D. Tamai, D. Jablonka, M. Gesquiere, R.G. Stout, D.G. Silverman, The effect of venous pulsation on the forehead pulse oximeter wave form as a possible source of error in Spo2 calculation: anesth, Analgesia 100 (2005) 743−747, https://doi.org/10.1213/01.ANE.0000145063.01043.4B.

[90] M. Proença, O. Grossenbacher, S. Dasen, V. Moser, D. Ostojic, A. Lemkaddem, D. Ferrario, M. Lemay, M. Wolf, J.-C. Fauchère, T. Karen, Performance assessment of a dedicated reflectance pulse oximeter in a neonatal intensive care unit, in: Presented at the 2018 40th Annual International Conference of the IEEE Engineering in Medicine and Biology Society (EMBC), IEEE, Honolulu, HI, 2018, pp. 1502−1505, https://doi.org/10.1109/EMBC.2018.8512504.

[91] L. Schallom, C. Sona, M. McSweeney, J. Mazuski, Comparison of forehead and digit oximetry in surgical/trauma patients at risk for decreased peripheral perfusion, Heart Lung 36 (2007) 188−194, https://doi.org/10.1016/j.hrtlng.2006.07.007.

[92] A. Guber, G. Epstein Shochet, S. Kohn, D. Shitrit, Wrist-sensor pulse oximeter enables prolonged patient monitoring in chronic lung diseases, J. Med. Syst. 43 (2019) 230, https://doi.org/10.1007/s10916-019-1317-2.

[93] F. Braun, P. Theurillat, M. Proenca, A. Lemkaddem, D. Ferrario, K.D. Jaegere, C.M. Horvath, C. Roth, A.-K. Brill, M. Lemay, S.R. Ott, Pulse oximetry at the wrist during sleep: performance, challenges and perspectives, in: EMBC 2020. Presented at the EMBC 2020, IEEE, Montréal, Canada, 2020, pp. 5115−5118, https://doi.org/10.1109/EMBC44109.2020.9176081.

[94] F. Braun, C. Verjus, J. Solà, M. Marienfeld, M. Funke-Chambour, J. Krauss, T. Geiser, S.A. Guler, Evaluation of a novel ear pulse oximeter: towards automated oxygen titration in eyeglass frames, Sensors 20 (2020) 3301, https://doi.org/10.3390/s20113301.

[95] J. Dieffenderfer, H. Goodell, S. Mills, M. McKnight, S. Yao, F. Lin, E. Beppler, B. Bent, B. Lee, V. Misra, Y. Zhu, O. Oralkan, J. Strohmaier, J. Muth, D. Peden, A. Bozkurt, Low-power wearable systems for continuous monitoring of environment and health for chronic respiratory disease, IEEE J. Biomed. Health Inform. 20 (2016) 1251−1264, https://doi.org/10.1109/JBHI.2016.2573286.

[96] O. Chételat, J. Oster, O. Grossenbacher, A. Hutter, J. Krauss, A. Giannakis, A highly in-tegrated wearable multi-parameter monitoring system for athletes, in: K. Dremstrup, S. Rees, M.Ø. Jensen (Eds.), 15th Nordic-Baltic Conference on Biomedical Engineering and Medical Physics (NBC 2011), IFMBE Proceedings, Springer Berlin Heidelberg, Berlin, Heidelberg, 2011, pp. 148–151, https://doi.org/10.1007/978-3-642-21683-1_37.

[97] O. Chételat, D. Ferrario, M. Proença, J.-A. Porchet, A. Falhi, O. Grossenbacher, R. Delgado-Gonzalo, N. Della Ricca, C. Sartori, Clinical validation of LTMS-S: a wearable system for vital signs monitoring, in: Presented at the 2015 37th Annual International Conference of the IEEE Engineering in Medicine and Biology Society (EMBC), IEEE, Milan, 2015, pp. 3125–3128, https://doi.org/10.1109/EMBC.2015.7319054.

Chapter 4

Textiles and smart materials for wearable monitoring systems

Rita Paradiso and Laura Caldani
Smartex srl, Prato, PO, Italy

Introduction

A new class of sensors, made by textile technology combining conductive and elastic yarns, cabling textile integration, industrial serigraphy, and embroidery processes, has produced in the last two decades a variety of systems and advanced platforms for application in the field of cardiopulmonary disease monitoring. The performances of these novel sensors have been studied and assessed to evaluate their use as plethysmography signal detector as well as electrodes for impedance and ECG monitoring.

Knitted and printed sensors have been evaluated, applying controlled small amplitude strains at low frequencies, in the range of the respiratory signal, and by measuring the resistance value variations to select the sensor features in terms of size, structure, and orientation compatible with pulmonary disease monitoring. Fabric sensors have been integrated in multiple configurations in several sensing textile platforms. Cardiopulmonary signals can then be acquired simultaneously in basal and movement conditions by means of fully integrated garments.

Fabric sensing functionality, a combination of conductivity and elasticity

Electrical conductivity is the main physical property that is capable to transform a textile material into a sensing material and that plays an important role in the development of e-textile apparels. Conductive fabrics can be used as bioelectrodes or (when combined with elastomers) as piezoresistive sensors that are capable to sense biomechanical variables. Several different methods can be used to build an electrically conductive fabric structure, starting from the integration of metal monofilaments into the yarn, the enrichment of the

Wearable Sensing and Intelligent Data Analysis for Respiratory Management
https://doi.org/10.1016/B978-0-12-823447-1.00005-1

fibers with conductive components, the coating of man-made fibers or fabrics with conductive layers, to the printing of conductive circuits onto the fabric surface.

In the conventional textile production, metal components in the form of fibers, filaments, or particles are typically used for technical applications such as shielding and antistatic protection, bacteriostatic applications, automotive, as well as for traditional manufacturing to create a wrinkle or shaping effect. Pure stainless-steel slivers can be blended with fibers such as polyamide, polyester, and cotton to obtain inert and stable electrically conductive yarns. Silver-plated fibers are better in terms of conductivity and manufacturability; however, poor resistance to strain and sweat oxidization problems make their lifetime shorter than fibers and yarns made of stainless steel.

Another important property required for sensing applications is the elastic recovery of the fabric that is the result of a combined use of elastomeric and functional fibers. Elastic components are used for the production of stretchable fabric capable to fit the body shape. This is an important property as a textile-sensing surface collects the information from the human body only if there is a good fitting of the sensors to the body shape. A system based on textile technology requires the construction of a fabric containing shaped regions where sensors are located, and these regions will correspond to a specific part on the body, alternatively a cut and sew process can be used to combine fabrics and layers with different mechanical properties. Sensors have to fit the body particularly in the area closed to the source of signal, to increase the accuracy of the data, and to reduce the movement artifacts. The sensing surface has to act as a second skin, stretchable, and comfortable. Flat knitting and seamless technology allow to confine specific yarns in defined regions of the fabric, working with yarn carries, at the same time it is possible to process different yarns together according to a desired topology.

Seamless technology provides comfortable, stretchable, fitting, and adherent garments, making this technology preferable for sensing applications where adherence, elasticity, and comfort are required. It is possible to realize seamless systems, where electrodes and sensors are knitted in the same production step. The main difference between these two technologies is the possibility for flat knitting technology to combine intarsia (a domain with a different yarn) and double knitting (two layers of fabric are knit simultaneously with two yarns), while seamless can only manage these two processes separately with different machineries. However, seamless technology is unique in combining elasticity, comfort with low production costs.

Textile materials for sensing

In the standard clinical practice, electrodes located on specific parts of the body are used to measure differences of electrical potential of biological origin. Biopotentials are related to the electrochemical activity of the cells, where the electrical activity is caused by differences in ion concentrations within the body. There are several examples of biopotential measurements: Electrocardiography (ECG), electromyography (EMG), and electrooculography (EOG), that have been successfully acquired by means of fabric electrodes [1], as well as to measure body impedance and skin conductance.

Biomechanical fabric sensors characterized by a piezo functionality have been realized in the past, by printing conductive elastomer (CE) onto an elastic fabric, through industrial serigraphy process [2,3]. These sensors compared with knitted piezoresistive fabric (KPF) [4] show lower performances in terms of transient time, hysteresis, lifetime, reliability, and robustness [5]. KPF sensors change the electrical resistance according to the strain; the variation in electrical properties is due to the modification of the electrical contacts inside the fabric structure. KPF sensors can be used either as strain sensors (single layer) or in a goniometric configuration (double layer) [6].

Respiratory monitoring, measurement methodologies compatible with wearable applications

As functioning units, the lung and heart are usually considered a single complex organ, but because these organs contain essentially two compartments, one for blood and one for air, they are usually separated in terms of the tests conducted to evaluate the heart or pulmonary functions. Two parameters are particularly important in monitoring the function of lungs:

The respiratory rate, expressed as the number of breaths per minute, is a parameter characterized by high sensitivity, also if it is not specific for any respiratory dysfunctions.

The inspiratory fraction, the ratio of inspiratory time to total respiratory time. Low values of this ratio, associated with an increase in the expiratory phase, may be an indication of an airway obstruction. Low values are also detected while a subject is talking, while high values can be measured during sleep especially if the subject has difficulty in breathing.

Multiple scientific papers [7−10] indicated respiratory rate as one of the most sensitive physiological parameters, although it is not considered as a specific indicator of respiratory dysfunction, the estimate of the tidal volume, the amount of air that moves in or out of the lungs with each respiratory cycle, instead provides useful information for diagnostic purposes. Knowing the respiratory rate and a volumetric indication, it is possible to derive the ventilation, the inspiratory fraction, and the expiratory and inspiratory flow.

Several methods and strategies have been used to measure continuously the kinematics of the chest wall that is associated with changes in thoracic volume, although the respiratory system is very complex, and its analysis requires identification of multitude of parameters, in practice only very few of them can be measured, especially if the methods are required to be noninvasive, like in the wearable applications. In particular, three measurement strategies are compatible with wearable applications; the first is based on the use of piezoresistive sensors for monitoring the movements correlated to the respiratory activity.

The second is based on the use of the inductance plethysmography. A conductor is integrated in an elastic band to form an inductor. A sinusoidal signal flows into the inductor and an inducted voltage is measured related to the changes in the cross-sectional area of the inductor. Using this information, the area enclosed by the inductor can be evaluated, hence the thoracic and abdominal cross-sectional area. When the inductance plethysmography is acquired by at least two coils located at abdominal and thoracic level, it is possible to estimate the tidal volume during normal breathing. In order to know an absolute measurement of the tidal volume the system must be calibrated once for each patient using a spirometer.

The third method is based on the impedance plethysmography. Body impedance is conventionally measured by collecting the potential drop caused by the movement of charges due to the injection of a high-frequency and low-intensity current through two electrodes. Measurement depends on body mass and the flow of fluids through the body. Impedance plethysmography allows monitoring the respiratory activity relying on the thorax impedance variations determined by the airflow through the lungs. The measurement can be made with two or four electrodes. The method with four electrodes is less affected by movement artifacts, the outer electrodes are used to inject a high-frequency and low-intensity current, the inner ones to detect the variation in potential due to the change in impedance correlated to the respiratory activity. The four electrodes measurement method, thanks to the high input impedance of the measurement amplifier, allows to remove the contribution of the resistance of the connection and the contact resistance between the electrodes and the skin, which could affect the value of the acquired measurement. Furthermore, the use of two electrodes to inject current, which are different from those used to measure the voltage, lets the spatial distribution of the current density between the measuring electrodes be approximately constant; this allows to reduce the effect of impedance variations in proximity of the injection electrodes due to the higher current density in their proximity. The value of the detected impedance varies according to the volume of air present in the lungs, for this reason the system can be calibrated with a spirometer. Impedance plethysmography signal is more robust compared to other indirect techniques with regard to motion artifacts; however, the signal quality is influenced by the value of the skin—electrode contact resistance.

When respiratory function is recorded separately from each hemithorax, it could be used to detect possible asymmetries in tidal volume and phase relations between right and left sides. Starting from rate and volume indexes, a number of other respiratory parameters can be computed, such as ventilation, fractional inspiratory time, both inspiratory and expiratory flows over mean inspiratory flow, as well as the phase and amplitude relation between abdominal and rib cage respiratory patterns. In addition, the same set of sensors will yield a wide series of respiratory indexes useful for diagnosing several clinical conditions affecting respiratory dynamics.

Respiratory drive signifies neural respiratory center activity. The gold standard for measuring this drive is the diaphragmatic electromyogram, an invasive, technically difficult procedure. Peak aspiratory flow and peak aspiratory acceleration, which can be easily obtained from the pneumogram, have been found to be a good substitute for diaphragmatic electromyogram in measuring respiratory drive. Several clinical respiratory phenomena can be characterized by the ratio of breath-by-breath ventilation divided by respiratory drive. For example, the ratio of ventilation to respiratory drive distinguishes psychogenic from organic causes of breathlessness and this may help to distinguish between a patient with chronic pulmonary and cardiac diseases from a patient with chronic anxiety.

In addition, ECG signal can be used to evaluate the time variations of the RR interval of the QRS component and respiratory sinus arrhythmia to study the sympathovagal balance [11]. RR intervals are commonly used to provide the power spectral density using an FFT-based approach. Two major oscillatory components are usually detectable in RR variability, one of which, synchronous with respiration and related to parasympathetic activity, is described as HF (high frequency, about 0.25 Hz and varying with respiration), whereas the other, corresponding to the slow waves of arterial pressure and mainly related to sympathetic activity, is described as LF (low frequency, about 0.1 Hz). The LF-to-HF ratio can thus be calculated to provide an indication of the sympathovagal balance.

Fabric sensors and electrodes for respiratory monitoring

Cardiopulmonary vital signs can be acquired and monitored by means of textile platforms with integrated fabric electrodes and piezoresistive sensors [12,13]. The implementation of appropriate techniques allows acquisition of both ECG and respiratory activity signals using the same type of electrode integrated in a garment that can be easily used in a remote setting.

Moreover, two piezoresistive fabric sensors allow the acquisition of plethysmography respiratory signal at abdominal and thoracic level [14].

A typical signal of impedance plethysmography compared with Biopac spirometer (used as gold standard) is shown in Fig. 4.1. During the first 5 s of the trial, the subject performs forced breath, as can be seen from the graphs, it's possible to discriminate between high and low rate breathing.

The two signals are comparable; the variation in peak-to-peak amplitude is detectable with both systems. In particular during the apnea phases, periodic fluctuations are observable on the impedance pneumography signal, with a frequency comparable with the cardiac periodicity.

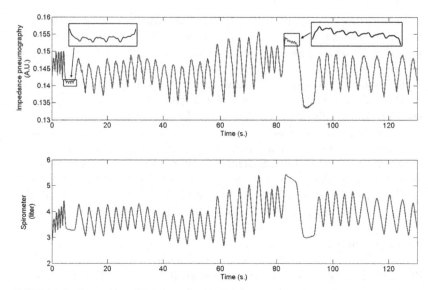

FIGURE 4.1 Comparison of signals obtained by impedance pneumography collected with fabric electrodes (upper) and Biopac system (lower).

Within the 82−97 s interval, two types of apnea have been registered: the first after a full lung inspiration, and the second after an expiration to the empty lung.

Signals acquired in basal condition by fabric sensors and fabric electrodes integrated in a wearable system are shown in Fig. 4.2 and compared with BIOPAC spirometer. Despite a different morphology of the signals due to the different physical principles of measurement (spirometer, impedance pneumography, Knitted Piezoresistive Fabric and Printed Piezoresistive Fabric thorax plethysmography, KPF abdominal plethysmography), it is possible to identify the various respiratory phases (inspiration and expiration). In fact, the maximum and the minimum peaks of the curves are perfectly synchronized.

Plethysmography measure is intrinsically subjected to movement artifact, while at rest this methodology can provide information that is impossible to acquire using impedance pneumography, such as difference in the movement of the body during abdominal and thoracic respiration.

In fact, in Fig. 4.3 are shown examples of differences between the abdominal and thoracic respiration acquired during a different breathing exercise. In Section 1 Introduction respiration is normal; in Section 2 Fabric sensing functionality, a combination of conductivity and elasticity respiration is predominantly abdominal, while in the third it is thoracic. This is regularly pointed out by the comparison of the two signals.

The possibility to discriminate between abdominal and thoracic movements permits to discriminate the paradoxical respiration, in which the chest and abdominal functions oppose each other; the patient exhales with the diaphragm

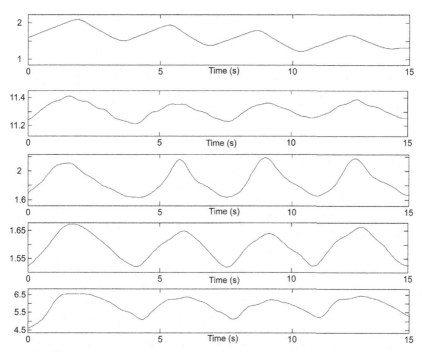

FIGURE 4.2 Comparison of the respiratory signal obtained in resting condition by means of Biopac spirometer and different fabric-based sensing methodologies. From the top: Biopac spirometer, impedance pneumography, KPFT and PPFT thorax plethysmography, KPFA abdominal plethysmography.

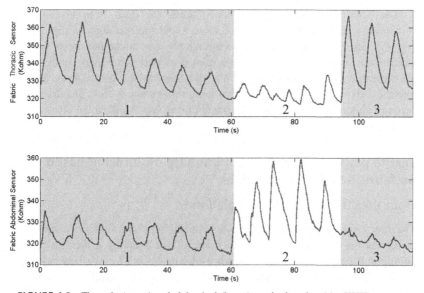

FIGURE 4.3 Thoracic (upper) and abdominal (lower) respiration signal by KPFT sensors.

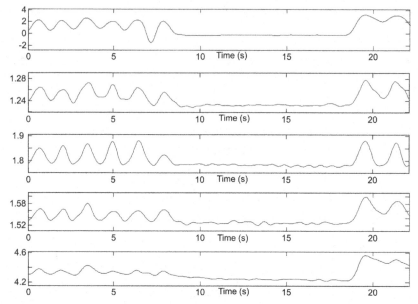

FIGURE 4.4 Comparison of the respiratory signal obtained in basal condition by means of the different fabric sensors and the Biopac spirometer. From the top: Biopac pneumography transducer, impedance pneumography, KPF_T and PPF_T thorax plethysmography, KPF_A abdominal plethysmography.

while inhaling via the thoracic muscles, and vice versa. An example of a trial of such simulated respiration is shown in Fig. 4.3, where the upper and lower signals are in phase opposition.

The possibility to detect the apnea phase has also been investigated; comparison between different fabric sensors and methodology is reported in Fig. 4.4. In this case a Biopac pneumograph transducer has been used to validate the textile sensors. The Biopac signal as well as the ones coming from the wearable system allows to identify six inspiration/expiration cycles followed by about 15 s of apnea and two more respiratory cycles. The difference between abdominal and thorax movement during breathing can be observed on the graph.

Textile platforms for cardiopulmonary monitoring based on fabric sensor components

From Wealthy to HealthWear platforms

The first e-textile platform addressing clinical rehabilitation for cardiac patients has been implemented in the frame of the Wealthy project, a project aiming at the implementation of a proof-of-concept platform, where a knitted textile infrastructure was designed and manufactured to acquire, process, and

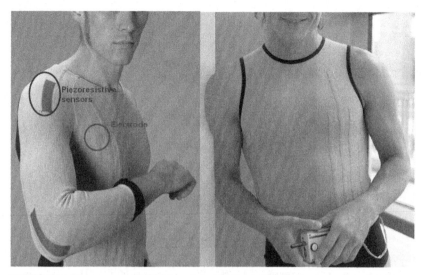

FIGURE 4.5 Two versions of Wealthy sensing platform, the system combines six fabric ECG electrodes, a double fabric layer with the conductive part in contact with the skin, three fabric impedance electrodes, bigger than the ECG electrodes, nine fabric insulated connections, a multilayer structure with the conductive fabric sandwiched between two insulating layers of textile, four fabric piezoresistive sensors, double fabric layer to insulate the sensor from the skin, two embedded temperature sensors.

transmit a selected set of physiological signals. The Wealthy project resulted in a wearable fully integrated system, capable to acquire, simultaneously and in a natural environment, a set of physiological parameters like electrocardiogram, respiration, posture, temperature, and movement index. Sensors and connections are integrated into the fabric structure in one production step (details about materials and processes of Wealthy shirt design and implementation have been given in Refs. [12,15]); conductive filaments are combined with stretchable yarns, and the use of flat knitting technology allows the integration of sensing components in the knitted interlaced structure during the manufacturing process (see Fig. 4.5).

The garment is connected with an electronics device: the PPU (Portable Patient Unit), shown in Fig. 4.6, the core of a two-way communication system, where local processing as well as communication with the network is performed. The project successfully established the first textile-based monitoring system capable of acquiring vital signs and transmitting them in real time to a monitoring center for immediate analysis of data. After initial ECG signal processing, data are transmitted to a monitoring center where they can be visualized, processed, and recorded in a database. Feedback from doctors to patients is possible at all times, and in case of alarm the doctor can take immediate action.

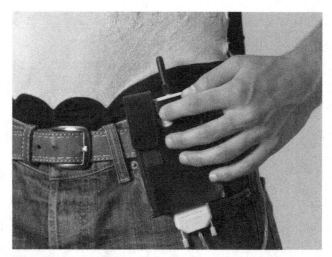

FIGURE 4.6 Wealthy Patient Portable Unit (PPU), realized by CSEM Switzerland. ECG signals are sampled on the PPU at 250 Hz where a local processing is applied in order to extract parameters with the highest sampling rate, in a way to compute ECG parameters such as heart rate (HR) value and QRS1 duration. The system is allowed to detect: six ECG electrodes configurable in Einthoven configuration (lead I, II, and III) and Wilson configuration (V2 and V5). Only one lead is transmitted at a time (for the General Packet Radio Service (GPRS) bandwidth limitation reasons). The ECG lead to be transmitted can be selected remotely by the monitoring center, respiration by impedance measurement, up to four skin temperature sensors (monolithic circuits), one 3D accelerometer (integrated in the unit), four piezoresistive strain sensors, SpO2 (pulse oximetry) from a commercial device provided by Nonin Medical Inc.

The garment was used as starting platform to develop a more comfortable system in the frame of the project: HealthWear (e TEN C029402), a collaborative EU-funded project aiming at the provision of a service that offers uninterrupted and ubiquitous monitoring of the health condition of individuals in the rehabilitation phase [16]. To achieve this purpose, the service has been validated for several use-cases, involving volunteers with COPD and heart diseases. The PPU was used to transmit the signals to a central processing site through the use of GPRS wireless technology. This service was applied to three distinct clinical contexts: rehabilitation of cardiac patients, following an acute event; early discharge programs in chronic respiration patients; promotion of physical activity in ambulatory stable cardiorespiratory patients.

Garment design improvement

In order to increase comfort and to explore the best solution for the beneficiaries of HealthWear service, a new model has been implemented according to the following criteria: the garment has to be easy to wear and comfortable from the thermal aspect and from the ergonomics aspect, washable and good enough from the look and feel perspectives. Males in the age range of 40–60

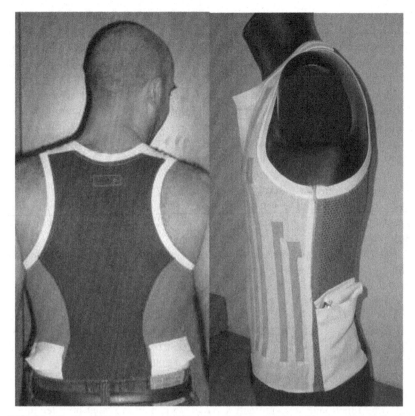

FIGURE 4.7 HealthWear new model; left: view of the back side, right: view of the left side.

constituted the target subjects; most of them were overweight; comfort and functionality were key issues that have been considered. Fig. 4.7 shows the HealthWear platform.

As can be seen from Fig. 4.7, there is a pocket on the back of the garment for the portable electronics and net-fabric inserts used to give a freshness feeling; the garment foresees a zip on the front to facilitate the wearing process.

Textile platform for cardiopulmonary monitoring based on hardware components

From WELCOME to WELMO monitoring system

WELCOME was a European-funded project, which aimed to create a technology solution enabling step-change in the integrated care of, and self-management by, patients suffering from chronic obstructive pulmonary disease (COPD) and its comorbidities.[1] The project foresaw the development of a light, easy, comfortable-to-wear, and washable vest integrating the electrical connections among 20 standalone, noninvasive chest sensors working in concert for measuring and monitoring various parameters of COPD and comorbidities.

More specifically, the parameters provided by the measuring sensors were the electrical impedance tomography (EIT), 12-lead ECG configuration, and chest sound (used to identify crackles, rhonchi, or wheezing); the sound measurements foresaw the placement of some sensors also on the back. The EIT was monitored to evidence if the findings were consistent with increased degree of ventilation heterogeneity associated with exacerbation of COPD, inadequate therapy or natural progression of the disease, pulmonary edema, and/or pleural effusion (especially in the COPD). The ECG was monitored to detect the onset of a not previously reported tachycardia or arrhythmia like atrial fibrillation.

Moreover SpO2 (oxygen saturation) and body movements were measured with a special reference sensor.

Vest design

The aim of the project WELCOME was the development of a textile platform where sensors were easily integrated in a vest that could be worn and used during the daily life routine of the COPD patients [17].

To reach this goal the vest was designed considering the user requirements in terms of wearing comfort, body size, gender, and age. But primarily the vest had to be functional: to manufacture a system capable to meet both technical and wearable requirements a multilayered structure was implemented, realized with different kinds of fabrics and textile materials afferent to different technologies. The first challenge was related to the handling of a large number of electronic devices in strict contact with the body, considering that the vest has to be comfortable and easy to wear, so an accurate selection of materials was done. Fabrics with different weight were selected using them according to the function and the part of the body to be covered. To enhance the adhesion of sensing regions (where the sensors were placed) a more structured fabric was used, and a frame was created to keep the electrodes in the right location; as the final weight of a large number of devices all close together was remarkable, even if the weight of a single device was small; the lightweight fabric was used to manufacture the rest of the vest to reduce as much as possible all the discomfort given by the combination of many rigid cases and the heavy holding fabric. The male model of the vest size L is shown in Fig. 4.8.

---

1. http://www.welcome-project.eu.

FIGURE 4.8 On the left the outer of the male model, on the right the inner (back and front); inside the vest there are 21 labels with impressed letters to indicate the location of each sensor; the color is different based on the type of the sensors; the bigger label (Ref) indicates the position of the Reference electrodes which collects the data from the other electrodes.

Differences between the male and the female models

One of the main requirements for the sensor's location concerned the configuration of the electrodes in terms of position and symmetry that had to be the same for the male and the female models, to apply the same model for the software image reconstruction. Unfortunately, considering the anatomic difference due to the gender, this request was not totally satisfied: to reduce the differences the vest for women was optimized for an androgynous body shape, while for women with a large breast, the electrodes weren't perfectly aligned on the same plane for a male or female subject. For the electrodes that are located slightly above the diaphragm, to increase the adherence in the female model and to maintain the symmetry with the male model, an elastic belt was added in the inner side of the garment: in this way the belt held the sensors under the breast, as shown in Fig. 4.9.

Conductive interconnections

A textile two-wire bus was foreseen to comply with the functional electrical requirements. The bus realized was done by two wires to connect the two points that were used to recharge the system (positive and negative contacts). The path followed to electrically connect all the measuring sensors is shown in Fig. 4.10.

From the textile point of view the complexity was to find a conductive wire that was compatible with textile technology to be integrated in the vest and highly conductive to guarantee the functionality of all the measuring

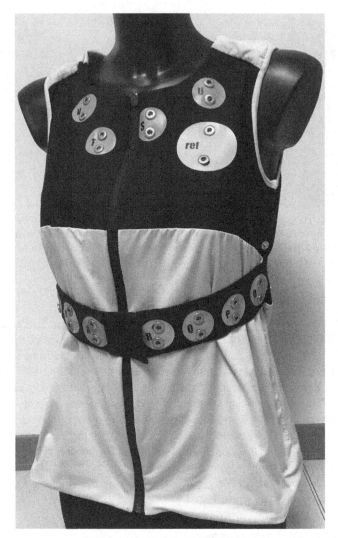

FIGURE 4.9 Female model inner side.

sensors; the selected wire was an insulated stainless-steel multifilament with 3.6 Ohm/m of resistance value. The multifilament's structure provides flexibility to this solution, and the quantity of conductive monofilaments was consistent with the conductivity requirement of the electronics functionality.

The global cabling path (Fig. 4.10) was divided into four subpaths to facilitate the manufacturing process, as shown in Fig. 4.11.

Since the chest circumference changes with the size, the distance between consecutive sensors is different and therefore the estimation of the length of

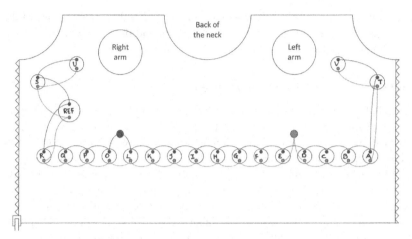

FIGURE 4.10 Cabling path for the connection of 20 electrodes plus the reference.

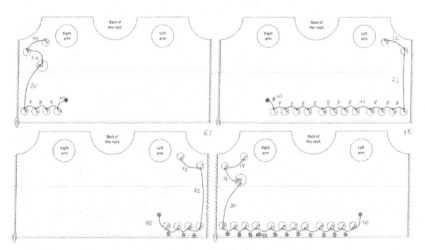

FIGURE 4.11 The *blue lines* (black in print version) in the drawings connect the positive poles; the red (gray in print version) ones join the negative poles.

the cable between the sensors' positions was adapted to the size development. In Fig. 4.11 the sensors connections of the male vest are depicted; in each drawing the length of the cable between two consecutive positions is reported; the global length was evaluated to optimize the consumption and to reduce the resistance value of the two-wire bus. For each size different estimations have been done considering the dimensions of the garment; for the whole cabling at least 5 m of cable were needed.

For each connection, the cable was soldered on the spring of the snap and then located in the inner side of the vest that is in contact with the skin when the garment is worn. All the cables were embedded in the internal space of the vest, between the fabric in contact with the skin and the one external, to limit any access from the user.

Interconnection through snap buttons

Snap buttons of medium size have been selected to hold the sensors in place, to be compatible with the contacts of the measuring electrodes represented by molded male snap buttons (Fig. 4.12). From the textile point of view this choice was not ideal because the high number of snap buttons made the garment heavy, affecting the wearing comfort of the vest. Lighter buttons with the same size weren't acceptable from the electronic point of view because this implicated a redesign of the electronics; furthermore, stainless-steel buttons with a smaller diameter weren't available for sale in stock service; therefore this solution was not applicable in the time frame of the project.

The electronic connection of the snap with the wire was a technological challenge: to increase the functionality of the system, from the electronic point of view, the conductive wire was soldered as close as possible to the contact with the electronics; therefore the better place to solder the conductive cable was close to the spring that is inside the female snap button (Fig. 4.13); this

FIGURE 4.12 Molded male snap buttons for the electrical contacts with the sensors.

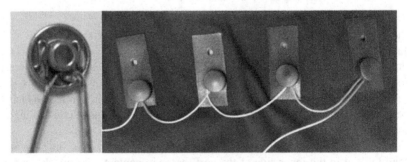

FIGURE 4.13 Welding on the snap spring.

solution was optimal to fulfill the electrical requirements in terms of connectivity but the welding reduced the movement of the spring of the female snap; moreover, during the attachment and detachment of the sensor time, the soldering underwent to a mechanical solicitation that deteriorates the quality of the electrical contact with the sensor and the robustness of the holding structure.

Standard textile tests

The textile vest was tested to verify the washing capability (after removing the electronics) and the capability to resist to high temperature and humidity following the standard procedure described in the norm UNI EN ISO 6330:2012 and UNI EN ISO 1419:1995. The result of the test showed that the lifetime of the garment was long enough to be used in a daily practice, in particular the welding was strong enough to resist to high values of temperature and humidity. Several vest prototypes with mock sensors were developed. A structured interview was designed to explore COPD patient perceptions pertaining to vest comfort, easiness of wearing and handling, willingness to use. Interviews have taken place at the clinic (England and Germany) where patients were provided with a suitably sized vest to try on followed by the structured interviews [18–20].

WELMO

WELMO[2] project aims at developing and validating a new generation of low-cost and low-power miniaturized sensors, integrated in a comfortable vest, enabling the effective and accurate monitoring of the lungs, through the simultaneous collection of sound and EIT signals with the same sensors that can be combined, processed, and linked with specific clinical outcomes making the systematic, accurate, and real-time evaluation of respiratory conditions possible. The project design is based on the experience acquired within WELCOME project [21,22].

The new vest incorporates secure wireless transmission capabilities through which the data collected are transferred to the WELMO central node where the major processing and results visualization takes place.

The concept of the vest

The wearable system merges three main functional modules. This structure is designed to maintain the sensors in close contact with the skin, to guarantee their functionality and location, as well as the easiness of use and wearing comfort.

---

2. http://www.welmo-project.eu/.

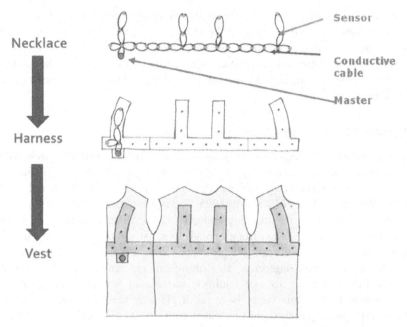

FIGURE 4.14   The main functional components of the WELMO Vest.

The sensing functionality is given by the sensors and the interconnection, each sensor is electrically linked to the neighbor through two wires, one side of the sensors has to be in contact with the skin in the right position without moving for the whole duration of the monitoring session. This has been conceived as a sort of Necklace whose chain is made of conductive cables and whose pendants are the sensors (see Fig. 4.14). The Necklace is embedded in an elastic multilayered structure, the Harness, which represents the middle layer, and the functional core of the entire system. The use of an elastic structure allows to push the electrodes on the body, for the correct functionality of the system. The harness is connected to the Master that is the device to which all the sensors are connected, and that has to be located on the external side to be used by the patient during the monitoring session.

Finally, the outer layer consists of a 'real' garment, the Vest: the main function of the vest is to locate the sensors on the right position, to increase the adherence of the sensors to the skin, to hold the Master, and the reference fabric electrodes; the design of the vest conceives anchor mechanisms (i.e., zippers, velcro, buttons, etc.) to integrate the harness. Two textile electrodes, Guard and Reference, are placed on the inner side of the vest, in contact with the skin. These two electrodes guarantee the correct functionality of the system and are electrically connected to the Master. Different areas of the external layer are filled to improve the adhesion of the sensors and the textile electrodes with the body.

Vest design

The need for adherence of the electronic devices is one of the main constraints for the study of the design of the garment, combined with the usability requirements, as we have to consider that the vest will be used by patients suffering from chronic diseases. For this reason, it is important to address the easiness of wearing with a zip closure that can be easily used: the need for a front opening has a strong impact on the whole symmetry of the vest design and circuit arrangement.

Another essential aspect is the integration of the whole set of the electronic devices and the electrical interconnections. With the exception of the sensors' sensing side, the skin of the wearer has to be protected from the electrical components that must be inaccessible during the monitoring session for safety reasons. The harness was then insulated in a double layered structure. The internal layer is equipped with a series of "loops" from which the sensors are coming out (Fig. 4.15), to contact the skin at the desired positions. The area of the fabric supporting the loops is reinforced. The external layer holds the Master, through a contact's base that is connected to the inner cables placed on the harness.

The architecture of the inner layer is the functional core of the vest; it is the interspace between the external and the internal side, from the source of body signals (sensors) to the management of the acquisition (master).

The sensors come out through the loops, while the harness circuit always remains in the interspace. The structure provides electrical, the insulated cables are not accessible, and mechanical protection, sensors are covered with multiple textile layers that can adsorb any mechanical inputs.

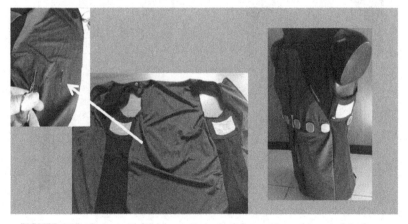

FIGURE 4.15 Inner part of the garment; on the left top side, a close up of the loops.

In order to access the interspace between the two layers of fabric and to be able to place the harness in the vest, a passage large enough to work comfortably in the interspace was implemented.

The design foresees the use of two jersey fabrics, with different weight, the back of the vest is realized with the lighter one, to smoothen the compressive feeling of the sensors regions and to increase the wearing comfort for the user. The parts that have to compress the sensors on the body are with the fabric with a higher elastomeric components and weight: to improve the contact of the sensors with the skin.

In Fig. 4.15 the inner side of the vest is shown with a detail on the loop for the sensor and finally the complete system, with all the integrated components. This prototype was used to verify the sensors placement and to perform ergonomic tests.

In Fig. 4.16 the distribution of the sensors is represented, the harness connects 18 sensors and 1 master: 16 sensors are distributed around the thorax; the remaining 2 chest sound sensors (number 21 and 22) are not aligned to the others, and placed at the front top part of the chest.

In order to place the sound sensors in contact with the skin, a dedicated structure (Fig. 4.17) has been foreseen. This structure consists of two nonelastic strips sewn into the harness in correspondence with the positions of

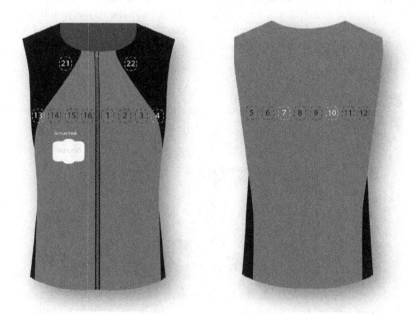

FIGURE 4.16 Distribution of the 18 sensors and the master on the front and back parts of the vest.

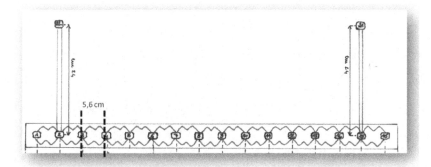

FIGURE 4.17 Drawing of the harness, 16 sensors for L size (thorax length 90 cm); the distance between two adjacent sensors is 5.6 cm.

the electrodes 2 and 15. Two male snaps buttons are attached to the ends of these strips: once the male buttons are clipped to the corresponding female ones (that are placed in the inner part of the vest) each sound sensor (21 and 22) goes into its correct position.

Another constraint to consider is related to distance between the sensors that has to be constant along the whole circumference, in order to allow the EIT image reconstruction. Each vest size will have its own intersensor distance. This is considered in the size development to guarantee a correct distribution of the sensors around the chest.

A first prototype was manufactured to fit the body of the engineer that did the first ergonomics trials. The following sketch reports the measures used for this first harness, related to a thorax circumference of 90 cm (Fig. 4.17).

The distance between one sensor and the neighbors is related to the subject thorax size, for instance, for an L size, a total of 16 sensors has to be placed in 90 cm: 5.6 cm is the distance between one sensor and the next (Fig. 4.18).

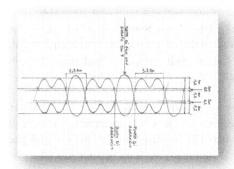
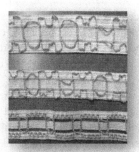

FIGURE 4.18 Arrangements of the cables on the harness.

FIGURE 4.19 The elastic cabling structure connected to one sensor.

The design of the elastic core of the harness foresees two lines with wider meshes to facilitate the attachment of the sensors to it. Each sensor has a plastic plate to fix the sensor to the harness. The plate and the sensor are held together with two screws; the two lines in the structure allow the screws to cross the elastic structure (Fig. 4.19).

The harness pushes the sensors on the skin due to its elastic nature; the strips for the sound sensors 20 and 21 instead are not elastic and extra fillers are foreseen to increase the adhesion of these sensors on the skin. The two conductive cables are arranged on the elastic structure following a sinusoidal arrangement, since one of the main electrical requirements is related to the stability of cable conductivity, i.e., the cable cannot be stretchable to avoid the presence of piezoresistive behavior. This design solution allows the elongation of the harness in the horizontal direction without stressing the cable (this occurs when the user wears the vest). The first version of the elastic structure integrating the conductive cables was produced with an industrial equipment, the structure is a 4 cm wide elastic band with the cables that realizes eight waves in the space of 3.3 cm. Each wave is anchored to the elastic at four points (see Fig. 4.20).

The master has four connections: two connections for the upper and lower lines of the 18 sensors, and the other two for the textile electrodes (Guard and Reference). In Fig. 4.21 the first harness manufactured is shown with the sensors mounted on it (16 EIT plus 2 sounds). Two snap buttons placed at the ends of the vertical strips allow to "hang" the structure to the vest. The elastic band can then be fixed around the chest with a closure.

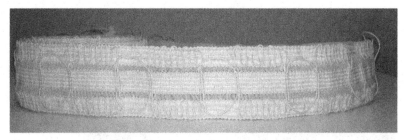

FIGURE 4.20 The final elastic cabling structure.

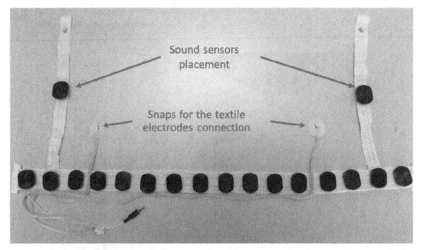

FIGURE 4.21 First prototype of the harness. Two knitted ribbons have been realized as a Faraday's cage (with a cotton yarn outside and a metal yarn inside), to protect and insulate the conductive cable for the Master electrodes (Reference and Guard); these two ribbons are stitched in some points to the belt; the electronics connections with the two textile sensors, that are integrated in the vest, are done by the two snap buttons.

In Fig. 4.22 the overall view of the WELMO system is shown: it consists of the harness that is between the two layers of the vest and the two textile electrodes Master and Guard that are integrated in the inner side of the vest, under the armpits. This location was chosen to guarantee a constant contact of the sensing areas with the body and to improve the conductivity of the fabric, thanks to the natural humidity of this part of the body. The electrodes work better if the contact with the skin is improved by the presence of sweat due to the presence of ions. The electrodes are filled with neoprene to increase the pressure of the conductive areas against the skin and thus, the positive effect of sweating.

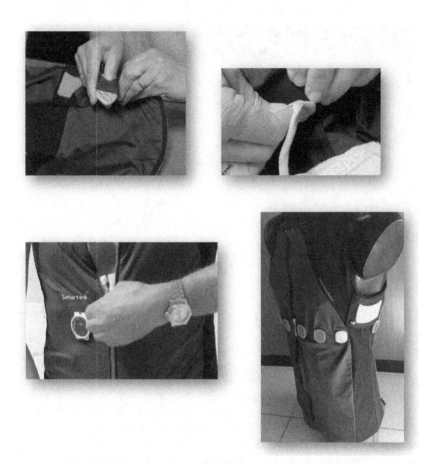

FIGURE 4.22   Overall view of the first prototype of WELMO vest.

Conclusions

Wearable monitoring systems capable of acquiring physiological signals and providing health information from a remote setting have been designed, implemented, and used in numerous research projects starting from the integration of fabric sensing components to the design of sophisticated collaborative architecture of miniaturized devices for imaging and sound sensing.

Textile technology and knowledge have been exploited to build a new class of tools specifically designed to be used without any additional assistance in a home environment, to provide a remote efficient service of diagnosis and assistance.

Any wearable application requires a very good fit of the sensors on the body, as well as efficient interconnections structure, which must be highly conductive, robust, and reliable.

To overcome these constraints, a system has to combine performing fabrics in terms of shaping, softness, and comfort to feel natural and confident, while wearing a monitoring garment. Supported by advanced acquisition, powering and communication technology. The paradigm is once again to develop a complex pipeline that can be perceived as natural as a traditional technical cloth, easy to use, beautiful to wear.

References

[1] E.P. Scilingo, A. Gemignani, R. Paradiso, N. Taccini, B. Ghelarducci, D. De Rossi, Performance evaluation of sensing fabrics for monitoring physiological and biomechanical variables, IEEE Trans. Inf. Technol. Biomed. 9 (3) (Sept. 2005) 345−352, https://doi.org/10.1109/TITB.2005.854506.

[2] P.T. Gibbs, H. Asada, Wearable conductive fiber sensors for multi-axis human joint angle measurements, J. NeuroEng. Rehabil. 2 (2005) 7, https://doi.org/10.1186/1743-0003-2-7.

[3] A. Tognetti, F. Lorussi, R. Bartalesi, et al., Wearable kinesthetic system for capturing and classifying upper limb gesture in post-stroke rehabilitation, J. NeuroEng. Rehabil. 2 (2005) 8, https://doi.org/10.1186/1743-0003-2-8.

[4] M. Pacelli, L. Caldani, R. Paradiso, Performances evaluation of piezoresistive fabric sensors as function of yarn structure, in: 2013 35th Annual International Conference of the IEEE Engineering in Medicine and Biology Society (EMBC), 2013, pp. 6502−6505, https://doi.org/10.1109/EMBC.2013.6611044.

[5] M. Pacelli, L. Caldani, R. Paradiso, Textile piezoresistive sensors for biomechanical variables monitoring, in: 2006 International Conference of the IEEE Engineering in Medicine and Biology Society, 2006, pp. 5358−5361, https://doi.org/10.1109/IEMBS.2006.259287.

[6] A. Tognetti, F. Lorussi, G.D. Mura, et al., New generation of wearable goniometers for motion capture systems, J. NeuroEng. Rehabil. 11 (2014) 56, https://doi.org/10.1186/1743-0003-11-56.

[7] J.F. Fieselmann, M.S. Hendryx, C.M. Helms, D.S. Wakefield, Respiratory rate predicts cardiopulmonary arrest for internal medicine inpatients, J. Gen. Intern. Med. 8 (7) (1993) 354−360.

[8] R. Strauß, S. Ewig, K. Richter, T. König, G. Heller, T.T. Bauer, The prognostic significance of respiratory rate in patients with pneumonia: a retrospective analysis of data from 705,928 hospitalized patients in Germany from 2010-2012, Dtsch. Arztebl. Int. 111 (29−30) (July 21, 2014) 503−508, https://doi.org/10.3238/arztebl.2014.0503, i−v, PMID: 25142073; PMCID: PMC4150027.

[9] J.B. West, Respiratory Physiology: The Essentials, fourth ed., Williams and Wilkins, Baltimore, 1990.

[10] D.R. Goldhill, A.F. McNarry, Physiological abnormalities in early warning scores are related to mortality in adult inpatients, Br. J. Anaesth. 92 (2004) 882−884.

[11] Task Force of the European Society of Cardiology and the North America Society of Pacing and Electrophysiology, Heart rate variability standards of measurement, physiological interpretation and clinical us, Circulation 93 (5) (1996) 1043−1065.

[12] R. Paradiso, G. Loriga, N. Taccini, A wearable health care system based on knitted integrated sensors, IEEE Trans. Inf. Technol. Biomed. 9 (3) (Sept. 2005) 337−344, https://doi.org/10.1109/TITB.2005.854512.

[13] G. Loriga, N. Taccini, M. Pacelli, R. Paradiso, Flat knitted sensors for respiration monitoring [From mind to market], IEEE Ind. Electron. Mag. 1 (2007) 4—7.

[14] N. Taccini, G. Loriga, M. Pacelli, R. Paradiso, Wearable monitoring system for chronic cardio-respiratory diseases, in: 2008 30th Annual International Conference of the IEEE Engineering in Medicine and Biology Society, 2008, pp. 3690—3693, https://doi.org/10.1109/IEMBS.2008.4650010.

[15] R. Paradiso, L. Caldani, Electronic textile platforms for monitoring in a natural environment, Res. J. Text. Appar. 14 (4) (2010) 9—21, https://doi.org/10.1108/RJTA-14-04-2010-B002.

[16] R. Paradiso, A. Alonso, D. Cianflone, A. Milsis, T. Vavouras, C. Malliopoulos, Remote health monitoring with wearable non-invasive mobile system: the Healthwear project, in: 2008 30th Annual International Conference of the IEEE Engineering in Medicine and Biology Society, 2008, pp. 1699—1702, https://doi.org/10.1109/IEMBS.2008.4649503.

[17] R. Paradiso, L. Caldani, G. De Toma, From the design to real e-textile platforms for Rehabilitation and chronic obstructive pulmonary diseases care, in: 2015 37th Annual International Conference of the IEEE Engineering in Medicine and Biology Society (EMBC), 2015, pp. 446—449, https://doi.org/10.1109/EMBC.2015.7318395.

[18] R. Kayyali, R. Siva, R.E. Kaimakamis, M.A. Spruit, A. Vaes, J. Chang, R. Costello, N. Davies, N. Philip, B. Pierscionek, E. Perantoni, R. Paradiso, A., Raptopoulos, S. Nabhani-Gebara, Wearable smart technology for monitoring COPD with co-morbidities — patients' perceptions, Eur. Respir. J. 46 (Suppl. 59) (September 2015) OA3278, https://doi.org/10.1183/13993003.congress-2015.OA3278.

[19] R. Kayyali, B. Odeh, I. Frerichs, N. Davies, E. Perantoni, S. D'arcy, A.W. Vaes, J. Chang, M.A. Spruit, B. Deering, N. Philip, R. Siva, E. Kaimakamis, I. Chouvarda, B. Pierscionek, N. Weiler, E.F. Wouters, A. Raptopoulos, S. Nabhani-Gebara, COPD care delivery pathways in five European Union countries: mapping and health care professionals' perceptions, Int. J. Chronic Obstr. Pulm. Dis. 11 (1) (2018), https://doi.org/10.2147/COPD.S104136.

[20] R. Kayyali, V. Savickas, M.A. Spruit, et al., Qualitative investigation into a wearable system for chronic obstructive pulmonary disease: the stakeholders' perspective, BMJ Open 6 (2016) e011657, https://doi.org/10.1136/bmjopen-2016-011657O.

[21] E. Kaimakamis, E. Perantoni, E. Serasli, V. Kilintzis, I. Chouvarda, R. Kayyali, S. Nabhani-Gebara, J. Chang, R. Siva, R. Hibbert, N. Philips, D. Karamitros, A. Raptopoulos, I. Frerichs, J. Wacker, N. Maglaveras, Experience of using the WELCOME remote monitoring system on patients with COPD and comorbidities, © Springer Nature Singapore Pte Ltd, in: N. Maglaveras, et al. (Eds.), Precision Medicine Powered by pHealth and Connected Health, IFMBE Proceedings 66, 2018, https://doi.org/10.1007/978-981-10-7419-6_1797.

[22] I. Frerichs, et al., Multimodal remote chest monitoring system with wearable sensors: a validation study in healthy subjects, Physiol. Meas. 41 (1) (February 2020) 015006, https://doi.org/10.1088/1361-6579/ab668f.

Part III

Data analysis and management

Chapter 5

Automated respiratory sound analysis

Diogo Pessoa[a], Bruno Machado Rocha[a], Paulo de Carvalho and
Rui Pedro Paiva
Department of Informatics Engineering, Centre for Informatics and Systems of the University of Coimbra, Coimbra, Portugal

Introduction

Respiratory diseases are among the most significant causes of morbidity and mortality worldwide [1] and are responsible for a substantial strain on health systems [2]. Fig. 5.1 shows the comparison between the top 10 causes of death globally in 2000 and in 2019. While the number of deaths caused by lower respiratory infections dropped, deaths due to chronic obstructive pulmonary disease (COPD) increased, as well as the number of cancers in organs of the respiratory tract. Therefore, significant research efforts have been dedicated to improving early diagnosis and routine monitoring of patients with respiratory diseases to allow for timely interventions [3].

Spirometry and auscultation are typically used to assess respiratory function and the progress of respiratory diseases, the former measuring the volume of air circulated during the inspiratory phase and the latter assessing airflow through the tracheobronchial tree, via the sounds produced [4]. Spirometry is effective and well validated for the diagnosis and monitoring of upper and lower airway abnormalities [5], but its dependence on the patient's motivation and cooperation is a major limitation for its use [3]. Concurrently, the existence of respiratory conditions may be assessed through the auscultation of respiratory sounds [4]. The stethoscope is the main tool for lung auscultation in clinical practice and auscultation is typically performed on the anterior and posterior chest [6]. Expert physicians are trained to recognize the presence of anomalous findings, such as adventitious sounds. While the technological advances in stethoscopes have enabled the visualization and analysis of respiratory sounds in computers, digital auscultation is not yet entirely

---

a. These authors contributed equally to this work.

Wearable Sensing and Intelligent Data Analysis for Respiratory Management
https://doi.org/10.1016/B978-0-12-823447-1.00003-8
123

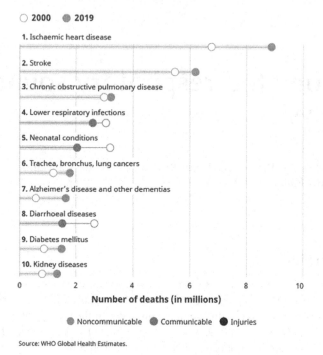

FIGURE 5.1 Top 10 causes of death in 2000 and in 2019 [1].

computational [7]. Some handicaps limit the use of conventional auscultation in research: (1) the presence/absence and clinical meaning of respiratory sounds need to be evaluated by an expert [8]; (2) continuous monitoring is impractical; (3) human audition and memory limitations [9]; (4) interlistener variability [10]; (5) stethoscopes can be a source of infection and need to be constantly sterilized, as demonstrated during the COVID-19 crisis [11]. These drawbacks hinder the effectiveness of conventional auscultation as a way of monitoring and managing respiratory conditions. Automated respiratory sound analysis could potentially overcome these limitations [4].

Description of respiratory sounds

In general terms, respiratory sounds refer to the specific sounds generated by the movement of air through the respiratory airways. These may be easily audible or identified through auscultation of the respiratory system over the lung fields with a stethoscope, as well as from the spectral characteristics of lung sounds.

Respiratory sounds exhibit different acoustic properties depending on each person's characteristics (e.g., age, gender, height, position, and airflow), location of sound acquisition, and position in the respiratory phase [12]. The definition of such characteristics and the mastering of the pulmonary auscultation technique, by health professionals, allows objective interpretations of respiratory sound alterations and, potentially, enhances the early detection, treatment, and monitoring of respiratory diseases. Moreover, the clear characterization of such sounds boosts the development of computerized methods to analyze respiratory sounds, since it creates a set of objective properties that identify a certain type of sound. Today, most of the characteristics that describe respiratory sounds are obtained from data gathered through computerized acoustic devices, such as electronic stethoscopes and microphones. Notwithstanding, the methods for recording, analyzing, and reporting respiratory sounds still differ significantly among studies, which weakens our understanding of the clinical meaning of these sounds and the changes in their acoustics. Therefore, keeping updated guidelines on the recording and further characterization of respiratory sounds is extremely important to compare results among studies and will ease the integration of the information provided by respiratory sound analysis. This development is essential to solidify and enhance the utility of auscultation at patients' bedside and further develop auscultation-based diagnosis and monitoring applications.

In 2000, a task force of the European Respiratory Society (ERS) published a set of guidelines for research and clinical practice in the field of computerized respiratory sound analysis (CORSA) [13]. These guidelines included an extensive list of term definitions, namely those of respiratory diseases, pulmonary physiology, acoustics, automatic data handling, and instrumentation. In Fig. 5.2, a schematic with the hierarchy proposed in Ref. [13] for the categorization of respiratory sounds is presented. According to this figure, respiratory sounds can be subdivided into three main groups: breath sounds, adventitious sounds, and lung sounds. Therefore, those three main groups include breath sounds, adventitious sounds, cough sounds, snoring sounds, sneezing sounds, and sounds from the respiratory muscles. Voiced sounds during breathing are not included in respiratory sounds [13]. Moreover, the sounds can also be divided according to the region from which they are heard. In general, the current trend is to reduce the number of categories and standardize the nomenclature [14].

Breath sounds

According to Ref. [13], breath sounds are those arising from breathing, excluding adventitious sounds, heard or recorded over the chest wall, the trachea, or at the mouth. The generation of breath sounds is related to airflow in the respiratory tract. Acoustically, they are characterized by broad spectrum noise with a frequency range depending on the pick-up location.

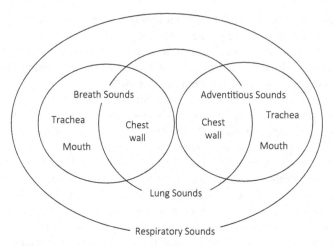

FIGURE 5.2 Automated respiratory sound analysis. *Adapted from A.R. Sovijärvi, F. Dalmasso, J. Vanderschoot, L.P. Malmberg, G. Righini, S.A. Stoneman, Definition of terms for applications of respiratory sounds, Eur. Respir. Rev. 10 (77) (2000) 597–610. Acknowledgement Wording: Reproduced with permission of the © ERS 2022: Eur Respir Rev 2000; 10: 77, 597–610.*

Adventitious sounds

Adventitious respiratory sounds (ARSs) are additional respiratory sounds superimposed on normal respiratory sounds [9,12,13]. There are two main types of adventitious sounds: continuous and discontinuous.

Lung sounds

Lung sounds concern the respiratory sounds heard or detected over the chest wall, or within the chest, including breath sounds and adventitious sounds detected at this location [9,13].

Normal respiratory sounds

Normal respiratory sounds are the sounds produced from breathing and can be categorized based on the location where they are heard or generated, such as the trachea, chest wall, and mouth [8,9,12,13]. Depending on the auscultation location, different types of respiratory sounds have distinct characteristics such as duration, pitch, and sound quality [8]. Characteristics of normal respiratory sounds also change from individual to individual because airway dimensions are a function of body height [15]. However, other factors such as gender, chest location where they are heard, body position, and airflow can also affect normal respiratory sounds [12].

Lung or vesicular sounds

The sound of normal breathing heard over the surface of the chest is markedly influenced by the anatomical structures between the site of sound generation and the site of auscultation. The sounds heard over the chest wall, normal or vesicular lung sounds, are soft and nonmusical, and generally only heard during inspiration and early part of expiration [8,12,16]. It is believed that normal lung sounds at frequencies above 300 Hz are generated by turbulent airflow vortexes; however, the generation of sounds heard below 300 Hz is not clear [17]. It seems well established that the sound heard during inspiration is produced primarily within the lobar and segmental airways, whereas the sound produced during expiration seems to be originated in more proximal airways. Several mechanisms of vesicular sounds have been suggested, including turbulent flow, vortexes, and other, hitherto unknown mechanisms [16]. Clinically, a decrease in sound intensity is the most common abnormality. Sound generation can be decreased when there is a drop in inspiratory airflow, which can result from several conditions, ranging from poor cooperation to depression of the central nervous system [16]. Airway conditions include blockage (e.g., by a foreign body or tumor) and the narrowing that occurs in obstructive airway diseases (e.g., asthma and COPD). The decrease in the intensity of breath sounds may be permanent, as in cases of pure emphysema, or reversible, as in asthma (e.g., during a bronchial provocation test or an asthma attack).

Tracheal sounds

Tracheal auscultation is not frequently performed, but, in certain situations, can convey important clinical information. When heard at the suprasternal notch or the lateral neck, normal tracheal sounds characteristically contain a large amount of sound energy and are easily heard during the two phases of the respiratory cycle [16]. Normal tracheal sounds are hollow and clearly heard in both phases of respiratory cycle, mostly on inspiration and on early expiration [8,12,16]. Their intensity and their pitch on the inspiration phase are higher than on the expiration phase [18]. Typically, they are harsh, loud, and high-pitched sounds covering a wide range of frequencies, from less than 100 Hz up to 5000 Hz, with a sharp drop in power at frequencies above 800 Hz and little energy beyond 1500 Hz [12,18]. As mentioned before, these sounds can be heard at the suprasternal notch and are generated by turbulent airflow in the upper airways, including pharynx, glottis, and subglottic regions [19]. Listening to tracheal sounds can be useful in a variety of circumstances [16]. First, the trachea carries sound from within the lungs, allowing auscultation of other sounds without filtering from the chest cage. Second, the characteristics of tracheal sounds are similar in quality to the abnormal bronchial breathing heard in patients with lung consolidation. Third, in patients with upper airway obstruction, tracheal sounds can become frankly musical, characterized as either a typical stridor or a localized, intense wheeze.

Bronchial sounds

Normal bronchial sounds are heard over the large airways on the chest, specifically near the second and third intercostal space. Bronchial sounds are hollower and more high-pitched than vesicular sounds [18]. Bronchial sounds are audible during both inspiratory and expiratory phases [8]. In contrast with vesicular sounds, due to the sounds being originated in larger airways, the expiratory phase sounds are normally audible for longer than the inspiratory phase ones. The intensity of expiration phase sounds is also higher than the intensity in the inspiration phase. Unlike vesicular sounds, there is a short pause in-between each cycle of breathing.

Mouth sounds

Breath sounds heard from the mouth are produced by central airways and caused by turbulent airflow below the glottis. Breath sounds from the mouth have a wide frequency range of 200—2000 Hz [20]. The energy distribution is like that of white noise. For a healthy person, breath sounds heard from the mouth should be silent.

Adventitious respiratory sounds

As already mentioned, ARSs are additional respiratory sounds superimposed on normal respiratory sounds [9,13,16,18]. These sounds are related to changes within lung morphology, in all airways. They are divided into two main groups: discontinuous adventitious sounds (DASs) (e.g., crackles) and continuous adventitious sounds (CASs) (e.g., wheezes). Usually, the presence of these types of sounds indicates the existence of a pulmonary disorder [9,13]. Therefore, they can be characterized based on the underlying conditions and, hence, be very useful in helping the diagnosis of several diseases [8]. In Table 5.1, a summary table with the main characteristics of the different types of adventitious sounds is presented.

Continuous adventitious sounds

CASs typically last more than 250 ms [8]. CASs can be further divided according to their pitch, namely as high-pitched CASs (wheeze, stridor, and gasp) or low-pitched CASs (rhonchi and squawk) [8].

Wheezes and rhonchi

The wheezing sound is probably the most easily discernible adventitious sound. Wheezes and rhonchi can be heard during inspiration, expiration (most common), or during both phases. However, rhonchi are low-pitched CASs and wheezes are high-pitched [8]. Wheezes have a typical dominant frequency above 400 Hz, while rhonchi have a dominant frequency typically below 300 Hz [8,9]. Although many physicians still use the term rhonchi, some

TABLE 5.1 Summary table of adventitious sounds main characteristics.

Name	Typical duration	Timing	Pitch
Wheeze	>80–100 ms	Inspiration, expiration, biphasic	High (>400 Hz)
Rhonchi	>80–100 ms	Inspiration, expiration, biphasic	Low (<200 Hz)
Stridor	>250 ms	Inspiration, expiration, biphasic	High (>500 Hz)
Squawk	±200 ms	Inspiration	Low (200–300 Hz)
Gasp	>250 ms	Inspiration	High
Fine crackle	±5 ms	Mid-to-late inspiration and occasionally on expiration	High (650 Hz)
Coarse crackle	±15 ms	Early inspiration, expiration, biphasic	Low (350 Hz)
Pleural rub	>15 ms	Biphasic	Low (<350 Hz)

Based on R.X.A. Pramono, S. Bowyer, E. Rodriguez-Villegas, Automatic adventitious respiratory sound analysis: a systematic review, PLoS One, 12 (2017) e0177926; A. Bohadana, G. Izbicki, S.S. Kraman, Fundamentals of lung auscultation, N. Engl. J. Med. 370 (2014) 744–751.

prefer to simply refer to these sounds as low-pitched wheezes [16]. Whilst wheezes are caused by the airway narrowing which then causes an airflow limitation, rhonchi are related to the thickening of mucus in the larger airways [8]. Both sinusoid-like signals can have up to three harmonic frequencies, with a typical frequency range between 100 and 1000 Hz [8].

Wheezes are usually louder than the underlying breath sounds and are often audible at the patient's open mouth or by auscultation over the trachea, chest wall, and larynx [12,19]. On the other hand, rhonchi, being low-pitched, are best heard over the chest wall [16]. Moreover, the presence of wheezes can reflect acoustical symptomatic characteristics, not only due to the presence of abnormalities in the respiratory system, but also due to the severity and the location of the most frequently found airway obstructions in asthma and respiratory stenoses [9,13]. Both wheezes and rhonchi are sounds commonly associated with obstructive diseases, such as asthma, COPD, and bronchitis [8].

Figs. 5.3–5.5 illustrate examples of the previously described sounds, both in time and time–frequency domains.

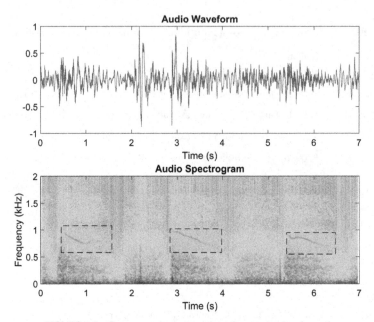

FIGURE 5.3 Spectrogram representation of monophonic wheezes.

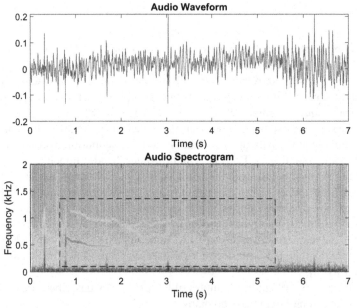

FIGURE 5.4 Spectrogram representation of a polyphonic wheeze.

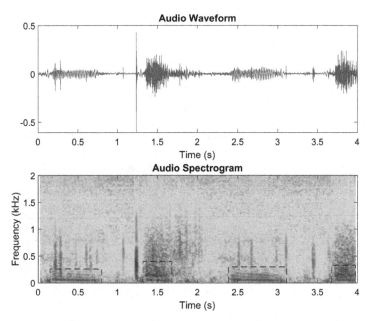

FIGURE 5.5   Spectrogram representation of rhonchi.

Stridors

Stridors are sibilant and musical wheeze-like CASs, with a very loud low frequency originating in the larynx or trachea [8,13]. Stridors can mainly be heard on the inspiration phase although, on some occasions, they can also be heard on expiration or even in both phases [16]. Unlike wheezes, the stridor sound is generated by turbulent airflow in the larynx or bronchial tree and is related to an upper airway obstruction of the upper respiratory tract, being clearly heard without the aid of a stethoscope, due to its intensity [8,16,18]. Thereby, stridor can mostly be audible on the trachea, while wheezing can also be heard clearly by chest auscultation. Stridor sounds typically range between 500 and 1000 Hz, and they are also normally harsher and louder than wheeze sounds [8,12,16]. As a type of CASs, stridor sounds have a typical duration of more than 250 ms.

The differential diagnoses for stridor are epiglottitis, croup, and laryngeal edema. All of these conditions are related to upper airway obstructions [8]. Stridor sounds can also be heard when there is a foreign body, such as a tumor, in the upper airway tract [8].

Fig. 5.6 illustrates examples of the previously described sounds, both in time and time−frequency domains.

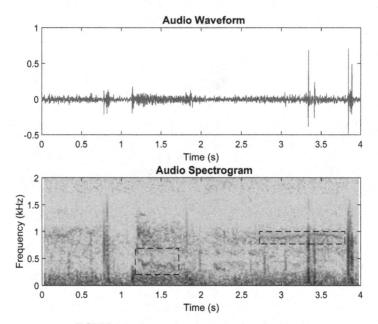

FIGURE 5.6 Spectrogram representation of stridors.

Squawk

Squawks are relatively short inspiratory CASs with a mixed constitution, presenting both musical and nonmusical components [8]. A squawk typically appears as a short wheeze preceded by a crackle [19]. The duration of squawks generally varies between 50 and 400 ms, while their fundamental frequency is usually between 200 and 300 Hz [8,9,18]. Squawks are typically heard from the middle to the end of inspiration in patients with restrictive (e.g., allergic alveolitis and diffuse interstitial fibrosis) or acute (e.g., severe pneumonia and bronchiolitis obliterans) lung pathologies [16,21,22].

Fig. 5.7 illustrates examples of the previously described sounds, both in time and time−frequency domains.

Gasp

Inspiratory gasps are high-pitched, long duration sounds that can be usually heard after a bout of coughing when a patient finally tries to inhale. The whoop sound of an inspiratory gasp is caused by fast moving air through the respiratory tract [8]. The whooping sound is a pathognomonic symptom of whooping cough (pertussis), with this being the only disease that is associated with this CAS.

Discontinuous adventitious sounds

DASs are abnormal sounds with a short duration, typically less than 25 ms [8]. This category of adventitious sounds is composed by crackles (fine and coarse)

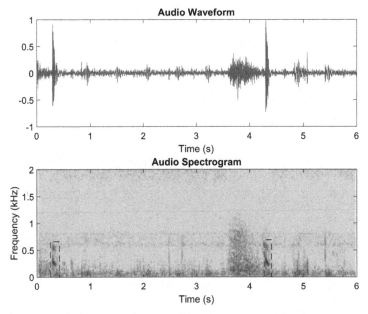

FIGURE 5.7 Spectrogram representation of squawks.

and pleural rubs. Crackles are the main type of DASs, and their evaluation is important, since it can help with the differential diagnosis of several diseases [16,23]. They are explosive and nonmusical sounds, usually classified as fine and coarse crackles based on their duration, loudness, pitch, timing in the respiratory cycle, and relationship to coughing and changing body position [9,18]. Besides crackles' frequency, duration, gravity (in)dependence, (non) response to forced expiratory maneuvers, and timing within the respiratory cycle, their number, and the anatomical place from where they are auscultated should also be considered. As several diseases progress, crackles tend to increase in number and occur first in the basal areas and later in the upper areas of the lungs [12,19]. In addition to the described crackles' characteristics, some studies have also used crackles' initial deflection width (IDW), largest deflection width (LDW), and two-cycle duration (2CD) for diagnostic and monitoring purposes [9,12].

Fine crackle

Fine crackle sounds are caused by explosive openings of the small airways and are produced within the small airways [8,18]. The sound is high pitched (around 650 Hz) and has a short duration (around 5 ms) [8]. Fine crackles are usually audible only at the late stages of inspiratory phases (mid-to-late), and are usually associated with pneumonia, congestive heart failure, and lung fibrosis [8].

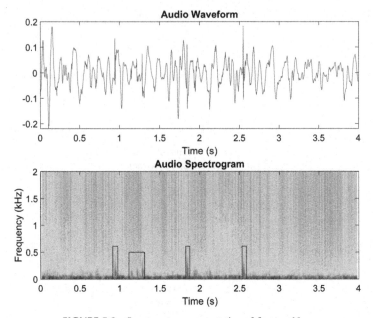

FIGURE 5.8 Spectrogram representation of fine crackles.

Fig. 5.8 illustrates examples of the previously described sounds, both in time and time—frequency domains.

Coarse crackle

Coarse crackle sounds are generated by air bubbles in large bronchi [8]. The sounds can be heard mostly during the early stages of inspiration but are also audible at the expiratory stage. Coarse crackles have a "popping" quality with a low pitch, around 350 Hz, and a duration of around 15 ms [8]. They may be heard over any lung region, are usually transmitted to the mouth, can change or disappear with coughing, and are not influenced by changes in body position [16]. They are commonly associated to patients with chronic bronchitis, bronchiectasis, COPD, and severe pulmonary edema [18].

Fig. 5.9 illustrates examples of the previously described sounds, both in time and time—frequency domains.

Pleural rub

In healthy persons, the parietal and visceral pleura slide over each other silently [16]. However, in persons with various lung-related diseases, the visceral pleura can become rough enough that its passage over the parietal pleura produces crackling sounds heard as a friction rub [16]. Pleural rub sounds are coarse crackles arising from parietal and visceral pleura rubbing

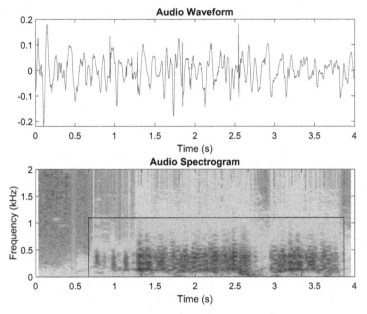

FIGURE 5.9 Spectrogram representation of coarse crackles.

against each other [13]. The sounds are heard on both respiratory phases (biphasic) and are nonmusical and explosive sounds [16]. They have a low pitch, normally below 350 Hz, and are mostly caused by inflammation of the pleural membrane [8,18].

Fig. 5.10 illustrates examples of the previously described sounds, both in time and time—frequency domains.

Diagnostic value of respiratory sounds

Respiratory complaints are one of the most common cause that brings a patient to the doctor [24]. Consequently, respiratory sounds, either heard at a distance or auscultated over the chest, are an integral part of the patients' evaluation and may provide valuable clues regarding their pathology [24]. Thus, these sounds are of significance as they provide valuable information regarding the health of the respiratory system, making auscultation valuable to extract information during physical examination [25]. Additionally, an auscultation-based diagnosis of pulmonary disorders relies heavily on the presence of adventitious sounds and on the altered transmission characteristics of the chest wall [24,26]. Thereby, ARSs provide crucial information on respiratory dysfunction, and changes in their characteristics (e.g., intensity, duration, and timing) might inform the clinical course of respiratory diseases and treatments [3].

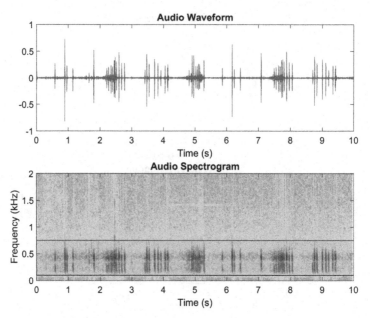

FIGURE 5.10 Spectrogram representation of pleural rub.

Breath sounds may be abnormal in certain pathological conditions of the airways or lungs. Breath sounds with abnormally high frequencies and intensity are typically observed in many diseases with airway obstruction, like asthma and chronic bronchitis [13]. In a population study [27], the authors have shown that changes in lung sounds spectra could help in the early detection of airways disease, enhancing the diagnostic value of respiratory sounds.

A model [28] showed that wheezes are always accompanied by flow limitation, but that flow limitation is not necessarily accompanied by wheezes. Wheezing is very common in asthma and a common clinical sign in patients with obstructive airway diseases. Another study [29] has demonstrated that there is an association between the degree of bronchial obstruction and the presence and characteristics of wheezes. The strongest association has been obtained when the degree of bronchial obstruction is compared to the proportion of the respiratory cycle occupied by wheezing.

Crackles occur frequently in cardiorespiratory diseases. The acoustic characteristics and timing of crackles depend on the disease being studied. When present, crackling sounds in patients with lung fibrosis are typically fine, repetitive, and end inspiratory, whereas those associated with chronic airways obstruction (e.g., COPD, emphysema, or bronchiectasis) are coarse, less repeatable, and occur early in inspiration. The appearance of crackles may be an early sign of respiratory disease [13]. When present, the number of crackles

TABLE 5.2 Associations between abnormal lung sounds and lung diseases [13,31].

Adventitious sounds	Associated respiratory diseases
Crackles	Alveolitis, pulmonary fibrosis, atelectasis, pneumonia, asbestosis, chronic bronchitis, bronchiectasis, congestive heart failure
Rhonchi	Chronic bronchitis, tumors, pneumonia, obstructive pulmonary disease
Squawks	Allergic alveolitis, pulmonary fibrosis, interstitial fibrosis
Stridors	Laryngitis, laryngomalacia, anatomic hypothesis, vocal cord paralysis, airway inflammation following extubation, tumors, tracheal stenosis
Wheezes	Obstructive lung diseases (e.g., asthma), cystic fibrosis, adults exposed to occupational hazards

per breath is associated with the severity of the disease in patients with interstitial lung disorders [30]. Moreover, the waveform and timing of crackles may have clinical significance in differential diagnosis of cardiorespiratory disorders.

Table 5.2 summarizes some of the most common respiratory diseases associated with different adventitious sounds.

Respiratory sound acquisition

Chest auscultation has long been considered a useful part of the physical examination, going back to the time of Hippocrates [16]. Notwithstanding, it was not until the 19th century that auscultation has become a widespread examination method, with the development of the stethoscope by Laënnec (see Fig. 5.11). The stethoscope can be regarded as the first of a long line of instruments that has given clinicians an ever-greater ability to examine the internal structure and function of the human body, whether to assess healthy or disease conditions.

As discussed before, auscultation presents significant drawbacks, namely its interuser variability. However, advances in technology have made it possible to digitally record respiratory sounds in a noninvasive way. This computerized approach aims at a more objective analysis of respiratory sounds, and, over the years, several techniques and methods have been developed in CORSA. Computational methods for the analysis of breath sounds offer significant advantages, such as digital storage, monitoring in critical settings, computer-supported analysis, comparison among different

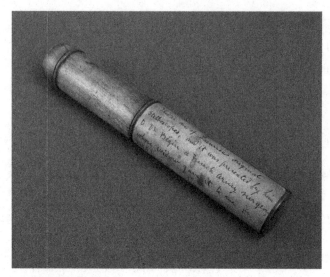

FIGURE 5.11 Early stethoscope developed by Laënnec [32].

recordings, and provision of objective parameters in their evaluation [33]. Having said that, the first step before developing any computerized technique is the recording of the lung sounds. This process typically comprises multiple phases of amplification, filtering, and digitization of the recorded sounds [33]. Multiple systems have been proposed and used to collect these types of data, being that the sound is typically recorded through either means of an electronic stethoscope or a microphone. In this section, we provide a broad overview of the most common sensors used to gather respiratory sound as well as the main difficulties associated with the process.

Sensors and placement

Listening to respiratory sounds (by means of a stethoscope) involves several physical phenomena: vibrations of the chest wall are converted into pressure variations of the air in the stethoscope and those pressure variations are then transmitted to the diaphragm of the ear [34]. Therefore, the basic principle in auscultation with a stethoscope is the transfer of vibrations of the chest wall to the air pressure variations that travel to the diaphragm of the ear throughout the stethoscope [33]. Subsequently, those air pressure variations cause sound waves and are recorded using microphones, which are transducers that convert sound signals to electrical signals [33].

One basic category of recording devices refers to microphones. Whatever the type of microphone, it always has a diaphragm, like the human ear, and the movement of the diaphragm is converted into an electrical signal. Two major microphone approaches exist: the "kinematic" approach, which involves the direct recording of chest wall movement ("contact sensor"), and the "acoustic" approach, which involves the recording of the movement of a diaphragm exposed to the pressure wave induced by the chest wall movement ("air-coupled sensor"). The chest wall movements are so weak that a free-field recording is not possible; therefore, it is essential to couple the diaphragm acoustically with the chest wall through a closed-air cavity [33,34]. Notwithstanding, whatever the approach, kinematic or acoustic, the captured vibrations have to be converted into electrical signals. To achieve that, three major principals of transduction are applied: (1) electromagnetic induction: a small movable induction coil's movement in the magnetic field induces a varying current in the coil through electromagnetic induction; (2) condenser principle: changing the distance between the two plates of a charged capacitor induces a voltage fluctuation; (3) piezoelectric effect: if a crystal (rod, foil) is bent, an electric charge on the surface is induced. Examples of "contact sensors" are piezoelectric accelerometers, and of "air-coupled sensors" are condenser microphones.

A piezoelectric accelerometer applies the piezoelectric effect in such a way that the output voltage is proportional to the acceleration of the whole sensor [34]. These sensors are commonly used as contact sensors since their diaphragm is directly coupled to the skin. They do not require an air coupler, but the pressure exerted on the skin through the sensor should be kept constant. Thereby, adhesive rings are suggested to be used for contact sensor attachments [33]. As for condenser microphones, the air that is captured in the air cavity of the coupler exerts enough force on the diaphragm to cause a displacement that produces measurable sound pressure.

Although both condenser microphones and piezoelectric contact sensors are displacement receivers, the waveforms that they deliver are different because of the coupling differences [35]. Both sensors have several disadvantages, that is, piezoelectric sensors are very sensitive to movement artifacts [36], whereas condenser microphones need mounting elements that change the frequency and amplitude of the sound transduction [34]. In summary, both types of sensors can be recommended to be coupled to the chest to record respiratory sound. Piezoelectric sensors have a higher sensitivity and are less influenced by ambient noise than condenser microphones, which show a larger frequency response in a free-field measurement [37].

Regarding the characteristics of the sensors that are used to perform respiratory sound recording, a set of recommendations for these devices has been described in Ref. [34]. Table 5.3 summarizes the CORSA recommendations for the acquisition of respiratory sounds, namely characteristics of sensors and microphone locations.

The CORSA task force has also settled some of the most relevant locations to perform the recording of respiratory sounds, such as the trachea (for tracheal sounds) and left and right posterior base of the lungs, usually 5 cm laterally from the paravertebral line and 7 cm below the scapular angle for adults [34].

TABLE 5.3 Summary of CORSA recommendations for the acquisition of respiratory sounds, namely, characteristics of sensors and microphone locations.

Sensor specification	
Frequency response	Flat in the frequency range of the sound. Maximum deviation allowed: 6 dB
Dynamic range	>60 dB
Sensitivity	Must be independent of frequency, static pressure, and sound direction
Signal-to-noise ratio	>60 dB (S = 1 mV Pa^{-1})
Directional characteristics	Omnidirectional
Coupling	
Piezoelectric contact	Immediate to the skin surface
Condenser air-coupled	Shape: Conical; depth: 2.5–5 mm; diameter at skin: 10–25 mm; vented
Fixing methods	
Piezoelectric contact	Adhesive ring
Condenser air-coupled	Either elastic belt or adhesive ring
Noise and interference	
Acoustic	Shielded microphones; protection from mechanical vibrations
Electromagnetic	Shielded twisted pair of coaxial cable
Amplifier	
Frequency response	Constant gain and linear phase in the band of interest
Dynamic range	>60 dB
Noise	Less than that introduced by the sensor
High-pass filtering	
	Cut-off frequency: 60 Hz; roll-off: >18 dB octave^{-1}; phase as linear as possible; minimized ripple
Low-pass filtering	
	Cut-off frequency: Above the higher frequency of the signal; roll-off: >24 dB octave^{-1}; minimized ripple

Adapted from L. Vannuccini, J. Earis, P. Helistö, B. Cheetham, M. Rossi, A. Sovijärvi, Vanderschoot, Capturing and preprocessing of respiratory sounds, Eur. Respir. Rev. 10 (2000) 616–620; H. Hadjileontiadis (2009); S.S. Kraman, G.R. Wodicka, Y. Oh, H. Pasterkamp, Measurement of respiratory acoustic signals: effect of microphone air cavity width, shape, and venting, Chest 108 (1995), 1004–1008.

Later, multichannel recording systems were developed in various laboratories, and multichannel analysis of respiratory sounds became increasingly more popular, since the simultaneous recording and analysis of sounds from multiple sites over the chest facilitated and accelerated an otherwise slow and tedious process [4,33,38,39]. Other optional locations have been recommended by CORSA for adults such as the right and left anterior area of the chest at the second intercostal space at the midclavicular line and right and left lateral area of the chest at the fourth or fifth intercostal space on the midaxillary line. In tracheal breath measurements, the sensor is usually attached to the neck at the anterior cervical triangle [33].

There are also several recommendations by the CORSA Task Force applicable to the environmental and subject conditions during breath sound recording namely [33]:

- The background noise level should preferably be less than 45 dB (A) and 60 dB (linear) with minimum ambient noise from the environment including voiced sounds. The sounds induced by airflow through the flow transducer should be considered.
- The recording should be performed at comfortable room temperature, humidity, lighting, and ventilation conditions.
- The sitting position is preferable for the subject for short-term recording, whereas supine position is preferable for long-term recording.
- For adults, tidal breathing of 7−10 respiratory cycles, with a peak expiratory and inspiratory flow of 1−1.5 L/s or 10%−15% of the predicted maximum peak flow, and tidal volume of 1 L or 15%−20% of predicted vital capacity are recommended for short-term recording, whereas tidal breathing without any voluntary effort is recommended for long-term recording.
- The airflow, volume, and flow/volume display should be, whenever possible, in front of the subject.
- For babies and young children, chest movement monitoring by strain gauge bands, pneumatic belts, or chest straps should be used.

Several commercial devices that can record respiratory sound are available on the market. These mostly include electronic stethoscopes. Among the most commonly used devices are the following ones: Littmann 3200 and 4000 electronic stethoscopes, ThinkLabs digital stethoscope, and Welch Allyn Meditron Master Elite Electronic Stethoscope, among others. Other types of devices have also been used in the literature to record respiratory sounds, such as the SONY ECM-77B microphone, AKG C417L Microphone, and KEC-2738 microphone, among others [8].

Challenges in data collection and future perspectives

Although respiratory sound recording is an active and growing area of research activity and development for more than 30 years, it is yet be established as a tool for clinical diagnosis. Despite the progressive interest in the collection of respiratory sounds, one of the main concerns is in the lack of standardization in the recording process, as different approaches have been reported in the literature [10,33]. This point is particularly relevant in regard to which positions should be used to record the respiratory sound or which type of maneuvers should be performed while recording. With the development of more efficient methods of recording breath sounds, multisensor arrays are also being designed to provide a better coverage and allow for multichannel algorithmic processing [4,38—42].

Wireless recording in wearable devices is another promising area for respiratory sound recording development [43,44]. This type of sensors is particularly suitable to be deployed in telemedicine in remote monitoring applications. These applications can be of two types, namely, asynchronous and synchronous telemedicine. In asynchronous telemedicine, data are collected and sent electronically to a medical professional for further analysis and diagnosis. Synchronous telemedicine is real time and can be interactive. Moreover, in synchronous telemedicine, distant examination may be enabled through audiovisual technology. There have been reports of self-monitoring of asthma and COPD patients at home using telemedicine tools, and the results were accurate and satisfactory, without affecting the quality of care received by these patients [45,46]. Similarly, with the advances in smartphone and Bluetooth technology, wireless stethoscopes working with mobile phones have also been developed [47]. As technological development progresses, more of these devices are likely to come into existence. The constant miniaturization of the sensors and the push toward personal health monitoring solutions are going to be two of the driving factors, alongside with the worldwide adoption of wearable monitoring systems.

In conclusion, electronic auscultation coupled with computerized lung sound analysis has the potential to improve diagnostic yield of pulmonary disorders, both in clinical and research settings [10]. However, more studies are required to address standardization of methodology and analytical methods across studies. This point is particularly important so that different studies can be fairly and appropriately compared. Therefore, these approaches face significant challenges before achieving clinical acceptance, mostly due to lack of standardized clinical trials and lack of diagnostic validity.

Respiratory sound databases

One of the core problems in the field of respiratory sound analysis is the lack of large publicly available databases that can serve to develop algorithms and benchmark results. Since most works in respiratory sound analysis and processing are based on machine learning approaches, having large and representative databases is an essential requirement for algorithmic development. Otherwise, it is harder to develop methods with good generalization capacity that work well in "real-world" conditions. In their systematic review, Pramono et al. identified the source of the data for all the 77 articles covered [8]. Even though most works have used private data collections, there were 13 publicly available databases among the data sources. These represent four online repositories and nine audio CD companion books. Table 5.4 presents a summary of the current publicly available databases, which include the ones listed by Pramono and others recently made available.

In the literature, the most referenced databases are the R.A.L.E. repository [52] and the audio CD from *Understanding Lung Sounds* [50]. As these repositories and CDs were designed for teaching purposes, they generally include a small number of examples of each type of respiratory sound.

TABLE 5.4 List of publicly available respiratory sound databases.

Database name and reference
The chest: its signs and sounds [48]
Understanding lung sounds, second edition and third edition [49,50]
Understanding Heart Sounds and Murmurs [51]
R.A.L.E. repository [52]
East Tennessee State University repository [53]
Fundamentals of lung and heart sounds [54]
Auscultation Skills: Breath and heart sounds, fourth edition [55]
Heart and lung sounds reference library [56]
Lung sounds: An introduction to the interpretation of the auscultatory findings [57]
Littmann repository [58]
Secrets Heart & lung sounds workshops [59]
SoundCloud lung sound repository [60]
ICBHI 2017/Respiratory Sound Database [4,61]
HF_Lung_V1 [62]
A dataset of lung sounds recorded from the chest wall using an electronic stethoscope [63]

Moreover, most of these sounds are clear and do not include environmental noise, commonly present in clinical practice. Thus, they are not suitable for the development of realistic classification models that are aiming to be deployed in the "wild".

In 2017, to overcome these difficulties, the RSD [4,61] was compiled for the scientific challenge at the International Conference on Biomedical and Health Informatics (ICBHI) 2017. The RSD represents one of the largest and most used databases in the area with data collected with multiple acquisition devices and under varied environmental setups with high levels of noise, mimicking the real hospital environment. This dataset was widely adopted as a considerable amount of works have been published using its data. Moreover, since the dataset contains a large number of patients and lung sound recordings, with more than 5.5 h of recordings, it has offered researchers the possibility to explore new approaches to analyze and process respiratory sounds, particularly with the use of data-driven approaches, such as deep learning. Since the RSD was made available, the number of works on respiratory sounds using deep learning has experienced a steady increase, leveraging the amount of data available. The RSD contains audio samples, collected independently by two research teams in two different countries, over several years, resulting in a total of 920 annotated audio samples from 126 participants [4].

More recently, in 2021, two new databases containing respiratory sounds were made available online [62−64]. Fraiwan et al. [63] have collected data from 112 subjects. However, unlike the RSD, each participant only has one recording, resulting in a total of 112 audio clips. The data were collected by placing the stethoscope (3M Littmann 3200) on various regions of the chest called zones (upper, mid, and lower). To date, the "HF_Lung_V1" [62] is largest public database containing respiratory sound recordings. It contains a total of 9765 audio files of lung sounds (duration of 15 s each) acquired from 279 patients [62]. The data were collected using two devices: 3M Littmann 3200 and a customized multichannel acoustic recording device (HF Type 1) that supports the connection of eight electret microphones. The breathing lung sounds were collected at the eight different auscultation points in hospital environment, also including intensive care units. An updated version of this database that includes more recordings, "HF_Lung_V2", has been announced [64].

Current methods

In the last decades, many researchers have developed methods for respiratory sound analysis. Most systems comprise two steps: (1) relevant features are extracted from the signal; (2) extracted features are used to detect or classify ARS events (i.e., crackles and wheezes).

Pioneering works in respiratory sound analysis

Urquhart et al. [65] discriminated between a group of healthy subjects and three groups of patients with different pulmonary diseases using spectral analysis. They divided the frequency range of 0–400 Hz into 20 intervals and produced a piecewise-constant approximation to the spectral envelopes. Subsequently, a Karhunen–Loeve transform was applied to extract the two most significant variables of the 20-dimensional feature space. A plot of these two variables was used to analyze the differences between the four groups. In another approach to discriminate between respiratory sounds of healthy and pathological subjects, autoregressive models of different orders were fed to a k-nearest neighbor classifier and their performance was compared [66].

Holford [67] used a balanced sample of 12 fine and coarse crackles to build a discriminant curve based on two features and the Mahalanobis distance. The discriminant curve was used to successfully classify all six samples of the test set.

The first validated method for the detection of crackles was introduced by Murphy et al. [68], but the details of its implementation were not revealed. Kaisia et al. [69] proposed a crackle detection method based on the analysis of the spectral stationarity of lung sounds. Du et al. [70] proposed a mathematical model for crackles, based on the criteria for an ideal crackle. Using this model as the mother wavelet, they used matched wavelet analysis to detect crackles in a sample of lung sounds from a teaching tape. Hadjileontiadis and Panas [71] proposed a wavelet-based filter for separating DASs from normal respiratory sounds.

Forkheim et al. [72] presented an analysis of the use of artificial neural networks (ANNs) to discriminate between isolated 100 ms segments containing wheezes and segments with no wheezes. They compared the performance of several ANNs on two small groups of patients, not revealing the number of patients.

Preprocessing

Respiratory sound recording is affected by environmental interference and by chest wall movement, muscle sounds, heart sounds [34], as well as external perturbations such as cough, throat clearing, and speech.

Heart sounds are one of the most studied interferences in respiratory sound recording. Gnitecki and Moussavi [73] reviewed 15 methods for filtering heart sounds from respiratory sounds, including linear adaptive, Short-time Fourier Transform (STFT) based, or wavelet-based techniques. However, the authors questioned the potential usefulness of those methods on a clinical setting, given that all the reviewed studies were based on data that had been acquired under ideal conditions, such as a quiet environment, and in known cardiac and respiratory states, with most subjects being healthy [73].

Another well-studied interference is cough. There are three main challenges in the automatic detection of cough: (1) differentiation from ambient noise; (2) differentiation from other patient sounds, especially speech, laughing, and sneezing; (3) variability in the acoustics of cough sounds both within and between individuals [74]. Several methods of automatic cough detection have been proposed in the literature [38,75−78]. While reported results are encouraging, all the algorithms were evaluated on different data, as there is no standard public dataset.

Various filtering techniques have been developed to denoise and to isolate different respiratory sound components. The most common filters applied in respiratory sound analysis are high-pass filters with cut-off frequencies between 60 and 100 Hz [34,79,80] and low-pass filters with cut-off frequencies depending on the sampling rate, considering that some components of respiratory sounds can reach 5000 Hz but most of their energy is below 1000 Hz [16]. Source separation techniques have also been used for respiratory sound analysis, namely nonnegative matrix factorization [81−83].

Several works have proposed Empirical Mode Decomposition (EMD) as a suitable tool for respiratory sound preprocessing, given the nonlinearity and nonstationarity properties of respiratory sounds [84]. The EMD algorithm decomposes a signal into intrinsic mode frequencies (IMFs) and a residual in an iterative sifting process [85], producing a local and fully data-driven separation of a signal in fast and slow oscillations [86]. For each IMF, instantaneous frequency and envelope can be defined at any point, achieving high temporal and spectral resolutions [87]. However, the local nature of the EMD may produce oscillations with very disparate scales in one mode, or oscillations with similar scales in different modes, originating the mode mixing effect [86]. This effect was observed by Lozano et al. [84] when applying EMD to respiratory sounds, resulting in poor separation of frequency scales. To overcome this problem, Wu and Huang [88] proposed a noise-assisted data analysis method, ensemble empirical mode decomposition (EEMD). EEMD performs the decomposition over an ensemble of the signal plus Gaussian white noise [89], taking advantage of the fact that EMD behaves a dyadic filter bank when applied to white noise [90]. Several improvements to the EEMD algorithm have been proposed since [89,91], one of which was applied to respiratory sound analysis [84]. Fig. 5.12 shows a respiratory sound signal containing wheezes and crackles. The waveforms of the first and second IMFs obtained using the EMD and the EEMD are shown below the original waveform.

Time−frequency representations

The most common time−frequency representation in the audio analysis literature is the spectrogram, obtained using the STFT. The spectrogram

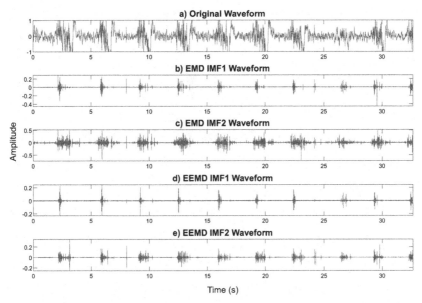

FIGURE 5.12 Waveforms of a) original sound file, b) first IMF obtained using EMD, c) second IMF obtained using the EMD, d) first IMF obtained using EEMD, and e) second IMF obtained using EEMD.

describes the evolution of the frequency components over time. The STFT representation (X) of a given discrete signal (x) is given by Ref. [92]:

$$F(n, \omega) = \sum_{i=-\infty}^{\infty} i\omega(n-i)e^{-j\omega} \tag{5.1}$$

where $\omega(i)$ is a window function centered at instant n. The spectrograms of the original signal and of the first and second IMFs obtained using the EMD and the EEMD are displayed in Fig. 5.13. The Hann window function was used to generate the spectrograms. Both wheezes (black dashed rectangles) and crackles (black closed rectangles) can be clearly seen in the original spectrogram. When decomposed using the EMD, mode mixing occurs, with higher wheeze harmonics appearing on IMF1 and lower wheeze harmonics appearing on IMF2. Applying the EEMD reduces the mode mixing effect.

Several works have proposed the Mel spectrogram as a time–frequency representation for respiratory sound analysis [7,93]. The Mel scale [94] is a perceptual scale of equally spaced pitches, aiming to match the human perception of sound. The conversion from Hz into Mel is performed using Eq. (5.2):

$$m = 2595 \cdot \log_{10}\left(1 + \frac{f}{700}\right) \tag{5.2}$$

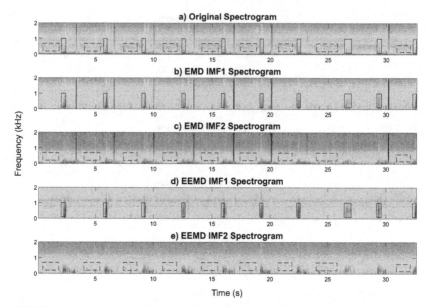

FIGURE 5.13 Spectrograms of a) original sound file, b) first IMF obtained with EMD, c) second IMF obtained using the EMD, d) first IMF obtained using the EEMD, and e) second IMF obtained using EEMD. Dashed black rectangles: wheezes. Closed black rectangles: crackles.

The Mel spectrogram displays the spectrum of a sound on the Mel scale. The Mel spectrograms of the original signal and of the first and second IMFs obtained using the EMD and the EEMD are shown in Fig. 5.14. As with spectrograms, both wheezes (dashed black rectangles) and crackles (closed black rectangles) can be clearly seen in the original Mel spectrogram, and the mode mixing effect is less pronounced when applying the EEMD instead of the EMD.

Wavelet analysis is also a popular alternative for respiratory sound analysis [71,95–97]. The typical time–frequency representation used in wavelet analysis is the scalogram, which can be obtained by employing the continuous wavelet transform (CWT). The CWT is good at detecting transients in nonstationary signals, and for signals in which instantaneous frequency grows rapidly [98].

The scalograms of the original signal and of the first and second IMFs obtained using the EMD and the EEMD are displayed in Fig. 5.15. The Bump wavelet was used to generate the scalograms [99]. Visualizing the scalogram of the original signal, wheezes (dashed white rectangles) and crackles (closed white rectangles) are not evident. After EMD or EEMD decomposition, crackles can be clearly seen in the IMF1 scalograms, but wheezes are still unclear.

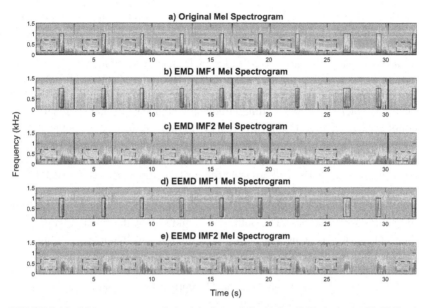

FIGURE 5.14 Mel spectrograms of a) original sound file, b) first IMF obtained with EMD, c) second IMF obtained using the EMD, d) first IMF obtained using the EEMD, and e) second IMF obtained using EEMD. Dashed black rectangles: wheezes. Closed black rectangles: crackles.

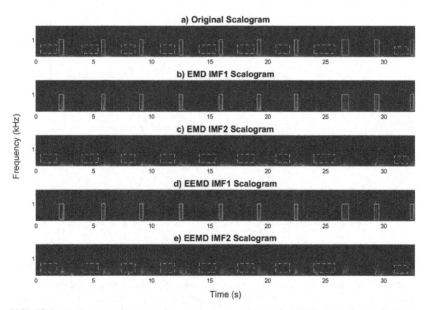

FIGURE 5.15 Bump scalograms of a) original sound file, b) first IMF obtained with EMD, c) second IMF obtained using the EMD, d) first IMF obtained using the EEMD, and e) second IMF obtained using EEMD. Dashed white rectangles: wheezes. Closed white rectangles: crackles.

Both the STFT and CWT have significant limitations. The STFT provides good frequency resolution but poor relative temporal resolution at high frequencies, whereas the CWT maintains a good relative temporal resolution throughout the spectrum but degrades in frequency resolution and becomes redundant with increasing frequency [100]. Huang et al. [101] proposed the Hilbert—Huang Transform (HHT) to overcome the limitations of other methods when performing time—frequency analysis of nonstationary and nonlinear data. The resulting Hilbert spectrograms of the original signal and of the first and second IMFs obtained using the EMD and the EEMD are shown in Fig. 5.16. As with scalograms, crackles (dashed black rectangles) can be seen in the IMF1 Hilbert spectrograms, while wheezes (closed black rectangles) are not easily observable.

A combination of EEMD and HHT was proposed by Lozano et al. [87] for the analysis of wheezes. The authors concluded that the main advantage of their approach was achieving both high temporal and frequency resolutions.

Several other time—frequency (TF) representations have been devised, but the ones presented in this section are the most prevalent in respiratory sound analysis accross the literature. All the TF representations can be useful if they are well suited for the required analysis.

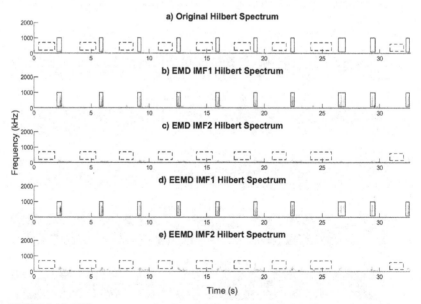

FIGURE 5.16 Hilbert spectrograms of a) original sound file, b) first IMF obtained with EMD, d) second IMF obtained using the EMD, d) first IMF obtained using the EEMD, and e) second IMF obtained using EEMD. Dashed black rectangles: wheezes. Closed black rectangles: crackles.

Feature extraction

Over the past 40 years, hundreds of hand-crafted features have been proposed for respiratory sound analysis. A systematic review [8] identified the most common features employed in the literature for respiratory sound analysis, which included Mel-frequency cepstral coefficients (MFCCs) [102], linear predictive coding coefficients (LPCCs) [103], spectral features [104], entropy [105], and wavelet coefficients [97]. Pramono et al. [79] evaluated 105 features for distinguishing between wheezes and normal respiratory sounds, concluding that features such as MFCCs, LPCCs, and tonality index were the most discriminative for that task. On a 3-class classification task (wheezes, crackles, and other), Rocha et al. [7] extracted 81 features from windows of varying length and applied feature selection to determine the most discriminative features, revealing that 7 of the first 10 features were obtained from MFCCs and that the other 3 were spectral entropy, spectral skewness, and spectral brightness ratio. The emergence of deep learning approaches in the past few years led many researchers in the respiratory sound analysis field to discard hand-crafted features and feed TF representations as inputs to ANNs. Table 5.5 provides a small description of most features employed by Pramono et al. [79] and by Rocha et al. [7].

MFCC features

The most common features used to describe the spectral shape of a sound are the MFCCs [106]. The MFCCs are calculated by converting the logarithm of the magnitude spectrum to the Mel scale and computing the Discrete Cosine Transform (DCT). As most of the signal information is concentrated in the first components, it is typical to extract the first 13 coefficients [107].

LPCC features

Linear predictive coding (LPC) is a time domain estimator of a signal based on linear combination of previous samples weighted with LPCCs [79].

Spectral features

Several features can be estimated from the spectrum. The first four standardized moments of the spectral distribution are commonly extracted: centroid, spread, skewness, and kurtosis. Other features that are commonly employed for characterizing the timbre of a sound are the zero-crossing rate, entropy, flatness, roughness, irregularity, and flux. Finally, the amount of high-frequency energy can be estimated in two ways: brightness, the high-frequency energy above a certain cut-off frequency; roll-off, which consists in finding the frequency below which a defined percentage of the total spectral energy is contained [107]. Brightness and roll-off can be computed at different frequencies and percentiles, respectively. Ratios between the brightnesses and the roll-offs can also be useful for respiratory sound analysis.

TABLE 5.5 Small description of features employed in respiratory sound analysis.

Type	Features	Description
MFCC	MFCC	13 Mel-frequency cepstral coefficients
LPCC	LPCC	8 Linear predictive coding coefficients
Spectral	Spectral centroid	Center of mass of the spectral distribution
	Spectral spread	Variance of the spectral distribution
	Spectral skewness	Skewness of the spectral distribution
	Spectral kurtosis	Excess kurtosis of the spectral distribution
	Zero-crossing rate	Waveform sign-change rate
	Spectral entropy	Estimation of the complexity of the spectrum
	Spectral flatness	Estimation of the noisiness of a spectrum
	Spectral roughness	Estimation of the sensory dissonance
	Spectral irregularity	Estimation of the spectral peaks' variability
	Spectral flux	Euclidean distance between the spectrum of successive frames
	Spectral brightness	Amount of energy above 100, 200, 400, and 800 Hz
	Brightness 400 ratio	Ratio between spectral brightness at 400 and 100 Hz
	Brightness 800 ratio	Ratio between spectral brightness at 800 and 100 Hz
	Spectral roll-off	Frequency at which 95%, 75%, 25%, and 5% of the total energy is contained below
	Roll-off Outlier ratio	Ratio between spectral roll-off at 5% and 95%
	Roll-off Interquartile ratio	Ratio between spectral roll-off at 25% and 75%
	ASE flux	Audio spectral envelope flux
	Tonality index	Ratio of the power in the dominant spectral bins to the total power of the input signal

Continued

TABLE 5.5 Small description of features employed in respiratory sound analysis.—cont'd

Type	Features	Description
Melodic	FF	Fundamental frequency estimation
	Inharmonicity	Partials nonmultiple of fundamental frequency
	Voicing	Presence of fundamental frequency

Melodic features

There are multiple ways to compute fundamental frequency [107,108], such as computing the cepstral autocorrelation of each frame to estimate each event's fundamental frequency curve. The inharmonicity and the voicing curves can then be computed based on the fundamental frequency curve. Rocha et al. [7] also computed these three features for a 400 Hz high-pass filtered version of the sound events. The rationale for this filter was the removal of the respiratory sounds, whose energy typically drops at 200 Hz [16], reaching insignificant levels at 400 Hz [109].

Classifiers

Various kinds of machine learning classifiers have been used for respiratory sound analysis. A systematic review [8] identified the most common classifiers proposed in the literature. They comprised empirical rule–based methods [110], support vector machines [84], discriminant analysis [111], ANNs [112], Gaussian mixture models [113], k-nearest neighbors [114], and logistic regression models [115]. More recently, several authors have used deep neural networks for respiratory sound analysis. Convolutional neural networks (CNNs) were the most common in the literature [92,93,116–118]; recurrent neural networks (RNNs) have also been tried [62,119]; as well as hybrid convolutional recurrent neural networks (CRNNs) [120,121].

Evaluation

In this section we introduce the most common tasks in RS analysis and summarize the best results found in the literature for each task. Additionally, a rating is given to each cited article (one to five stars) based on five criteria, i.e., a star is given for each criterion that is fulfilled: (1) number of patients is at least 30; (2) data source is public; (3) methodology is replicable and has no important flaws, e.g., data augmentation performed on the test set or lack of rationale for selection of a subset of patients; (4) a separate test set is used; (5) patient independence between training and test sets, i.e., a patient's recordings should be exclusively in the training or the test sets.

Evaluation metrics

Multiple evaluation metrics can be found in the literature. The most common are:

$$\text{Accuracy} = \frac{(\text{TP} + \text{TN})}{(\text{TP} + \text{TN} + \text{FP} + \text{FN})} \tag{5.3}$$

$$\text{Specificity} = \frac{\text{TN}}{\text{TN} + \text{FP}} \tag{5.4}$$

$$\text{Precision} = \frac{\text{TP}}{(\text{TP} + \text{FP})} \tag{5.5}$$

$$\text{Sensitivity} = \frac{\text{TP}}{(\text{TP} + \text{FN})} \tag{5.6}$$

$$\text{F1 Score(F1)} = \frac{(2 \times \text{Precision} \times \text{Sensitivity})}{(\text{Precision} + \text{Sensitivity})} \tag{5.7}$$

$$\text{MatthewsCorrCoef(MCC)} = \frac{((\text{TP} \times \text{TN}) - (\text{FP} \times \text{FN}))}{\sqrt{((\text{TP} + \text{FP})(\text{TP} + \text{FN})(\text{TN} + \text{FP})(\text{TN} + \text{FN}))}} \tag{5.8}$$

where TP (True Positives) are events of the relevant class that are correctly detected/classified; TN (True Negatives) are events of the other classes that are correctly detected/classified; FP (False Positives) are events that are incorrectly detected/classified as the relevant class; FN (False Negatives) are events of the relevant class that are incorrectly detected/classified.

Another commonly reported metric is the Area Under the ROC Curve (AUC). The AUC provides an aggregate measure of performance across all possible classification thresholds by measuring the two-dimensional area underneath the Receiver Operating Characteristic Curve (ROC) curve [122]. A graphical example of the AUC can be seen in Fig. 5.17.

Common tasks and current results

While many tasks can be performed in the field of RS analysis, the most common are: (1) ARS event detection or segmentation, i.e., segmentation of ARS events in a time series; (2) ARS event classification, i.e., classification of previously segmented ARS events; (3) respiratory cycle/phase detection or segmentation, i.e., segmentation of a time series into respiratory cycles or phases (inspiration and expiration); (4) respiratory cycle/phase classification, i.e., classification of previously segmented respiratory cycles or phases as containing one or more ARS events; (5) respiratory disease classification, i.e., classification of patients and/or audio files as one of several classes of respiratory diseases.

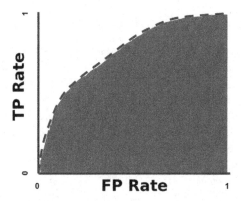

FIGURE 5.17 Area under the receiver—operator curve (AUC) [122].

Adventitious respiratory sounds event detection

Sound event detection or segmentation is the task of recognizing the sound events and their respective temporal start and end time in a recording [123]. A selection of works on ARS event detection is summarized in Table 5.6.

Adventitious respiratory sound event classification

Sound event classification is a task involving the recognition of active sound events in an audio recording [124]. A selection of works on ARS event classification is summarized in Table 5.7.

Respiratory phase segmentation

The timing of ARS in the respiratory cycle (inspiration or expiration) may have clinical significance for the assessment of patient respiratory status and for the differential diagnosis of respiratory disorders [125]. Therefore, respiratory phase detection is an important task that can be automated. A selection of works on respiratory phase detection is summarized in Table 5.8.

Respiratory cycle classification

While several databases have annotations at the event level, annotations at the level of the respiratory phase or cycle are also common. A selection of works on respiratory cycle classification is summarized in Table 5.9.

Respiratory disease classification

Physicians use various methods to diagnose respiratory diseases, such as spirometry or lung auscultation. Methods for the automatic diagnosis of respiratory diseases have been devised over the last decades. A selection of works on respiratory disease classification through automated auscultation is summarized in Table 5.10.

TABLE 5.6 Summary of selected works on ARS event detection.

Reference	Data	Classes	Best results (%)	Rating
Taplidou et al. [126]	*Participants*: 13; *Recordings*: 13; *Source*: Private	Wheezes; other	*Sensitivity*: 96; *Specificity*: 94	****
Riella et al. [127]	*Participants*: 28; *Recordings*: 28; *Source*: Public	Wheezes; other	*Accuracy*: 84; *Specificity*: 83; *Sensitivity*: 86	***
Pinho et al. [110]	*Participants*: 10; *Recordings*: 24; *Source*: Private	Crackles; other	*Precision*: 95; *Specificity*: 92; *Sensitivity*: 89	***
Mendes et al. [128]	*Participants*: 12; *Recordings*: 24; *Source*: Private	Wheezes; other	*Accuracy*: 98; *Specificity*: 99; *Sensitivity*: 91; *MCC*: 93	*
Mendes et al. [115]	*Participants*: 20; *Recordings*: 40; *Source*: Private	Crackles; other	*Precision*: 77; *Sensitivity*: 76; *Specificity*: 91	***
Messner et al. [129]	*Participants*: 15; *Recordings*: 480; *Source*: Private	Crackles; other	*Precision*: 73; *Sensitivity*: 73; *F1*: 72	***
Hsu et al. [62]	*Participants*: 261; *Recordings*: 9765; *Source*: HF lung V1	Wheezes; crackles; other	Wheezes *F1*: 52; Crackles *F1*: 71	*****

TABLE 5.7 Summary of selected works on ARS event classification.

Reference	Data	Classes	Best results (%)	Rating
Hadjileontiadis [111]	*Participants*: 25; *Recordings*: 25; *Source*: Public	Fine crackles; coarse crackles; squawks	*Accuracy*: 100	****
Chamberlain et al. [112]	*Participants*: NA; *Recordings*: 227; *Source*: Private	Wheezes; crackles; wheezes + crackles; other	Wheezes AUC: 86; Crackles AUC: 74	**
Lozano et al. [84]	*Participants*: 30; *Recordings*: 870; *Source*: Private	Wheezes; other	*Accuracy*: 94; *Precision*: 95; *Sensitivity*: 94; *Specificity*: 94	****

Continued

TABLE 5.7 Summary of selected works on ARS event classification.—cont'd

Reference	Data	Classes	Best results (%)	Rating
Gronnesby et al. [130]	*Participants:* NA; *Recordings:* 383; *Source:* Private	Crackles; other	*Precision:* 85; *Sensitivity:* 84; *F1:* 84	*
Jakovljevic et al. [131]	*Participants:* 126; *Recordings:* 920; *Source:* RSD	Wheezes; crackles; other	*Wheezes Sensitivity:* 52; *Crackles Sensitivity:* 56; *Other Sensitivity:* 52	****
Serbes et al. [132]	*Participants:* 126; *Recordings:* 920; *Source:* RSD	Wheezes; crackles; other	*Wheezes Sensitivity:* 79; *Crackles Sensitivity:* 95; *Other Sensitivity:* 91	****
Bardou et al. [116]	*Participants:* 15; *Recordings:* 15; *Source:* RALE	Wheezes; crackles; squawks; stridors; other	*Accuracy:* 96; *Precision:* 95; *Sensitivity:* 93	*
Pramono et al. [79]	*Participants:* 38; *Recordings:* 38; *Source:* Public	Wheezes; other	*F1:* 87; *MCC:* 73	*****
Rocha et al. [7]	*Participants:* 126; *Recordings:* 920; *Source:* RSD	Wheezes; crackles; other	*Accuracy:* 82; *Wheezes F1:* 73; *Crackles F1:* 88; *Wheezes MCC:* 69; *Crackles MCC:* 75	*****

TABLE 5.8 Summary of selected works on respiratory phase segmentation.

Reference	Data	Classes	Best results (%)	Rating
Chuah et al. [133]	*Participants:* 11; *Recordings:* 17; *Source:* Private	Inspiration; expiration	*Accuracy:* 97	***
Huq et al. [134]	*Participants:* 93; *Recordings:* 93; *Source:* Private	Inspiration; expiration	*Sensitivity:* 96; *Specificity:* 96	****

Continued

TABLE 5.8 Summary of selected works on respiratory phase segmentation.—cont'd

Reference	Data	Classes	Best results (%)	Rating
Messner et al. [129]	*Participants*: 15; *Recordings*: 480; *Source*: Private	Inspiration; expiration	*F1*: Inspiration: 87; *F1*: Expiration: 85	***
Jácome et al. [125]	*Participants*: 105; *Recordings*: 1954; *Source*: Private	Inspiration; expiration	*Sensitivity*: 97; *Specificity*: 84	****
Hsiao et al. [135]	*Participants*: 22; *Recordings*: 489; *Source*: Private	Inspiration; expiration	*Accuracy*: 92	*
Hsu et al. [62]	*Participants*: 261; *Recordings*: 9765; *Source*: HF lung V1	Wheezes; inspiration; expiration	*F1* Inspiration: 86; *F1* Expiration: 71	*****

TABLE 5.9 Summary of selected works on respiratory cycle classification.

Reference	Data	Classes	Best results (%)	Rating
Kochetov et al. [119]	*Participants*: 126; *Recordings*: 920; *Source*: RSD	Wheezes; crackles; wheezes + crackles; other	*Sensitivity*: 58; *Specificity*: 73	*****
Jakovljevic et al. [131]	*Participants*: 126; *Recordings*: 920; *Source*: RSD	Wheezes; crackles; wheezes + crackles; other	*Accuracy*: 40	*****
Serbes et al. [132]	*Participants*: 126; *Recordings*: 920; *Source*: RSD	Wheezes; crackles; wheezes + crackles; other	*Accuracy*: 50	*****
Chen et al. [136]	*Participants*: NA; *Recordings*: 240; *Source*: RALE and RSD	Wheezes; crackles; wheezes + crackles; other	*Accuracy*: 99; *Sensitivity*: 96; *Specificity*: 1000	**
Ma et al. [137]	*Participants*: 126; *Recordings*: 920; *Source*: RSD	Wheezes; crackles; wheezes + crackles; other	*Accuracy*: 50	*****
Perna et al. [138]	*Participants*: 126; *Recordings*: 920; *Source*: RSD	Wheezes; crackles; wheezes + crackles; other	*Accuracy*: 74; *Sensitivity*: 64; *Specificity*: 84	***
Jung et al. [139]	*Participants*: 22; *Recordings*: 489; *Source*: Private	Wheezes; crackles; other; unknown	*Accuracy*: 86; Wheezes *F1*: 89; Crackles *F1*: 82	***

TABLE 5.10 Summary of selected works on respiratory disease classification.

Reference	Data	Classes	Best results (%)	Rating
Aykanat et al. [140]	*Participants:* 1630; *Recordings:* 17,930; *Source:* Private	Healthy; Unhealthy	*Accuracy:* 85; *Precision:* 86; *Sensitivity:* 86; *Specificity:* 86	***
Perna et al. [138]	*Participants:* 126; *Recordings:* 920; *Source:* RSD	Healthy; chronic; nonchronic	*Accuracy:* 98; *Precision:* 93; *Sensitivity:* 90; *Specificity:* 82; *F1:* 91	***
Pham et al. [121]	*Participants:* 126; *Recordings:* 920; *Source:* RSD	Healthy; chronic; nonchronic	*Sensitivity:* 96; *Specificity:* 86	***
Garcia-Ordás et al. [141]	*Participants:* 126; *Recordings:* 920; *Source:* RSD	Healthy; chronic; nonchronic	*F1:* 99	***
Shuvo et al. [117]	*Participants:* 126; *Recordings:* 920; *Source:* RSD	Healthy; chronic; nonchronic	*Accuracy:* 99; *Sensitivity:* 99; *Specificity:* 100	****
Torre-Cruz et al. [142]	*Participants:* 208; *Recordings:* 208; *Source:* Public	Healthy; Unhealthy	*Accuracy:* 96; *Precision:* 100; *Sensitivity:* 93; *Specificity:* 100	*****
Fraiwan et al. [143]	*Participants:* 215; *Recordings:* 1484; *Source:* RSD and Jordan	Healthy; asthma; COPD; bronchiectasis; pneumonia; heart failure	*Accuracy:* 98; *Sensitivity:* 95; *Specificity:* 99; *F1:* 94	***

Conclusion

In this chapter we presented a global picture of the field of automated respiratory sound analysis in its current form. The most common methods for respiratory sound visualization were displayed, along with feature extraction and classification methods that are currently applied in the five most common tasks related to automated respiratory sound analysis: (1) ARS detection; (2) adventitious respiratory sound classification; (3) respiratory phase segmentation; (4) respiratory cycle classification; (5) respiratory disease classification. While current techniques are way more sophisticated than the simple rule-based methods used by the pioneers of automated respiratory sound analysis, the evaluation performed in most of the reviewed articles is far from ideal, restricting the use of these methods in clinical applications. However, the introduction, in recent years, of large public databases recorded in clinical settings is attracting a lot of attention to the field, and we believe the way algorithms are evaluated will become more objective soon, creating the basis for these techniques to be suitable in clinical practice.

List of Acronyms

2CD Two-cycle Duration
ANN Artificial Neural Network
ARS Adventitious Respiratory Sound
AUC Area Under the ROC Curve
CASs Continuous Adventitious Sounds
CNN Convolutional Neural Network
COPD Chronic Obstructive Pulmonary Disease
CORSA Computerized Respiratory Sound Analysis
CRNN Convolutional Recurrent Neural Network
CWT Continuous Wavelet Transform
DASs Discontinuous Adventitious Sounds
DCT Discrete Cosine Transform
EEMD Ensemble Empirical Mode Decomposition
EMD Empirical Mode Decomposition
ERS European Respiratory Society
HHT Hilbert–Huang Transform
IDW Initial Deflection Width
IMF Intrinsic Mode Frequency
LDW Largest Deflection Width
LPC Linear Predictive Coding
LPCC Linear Predictive Coding Coefficient
MFCC Mel-Frequency Cepstral Coefficient
RNN Recurrent Neural Network
ROC Receiver Operating Characteristic Curve
RSD Respiratory Sound Database
STFT Short Time Fourier Transform
TF Time–Frequency

References

[1] "The top 10 causes of death." www.who.int/news-room/fact-sheets/detail/the-top-10-causes-of-death. Accessed: 2021-07-15.

[2] G.J. Gibson, R. Loddenkemper, B. Lundbäck, Y. Sibille, Respiratory health and disease in Europe: the new European lung white book, Eur. Respir. J. 42 (3) (2013) 559−563.

[3] A. Marques, A. Oliveira, C. Jácome, Computerized adventitious respiratory sounds as outcome measures for respiratory therapy: a systematic review, Respir. Care 59 (May 2014) 765−776.

[4] B.M. Rocha, D. Filos, L. Mendes, G. Serbes, S. Ulukaya, Y.P. Kahya, N. Jakovljevic, T.L. Turukalo, I.M. Vogiatzis, E. Perantoni, E. Kaimakamis, P. Natsiavas, A. Oliveira, C. Jácome, A. Marques, N. Maglaveras, R.P. Paiva, I. Chouvarda, P. De Carvalho, An open access database for the evaluation of respiratory sound classification algorithms, Physiol. Meas. 40 (3) (2019).

[5] D. Hayes, S.S. Kraman, The physiologic basis of spirometry, Respir. Care 54 (December 2009) 1717−1726.

[6] S. Fleming, A. Pluddemann, J. Wolstenholme, C. Price, C. Heneghan, M. Thompson, "Diagnostic Technology : Automated Lung Sound Analysis for Asthma," Tech. Rep, July 2011.

[7] B.M. Rocha, D. Pessoa, A. Marques, P. Carvalho, R.P. Paiva, Automatic classification of adventitious respiratory sounds: a (un)solved problem? Sensors 21 (1) (2021) 1−19.

[8] R.X.A. Pramono, S. Bowyer, E. Rodriguez-Villegas, Automatic adventitious respiratory sound analysis: a systematic review, PLoS One 12 (May 2017) e0177926.

[9] S. Reichert, R. Gass, C. Brandt, E. Andrès, Analysis of respiratory sounds: state of the art, Clin. Med. Circ. Respir. Pulm Med. 2 (January 2008) p. CCRPM.S530.

[10] A. Gurung, C.G. Scrafford, J.M. Tielsch, O.S. Levine, W. Checkley, Computerized lung sound analysis as diagnostic aid for the detection of abnormal lung sounds: a systematic review and meta-analysis, Respir. Med. 105 (9) (2011) 1396−1403, 45.

[11] M.A. Marinella, COVID-19 Pandemic and the Stethoscope: Do Not Forget to Sanitize, 2020.

[12] A. Marques, A. Oliveira, Normal versus adventitious respiratory sounds, in: Breath Sounds, Ch. 10, Springer International Publishing, Cham, 2018, pp. 181−206.

[13] A.R. Sovijärvi, F. Dalmasso, J. Vanderschoot, L.P. Malmberg, G. Righini, S.A. Stoneman, Definition of terms for applications of respiratory sounds, Eur. Respir. Rev. 10 (77) (2000) 597−610.

[14] H. Pasterkamp, P.L.P. Brand, M. Everard, L. Garcia-Marcos, H. Melbye, K.N. Priftis, Towards the standardisation of lung sound nomenclature, Eur. Respir. J. 47 (3) (2016).

[15] F. Dalmay, M.T. Antonini, P. Marquet, R. Menier, Acoustic properties of the normal chest, Eur. Respir. J. 8 (10) (1995) 1761−1769.

[16] A. Bohadana, G. Izbicki, S.S. Kraman, Fundamentals of lung auscultation, N. Engl. J. Med. 370 (February 2014) 744−751.

[17] H. Pasterkamp, I. Sanchez, Effect of gas density on respiratory sounds, Am. J. Respir. Crit. Care Med. 153 (March 1996) 1087−1092.

[18] M. Sarkar, I. Madhavi, N. Niranjan, M. Dogra, Auscultation of the respiratory system, Ann. Thorac. Med. 10 (3) (2015) 158.

[19] H. Pasterkamp, S.S. Kraman, G.R. Wodicka, Respiratory sounds: advances beyond the stethoscope, Am. J. Respir. Crit. Care Med. 156 (3 I) (1997) 974−987.

[20] P. Forgacs, Crackles and wheezes, Lancet 290 (July 1967) 203−205.

[21] R. Paciej, A. Vyshedskiy, D. Bana, R.L. Murphy, Squawks in pneumonia, Thorax 59 (2004) 177—179.

[22] A. Sovijarvi, L.P. Malmberg, G. Charbonneau, J. Vanderschoot, F. Dalmasso, C. Sacco, M. Rossi, J.E. Earis, Characteristics of breath sounds and adventitious respiratory sounds, Eur. Respir. Rev. 10 (77) (2000) 591—596.

[23] H. Melbye, J.C. Aviles Solis, C. Jácome, H. Pasterkamp, Inspiratory crackles-early and late-revisited: identifying COPD by crackle characteristics, BMJ Open Resp. Res. 8 (1) (2021) 1—8.

[24] S. Fouzas, M.B. Anthracopoulos, A. Bohadana, Clinical Usefulness of Breath Sounds, Springer International Publishing, Cham, 2018, pp. 33—52.

[25] A. Abbas, A. Fahim, An automated computerized auscultation and diagnostic system for pulmonary diseases, J. Med. Syst. 34 (December 2010) 1149—1155.

[26] M. Yeginer, K. Ciftci, U. Cini, I. Sen, G. Kilinc, Y. Kahya, Using lung sounds in classification of pulmonary diseases according to respiratory subphases, in: 26th Annual International Conference of the IEEE Engineering in Medicine and Biology Society vol. 3, IEEE, 2004, pp. 482—485.

[27] N. Gavriely, M. Nissan, A.H. Rubin, D.W. Cugell, Spectral characteristics of chest wall breath sounds in normal subjects, Thorax 50 (December 1995) 1292—1300.

[28] J.B. Grotberg, N. Gavriely, Flutter in collapsible tubes: a theoretical model of wheezes, J. Appl. Physiol. 66 (5) (1989) 2262—2273.

[29] J.J. Marini, D.J. Pierson, L.D. Hudson, S. Lakshminarayan, The significance of wheezing in chronic airflow obstruction, Am. Rev. Respir. Dis. 120 (November 1979) 1069—1072.

[30] G.R. Epler, C.B. Carrington, E.A. Gaensler, Crackles (rales) in the interstitial pulmonary diseases, Chest 73 (3) (1978) 333—339.

[31] C.H. Chen, W.T. Huang, T.H. Tan, C.C. Chang, Y.J. Chang, Using K-nearest neighbor classification to diagnose abnormal lung sounds, Sensors 15 (6) (2015) 13132—13158.

[32] Stethoscope - Wikipedia. https://en.wikipedia.org/wiki/Stethoscope#/media/File:Laennecs_stethoscope,_c_1820._(9660576833).jpg. Accessed: 2021-07-15.

[33] Y.P. Kahya, Breath sound recording, in: Breath Sounds, Springer International Publishing, 2018, pp. 119—137.

[34] L. Vannuccini, J. Earis, P. Helistö, B. Cheetham, M. Rossi, A. Sovijärvi, J. Vanderschoot, Capturing and preprocessing of respiratory sounds, Eur. Respir. Rev. 10 (2000) 616—620.

[35] L. Hadjileontiadis, Lung sounds: An advanced signal processing perspective, in: J. Enderle (Ed.), Synthesis Lectures in Biomedical Engineering, Morgan & Claypool, 2009.

[36] M.J. Mussell, The need for standards in recording and analysing respiratory sounds, Med. Biol. Eng. Comput. 30 (2) (1992) 129—139.

[37] C.K. Druzgalski, R.L. Donnerberg, R.M. Campbell, Techniques of recording respiratory sounds, J. Clin. Eng. 5 (October 1980) 321—330.

[38] B.M. Rocha, L. Mendes, R. Couceiro, J. Henriques, P. Carvalho, R. Paiva, Detection of explosive cough events in audio recordings by internal sound analysis internal sound analysis, in: IEEE Engineering in Medicine and Biology Conference, No. July, 2017.

[39] E. Messner, M. Hagmüller, P. Swatek, F. Pernkopf, A Robust Multichannel lung sound recording device, in: Proceedings of the 9th International Joint Conference on Biomedical Engineering Systems and Technologies, SCITEPRESS—Science and and Technology Publications, 2016, pp. 34—39.

[40] M. Kompis, H. Pasterkamp, G.R. Wodicka, Acoustic imaging of the human chest, Chest 120 (4) (2001) 1309—1321.

[41] R. Murphy, Computerized multichannel lung sound analysis, in: IEEE Engineering in Medicine and Biology Magazine, 2007, pp. 16—19.

[42] I. Sen, Y. Kahya, A multi-channel device for respiratory sound data acquisition and transient detection, in: IEEE Engineering in Medicine and Biology 27th Annual Conference, No. January 2016, 2005, pp. 6658−6661.

[43] B.-Y. Lu, Unidirectional microphone based wireless recorder for the respiration sound, J. Bioeng Biomed. Sci. 6 (3) (2016).

[44] G. Yilmaz, M. Rapin, D. Pessoa, B.M. Rocha, A.M. de Sousa, R. Rusconi, P. Carvalho, J. Wacker, R.P. Paiva, O. Chételat, A wearable stethoscope for long-term ambulatory respiratory health monitoring, Sensors 20 (September 2020) 5124.

[45] D.S. Chan, C.W. Callahan, S.J. Sheets, C.N. Moreno, F.J. Malone, An Internet-based store-and-forward video home telehealth system for improving asthma outcomes in children, Am. J. Health Syst. Pharm. 60 (19) (2003) 1976−1981.

[46] A. Casas, Integrated care prevents hospitalisations for exacerbations in COPD patients, Eur. Respir. J. 28 (July 2006) 123−130.

[47] B. Reyes, N. Reljin, K. Chon, Tracheal sounds acquisition using smartphones, Sensors 14 (July 2014) 13830−13850.

[48] G. Druger, The Chest, its Signs and Sounds: The Examination and Interpretation of Physical Findings of the Chest : A New Teaching Program, Humetrics Corp, 1973.

[49] S. Lehrer, Understanding Lung Sounds, second ed., Saunders, Philadelphia, 1993.

[50] S. Lehrer, Understanding Lung Sounds, third ed., W.B. Saunders, Philadelphia, Pa, 2002.

[51] A.G. Tilkian, M.B. Conover, Understanding Heart Sounds and Murmurs With an Introduction to Lung Sounds, W.B. Saunders, Philadelphia, 2001.

[52] D. Owens, R.A.L.E. Lung sounds 3.0, Comput. Inf. Nurs. 5 (3) (2002) 9−10.

[53] East Tennessee State University 2002 Pulmonary breath sounds. http://faculty.etsu.edu/arnall/www/public_html/heartlung/breathsounds/contents.html. Accessed: 2021-07-15.

[54] R.L. Wilkins, J.E. Hodgkin, B. Lopez, Fundamentals of Lung and Heart Sounds, Mosby, St. Louis, 2004.

[55] L.W. Wilkins, Auscultation Skills : Breath & Heart Sounds, Wolters Kluwer/Lippincott Williams & Wilkins Health, Philadelphia, 2009.

[56] Diane Wrigley, Heart and Lung Sounds Reference Library CD-ROM, PESI HealthCare, 2011.

[57] S. Kraman, Lung sounds: an introduction to the interpretation of auscultatory findings, MedEdPORTAL 3 (January 2007) pp. mep_2374−8265.129.

[58] 3M Littmann Library. http://www.3m.com/healthcare/littmann/mmm-library.html. Accessed: 2021-07-15.

[59] S. Mangione, Secrets Heart & Lung Sounds Audio Workshop 2nd edition + Student Consult Online Access Companion to Physical Diagnosis Secrets, Elsevier, City, 2015.

[60] SoundCloud - Lung sounds. https://soundcloud.com/search?q=lung%20sounds. Accessed: 2021-07-15.

[61] B.M. Rocha, D. Filos, L. Mendes, I. Vogiatzis, E. Perantoni, E. Kaimakamis, P. Natsiavas, A. Oliveira, C. Jácome, A. Marques, R.P. Paiva, I. Chouvarda, P. Carvalho, N. Maglaveras, A respiratory sound database for the development of automated classification, IFMBE Proc. 66 (2017) 33−37.

[62] F.-S. Hsu, S.-R. Huang, C.-W. Huang, C.-J. Huang, Y.-R. Cheng, C.-C. Chen, J. Hsiao, C.-W. Chen, L.-C. Chen, Y.-C. Lai, B.-F. Hsu, N.-J. Lin, W.-L. Tsai, Y.-L. Wu, T.-L. Tseng, C.-T. Tseng, Y.-T. Chen, F. Lai, Benchmarking of eight recurrent neural network variants for breath phase and adventitious sound detection on a self-developed open-access lung sound database-Hf_lung_v1, PLos One (2021).

[63] M. Fraiwan, L. Fraiwan, B. Khassawneh, A. Ibnian, A dataset of lung sounds recorded from the chest wall using an electronic stethoscope, Data Brief 35 (2021) 106913.

[64] F.-S. Hsu, S.-R. Huang, C.-W. Huang, Y.-R. Cheng, C.-C. Chen, J. Hsiao, C.-W. Chen, F. Lai, An Update of a Progressively Expanded Database for Automated Lung Sound Analysis, arXiv, 2021.

[65] R.B. Urquhart, J. McGhee, J.E.S. Macleod, S.W. Banham, F. Moran, The diagnostic value of pulmonary sounds: a preliminary study by computer-aided analysis, Comput. Biol. Med. 11 (January 1981) 129–139.

[66] B. Sankur, Y.P. Kahya, E. Guler, T. Engin, Comparison of AR-based algorithms for respiratory sounds classification, Comput. Biol. Med. 24 (January 1994) 67–76.

[67] S.K. Holford, Discontinuous Adventitious Lung Sounds: Measurement, Classification, and Modeling, 1981. PhD thesis.

[68] R.L. Murphy, E.A. Del Bono, F. Davidson, Validation of an automatic crackle (rale) counter, Am. Rev. Respir. Dis. 140 (4) (1989) 1017–1020.

[69] T. Kaisia, A.R.A. Sovijärvi, P. Piirilä, H.M. Rajala, S. Haitsonen, T. Rosqvist, Validated method for automatic detection of lung sound crackles, Med. Biol. Eng .Comput. (1991). Physiological Measurement, No. September.

[70] M. Du, F. Chan, F. Lam, J. Sun, Multi-Resolution Decomposition Applied To Crackle Detection, CCS, 1997.

[71] L. Hadjileontiadis, S. Panas, Separation of discontinuous adventitious sounds from vesicular sounds using a wavelet-based filter, IEEE Trans. Biomed. Eng. 44 (12) (1997).

[72] K.E. Forkheim, D. Scuse, H. Pasterkamp, Comparison of neural network models for wheeze detection, in: IEEE WESCANEX Communications, Power, and Computing 1, 1995, pp. 214–219, no. 95.

[73] J. Gnitecki, Z.M.K. Moussavi, Separating heart sounds from lung sounds. Accurate diagnosis of respiratory disease depends on understanding noises, in: IEEE Engineering in Medicine and Biology Magazine 26, 2007, pp. 20–29, no.1.

[74] J. Smith, Ambulatory methods for recording cough, Pulm. Pharmacol. Therapeut. 20 (4) (2007) 313–318.

[75] S. Matos, S.S. Birring, I.D. Pavord, D.H. Evans, An automated system for 24-h monitoring of cough frequency: the leicester cough monitor, IEEE Trans. Biomed. Eng. 54 (8) (2007) 1472–1479.

[76] T. Drugman, J. Urbain, T. Dutoit, Assessment of audio features for automatic cough detection, in: 19th European Signal Processing Conference (Eusipco11), 2011, pp. 1289–1293.

[77] J. Amoh, K. Odame, Deep neural networks to identify cough sounds, IEEE Trans. Biomed. Circuits Syst. (2016) 1–9.

[78] R.X.A. Pramono, S.A. Imtiaz, E. Rodriguez-Villegas, A cough-based algorithm for automatic diagnosis of pertussis, PLoS One 11 (9) (2016) 1–20.

[79] R.X.A. Pramono, S.A. Imtiaz, E. Rodriguez-Villegas, Evaluation of features for classification of wheezes and normal respiratory sounds, PLoS One 14 (3) (2019) 1–21.

[80] F.G. Nabi, K. Sundaraj, C.K. Lam, Identification of asthma severity levels through wheeze sound characterization and classification using integrated power features, Biomed. Signal Process Control 52 (2019) 302–311.

[81] A.K. Kattepur, F. Jin, F. Sattar, Single channel source separation for convolutive mixtures with application to respiratory sounds, in: BIOSIGNALS 2010-Proceedings of the 3rd International Conference on Bio-inpsired Systems and Signal Processing, Proceedings, 2010, pp. 220–224.

[82] J.D.L.T. Cruz, F.J.C. Quesada, N.R. Reyes, P.V. Candeas, J.J.C. Orti, Wheezing sound separation based on informed inter-segment non-negative matrix partial co-factorization, Sensors 20 (9) (2020).

[83] F.J. Canadas-Quesada, N. Ruiz-Reyes, J. Carabias-Orti, P. Vera-Candeas, J. Fuertes-Garcia, A non-negative matrix factorization approach based on spectro-temporal clustering to extract heart sounds, Appl. Acoust. 125 (2017) 7−19.

[84] M. Lozano, J.A. Fiz, R. Jané, Automatic differentiation of normal and Continuous Adventitious Respiratory Sounds Using Ensemble Empirical Mode decomposition and instantaneous frequency, IEEE J. Biomed. Health Inform. 20 (2) (2016) 486−497.

[85] Empirical mode decomposition - MATLAB. www.mathworks.com/help/signal/ref/emd. html. Accessed: 2021-07-15.

[86] M.A. Colominas, G. Schlotthauer, M.E. Torres, Improved complete ensemble EMD: a suitable tool for biomedical signal processing, Biomed. Signal Process Control 14 (1) (2014) 19−29.

[87] M. Lozano, J. Antonio, R. Jané, Performance evaluation of the Hilbert − huang-transformforrespiratorysoundanalysisanditsapplicationtocontinuous adventitious sound characterization, Signal Process. 120 (March) (2016) 99−116.

[88] Z. Wu, N.E. Huang, Ensemble empirical mode decomposition: a noise-assisted data analysis method, Adv. Adapt. Data Anal. 1 (1) (2009) 1−41.

[89] M.E. Torres, M.A. Colominas, G. Schlotthauer, P. Flandrin, A complete ensemble empirical mode decomposition with adaptive noise, in: ICASSP, IEEE International Conference on Acoustics, Speech and Signal Processing Proceedings, 2011, pp. 4144−4147.

[90] P. Flandrin, G. Rilling, P. Goncalves, Empirical mode decomposition as a filter bank, IEEE Signal Process. Lett. 11 (February 2004) 112−114.

[91] J. Zhang, R. Yan, R.X. Gao, Z. Feng, Performance enhancement of ensemble empirical mode decomposition, Mech. Syst. Signal Process. 24 (7) (2010) 2104−2123.

[92] F. Demir, A.M. Ismael, A. Sengur, Classification of lung sounds with CNN model using parallel pooling structure, IEEE Access 8 (2020) 105376−105383.

[93] T. Nguyen, F. Pernkopf, Lung sound classification using snapshot ensemble of convolutional neural networks, in: 42nd Annual International Conference of the IEEE Engineering in Medicine & Biology Society, April, IEEE, July 2020, pp. 760−763.

[94] S.S. Stevens, J. Volkmann, E.B. Newman, A scale for the measurement of the psychological magnitude pitch, J. Acoust. Soc. Am. 8 (3) (1937) 185−190.

[95] A. Kandaswamy, C.S. Kumar, R.P. Ramanathan, S. Jayaraman, N. Malmurugan, Neural classification of lung sounds using wavelet coefficients, Comput. Biol. Med. 34 (2004) 523−537.

[96] M. Bahoura, Separation of crackles from vesicular sounds using wavelet packet transform, in: 2006 IEEE International Conference on Acoustics Speed and Signal Processing Proceedings, vol. 2, 2006. II1076−II1079.

[97] S. Ulukaya, G. Serbes, Y.P. Kahya, Overcomplete discrete wavelet transform based respiratory sound discrimination with feature and decision level fusion, Biomed. Signal Process Control 38 (September) (2017) 322−336.

[98] Continuous wavelet transform - MATLAB. www.mathworks.com/help/wavelet/ref/cwt. html. Accessed: 2021-07-15.

[99] Choose a wavelet - MATLAB. www.mathworks.com/help/wavelet/gs/choose-a-wavelet. html. Accessed: 2021-07-15.

[100] V.V. Moca, H. Bârzan, A. Nagy-Dăbâcan, R.C. Mures, Time-frequency super-resolution with superlets, Nat. Commun. 12 (1) (2021) 1−18.

[101] N.E. Huang, Z. Shen, S.R. Long, M.C. Wu, H.H. Snin, Q. Zheng, N.C. Yen, C.C. Tung, H.H. Liu, The empirical mode decomposition and the Hilbert spectrum for nonlinear and non-stationary time series analysis, in: Proceedings of the Royal Society A: Mathematical, Physical and Engineering Sciences, vol. 454, 1998, pp. 903−995, no. 1971.

[102] N. Nakamura, M. Yamashita, S. Matsunaga, Detection of patients considering observation frequency of continuous and discontinuous adventitious sounds in lung sounds, in: Proceedings of the Annual International Conference of the IEEE Engineering in Medicine and Biology Society, EMBS, 2016Octob, 2016, pp. 3457−3460.

[103] D. Oletic, B. Arsenali, V. Bilas, Towards continuous wheeze detection body sensor node as a core of asthma monitoring system, in: Lecture Notes of the Institute for Computer Sciences, Social-Informatics and Telecommunications Engineering, vol. 83, LNICST, 2012, pp. 165−172.

[104] P. Bokov, B. Mahut, P. Flaud, C. Delclaux, Wheezing recognition algorithm using recordings of respiratory sounds at the mouth in a pediatric population, Comput. Biol. Med. 70 (2016) 40−50.

[105] X. Liu, W. Ser, J. Zhang, D.Y.T. Goh, Detection of adventitious lung sounds using entropy features and a 2-D threshold setting, in: 2015 10th International Conference on Information, Communications and Signal Processing (ICICS), IEEE, December 2015, pp. 1−5.

[106] S. Davis, P. Mermelstein, Comparison of parametric representations for monosyllabic word recognition in continuously spoken sentences, in: IEEE Transactions on Acoustics, Speech, and Signal Processing 28, 1980, pp. 357−366, no.4.

[107] O. Lartillot, P. Toiviainen, Mir in matlab (II): a toolbox for musical feature extraction from audio, in: Proceedings of the 8th International Conference on Music Information Retrieval, ISMIR 2007, 2007, pp. 127−130.

[108] Pitch - MATLAB. https://www.mathworks.com/help/audio/ref/pitch.html. Accessed: 2021-07-15.

[109] G. Charbonneau, E. Ademovic, B. Cheetham, L.A. Malmberg, Basic techniques for respiratory sound analysis, Eur. Respir. Rev. 10 (2000) 625−635.

[110] C. Pinho, A. Oliveira, C. Jácome, J. Rodrigues, A. Marques, Automatic crackle detection algorithm based on fractal dimension and box filtering, Procedia Comput. Sci. 64 (2015) 705−712.

[111] L.J. Hadjileontiadis, A texture-based classification of crackles and squawks using lacunarity, IEEE Trans. Biomed. Eng. 56 (3) (2009) 718−732.

[112] D. Chamberlain, R. Kodgule, D. Ganelin, V. Miglani, R.R. Fletcher, Application of semi-supervised deep learning to lung sound analysis, in: Proceedings of the Annual International Conference of the IEEE Engineering in Medicine and Biology Society, EMBS, 2016-Octob, 2016, pp. 804−807.

[113] B.-S. Lin, B.-S. Lin, Automatic wheezing detection using speech recognition technique, J. Med. Biol. Eng. 36 (August 2016) 545−554.

[114] R. Naves, B.H. Barbosa, D.D. Ferreira, Classification of lung sounds using higher-order statistics: a divide-and-conquer approach, Comput. Methods Progr. Biomed. 129 (2016) 12−20.

[115] L. Mendes, I.M. Vogiatzis, E. Perantoni, E. Kaimakamis, I. Chouvarda, N. Maglaveras, J. Henriques, P. Carvalho, R.P. Paiva, Detection of crackle events using a multi-feature approach, in: 2016 38th Annual International Conference of the IEEE Engineering in Medicine and Biology Society (EMBC), 2016, IEEE, August 2016, pp. 3679−3683.

[116]   D. Bardou, K. Zhang, S.M. Ahmad, Lung sounds classification using convolutional neural networks, Artif. Intell. Med. 88 (June 2018) 58–69.

[117]   S.B. Shuvo, S.N. Ali, S.I. Swapnil, T. Hasan, M.I.H. Bhuiyan, A lightweight CNN model for detecting respiratory diseases from lung auscultation sounds using EMD-CWT-based hybrid scalogram, IEEE J. Biomed. Health Inform. XX (XX) (2020) 1–9.

[118]   S. Jayalakshmy, G.F. Sudha, Scalogram based prediction model for respiratory disorders using optimized convolutional neural networks, Artif. Intell. Med. 103 (January) (2020) 101809.

[119]   K. Kochetov, E. Putin, M. Balashov, A. Filchenkov, A. Shalyto, Noise masking recurrent neural network for respiratory sound classification, in: International Conference on Artificial Neural Networks, vol. 1, 2018, pp. 208–217.

[120]   J. Acharya, A. Basu, Deep neural network for respiratory sound classification in wearable devices enabled by patient specific model tuning, IEEE Trans. Biomed. Circuits Syst. 14 (3) (2020) 535–544.

[121]   L. Pham, I. McLoughlin, H. Phan, M. Tran, T. Nguyen, R. Palaniappan, Robust Deep Learning Framework for Predicting Respiratory Anomalies and Diseases, January 2020, pp. 90–93.

[122]   "Classification: Roc curve and auc." https://developers.google.com/machine-learning/crash-course/classification/roc-and-auc. Accessed: 2021-07-15.

[123]   S. Adavanne, T. Virtanen, "A Report on Sound Event Detection with Different Binaural Features," arXiv, November, 2017.

[124]   S. Adavanne, H. Fayek, V. Tourbabin, Sound event classification and detection with weakly labeled data, in: DCASE, October, 2019, pp. 15–19.

[125]   C. Jácome, J. Ravn, E. Holsbø, J.C. Aviles-Solis, H. Melbye, L.A. Bongo, Convolutional neural network for breathing phase detection in lung sounds, Sensors 19 (8) (2019) 1–10.

[126]   S.A. Taplidou, L.J. Hadjileontiadis, Wheeze detection based on time-frequency analysis of breath sounds, Comput. Biol. Med. 37 (August 2007) 1073–1083.

[127]   R. Riella, P. Nohama, andJ. Maia, Method for automatic detection of wheezing in lung sounds, Braz. J. Med. Biol. Res. 42 (July 2009) 674–684.

[128]   L. Mendes, I.M. Vogiatzis, E. Perantoni, E. Kaimakamis, I. Chouvarda, N. Maglaveras, V. Tsara, C. Teixeira, P. Carvalho, J. Henriques, R.P. Paiva, Detection of wheezes using their signature in the spectrogram space and musical features, in: Proceedings of the Annual International Conference of the IEEE Engineering in Medicine and Biology Society, 2015, EMBS, 2015, pp. 5581–5584.

[129]   E. Messner, M. Fediuk, P. Swatek, S. Scheidl, F.M. Smolle-Juttner, H. Olschewski, F. Pernkopf, Crackle and breathing phase detection in lung sounds with deep bidirectional gated recurrent neural networks, in: Proceedings of the Annual International Conference of the IEEE Engineering in Medicine and Biology Society, 2018, EMBS, 2018, pp. 356–359. July 2018.

[130]   M. Grønnesby, J.C.A. Solis, E. Holsbø, H. Melbye, andL.A. Bongo, Machine learning based crackle detection in lung sounds, arXiv (2017) 1–13.

[131]   N. Jakovljević, T. Lončar-Turukalo, Hidden markov model based respiratory sound classification, Int. Conf. Biomed. Health Inform. (2017) 39–43.

[132]   G. Serbes, S. Ulukaya, Y.P. Kahya, An automated lung sound preprocessing and classification system based on spectral analysis methods, Int. Conf. Biomed. Health Inform. vol. 66 (2018) 45–49.

[133] J.S. Chuah, Z.K. Moussavi, Automated respiratory phase detection by acoustical means, in: Engineering in Medicine and Biology Society, 2000. Proceedings of the 22nd Annual International Conference of the IEEE, 2000.

[134] S.H.Z. Moussavi, Acoustic breath-phase detection using tracheal breath sounds, Med. Biol. Eng. Comput. 50 (3) (2012) 297–308.

[135] C.H. Hsiao, T.W. Lin, C.W. Lin, F.S. Hsu, F.Y.S. Lin, C.W. Chen, C.M. Chung, Breathing sound segmentation and detection using transfer learning techniques on an attention-based encoder-decoder architecture, in: Proceedings of the Annual International Conference of the IEEE Engineering in Medicine and Biology Society, 2020-July, EMBS, 2020, pp. 754–759.

[136] H. Chen, X. Yuan, Z. Pei, M. Li, J. Li, Triple-classification of respiratory sounds using optimized S-transform and deep residual networks, IEEE Access 7 (2019) 32845–32852.

[137] Y. Ma, X. Xu, Q. Yu, Y. Zhang, Y. Li, J. Zhao, G. Wang, LungBRN: a smart digital stethoscope for detecting respiratory disease using bi-resnet deep learning algorithm, in: BioCAS 2019–Biomedical Circuits and Systems Conference, Proceedings, 2019, pp. 9–12.

[138] D. Perna, A. Tagarelli, Deep auscultation: predicting respiratory anomalies and diseases via recurrent neural networks, in: 2019 IEEE 32nd International Symposium on Computer-Based Medical Systems, 2019 June, IEEE, June 2019, pp. 50–55.

[139] S.Y. Jung, C.H. Liao, Y.S. Wu, S.M. Yuan, C.T. Sun, Efficiently classifying lung sounds through depthwise separable cnn models with fused stft and mfcc features, Diagnostics 11 (4) (2021).

[140] M. Aykanat, Ö. Kılıç, B. Kurt, S. Saryal, Classification of lung sounds using convolutional neural networks, EURASIP J. Image Video Process. 2017 (December 2017) 65.

[141] M.T. García-Ordás, J.A. Benítez-Andrades, I. García-Rodríguez, C. Benavides, H. Alaiz-Moretón, Detecting respiratory pathologies using convolutional neural networks and variational autoencoders for unbalancing data, Sensors 20 (4) (2020).

[142] J. Torre-Cruz, F. Canadas-Quesada, S. García-Galán, N. Ruiz-Reyes, P. VeraCandeas, J. Carabias-Orti, A constrained tonal semi-supervised nonnegative matrix factorization to classify presence/absence of wheezing in respiratory sounds, Appl. Acoust. 161 (2020) 107188.

[143] L. Fraiwan, O. Hassanin, M. Fraiwan, B. Khassawneh, A.M. Ibnian, M. Alkhodari, Automatic identification of respiratory diseases from stethoscopic lung sound signals using ensemble classifiers, Biocybern. Biomed. Eng. 41 (1) (2021) 1–14.

j

Chapter 6

Respiratory image analysis

Inéz Frerichs[1], Zhanqi Zhao[2,3], Meng Dai[3], Fabian Braun[4],
Martin Proença[4], Michaël Rapin[4], Josias Wacker[4], Mathieu Lemay[4],
Kostas Haris[5,6], Georgios Petmezas[5], Aris Cheimariotis[5], Irini Lekka[5],
Nicos Maglaveras[5], Claas Strodthoff[1], Barbara Vogt[1], Livia Lasarow[1],
Norbert Weiler[1], Diogo Pessoa[7], Bruno Machado Rocha[7],
Paulo de Carvalho[7], Rui Pedro Paiva[7] and Andy Adler[8]

[1]Department of Anaesthesiology and Intensive Care Medicine, University Medical Centre
Schleswig-Holstein, Campus Kiel, Kiel, Germany; [2]Institute of Technical Medicine, Furtwangen
University, Villingen-Schwenningen, Germany; [3]Department of Biomedical Engineering, Fourth
Military Medical University, Xi'an, China; [4]Swiss Center for Electronics and Microtechnology
(CSEM, Centre Suisse d'Electronique et de Microtechnique), Neuchâtel, Switzerland; [5]Lab of
Computing, Medical Informatics and Biomedical Imaging Technologies, Aristotle University of
Thessaloniki, Thessaloniki, Greece; [6]Department of Informatics and Computer Engineering,
University of West Attica, Athens, Greece; [7]Department of Informatics Engineering, Centre for
Informatics and Systems of the University of Coimbra, Coimbra, Portugal; [8]Systems and Computer
Engineering, Carleton University, Ottawa, Canada

Introduction

Medical imaging methods are an essential part of care and therapy of patients. Modern medicine uses medical imaging in diagnostics, monitoring, and treatment of patients. Albeit using different measuring principles the common feature of all medical imaging methods is that they seek to generate images of the internal structure (in some cases even function) of the examined body (or its part) that otherwise is not visible to the human eye.

Typical common medical imaging methods are radiography, computed tomography (CT), magnetic resonance imaging (MRI), scintigraphy, positron emission tomography (PET), single-photon emission computed tomography (SPECT), electrical impedance tomography (EIT), and medical ultrasound (sonography). Radiography and CT utilize X-rays, applied either in a single or multiple projections, and their spatially dissimilar attenuation by different biological tissues to produce images of the human body and its organs. MRI uses a strong magnetic field to polarize and excite hydrogen nuclei and a radiofrequency field to manipulate them to produce measurable signals that can be transformed into images. Scintigraphy, PET, and SPECT are all based on the application of radioactive isotopes to the body and on the detection of

Wearable Sensing and Intelligent Data Analysis for Respiratory Management
https://doi.org/10.1016/B978-0-12-823447-1.00001-4
169

emitted radiation that can be converted into images. EIT applies small imperceptible electrical alternating currents to a body section, measures the resulting voltages on the body surface, and calculates the internal distribution of electrical bioimpedance to produce an image. Medical ultrasound uses nonaudible sound waves (i.e., ultrasound) for image generation. All the medical imaging instruments mentioned so far probe the body from the outside. Endoscopic methods on the contrary examine the body from the inside. They are regarded as invasive methods because visual examination of body cavities or organs, like stomach or airways, is achieved by an insertion of an illuminated optical instrument.

In view of the important role of imaging methods in medicine and the recent developments and research toward wearable diagnostic and monitoring instruments the question arises whether any of the existing imaging methods could be meaningfully implemented into wearable instruments. It is obvious that conventional methods of radiology and nuclear medicine that utilize X-rays and radioactivity cannot be transformed into wearable instruments for multiple reasons. The radiation load and the size of the devices only allow patient examination at a specialized examination site. Although MRI does not use radiation, the operational principle of this technology precludes a wearable design. The invasiveness of endoscopic methods also does not permit remote use by patients without medical expertise. Thus, from the perspective of current medical knowledge, only ultrasound and EIT might be considered for potential transformation into wearable medical instruments. Both methods apply and measure signals that are not harmful either to the examined subject or to the surrounding environment; the devices are already portable and relatively low cost. Hence, wearable designs seem to be possible.

In this chapter we address in detail the medical imaging method of EIT [1]. EIT is currently used mainly in chest applications for monitoring of regional lung ventilation, typically in patients requiring ventilator therapy. Several EIT devices approved for clinical in-hospital use are commercially available. These are stand-alone portable instruments or they are integrated into mechanical ventilators. Another large group of patients that are expected to benefit from EIT examinations are the patients suffering from chronic lung diseases, like chronic obstructive pulmonary disease (COPD), asthma, or cystic fibrosis. In these patients EIT might provide valuable information on the natural history of the lung disease, assess the effects of pharmacological or physical therapy, support disease staging, and identify exacerbations of the disease [1,2]. The availability of wearable EIT, suitable for remote monitoring in home environment, would be of immense advantage in these patients.

We present here basic information on how EIT generates primary (raw) EIT images and how these images can be used to create functional EIT (fEIT) images as well as quantitative measures, characterizing regional lung ventilation suitable for clinical decision-making. We describe the challenges and possible pitfalls arising from the attempt to transfer EIT into a fully wearable medical instrument, and we provide information on the current research status regarding the development of wearable EIT. In addition, although chest EIT is primarily applied for lung ventilation monitoring, other potentially clinically relevant data regarding cardiac function and blood circulation can also be derived from chest EIT examinations. We present these novel fields of EIT research too. Finally, we address the issue how EIT could be meaningfully combined with additional biosignals to create a complex, but still easy-to-use and reliable, monitoring system for patients suitable for routine use.

Primary image reconstruction

EIT makes electrical stimulations and measurements at body-surface electrodes to calculate images of the internal distribution of tissue electrical properties. Like other tomographic imaging modalities, EIT is named for the source data, which are transfer-impedance values made using low-frequency (10−250 kHz) and low-amplitude (<10 mA) sinusoidal electrical current stimulation. Fig. 6.1 illustrates EIT data measurement: electrical current flows between a pair of electrodes and the resulting voltage distribution is measured before and after a change (here, an increase in conductive blood in the heart). EIT images are reconstructed using voltages measured at several different applied currents.

Most biological tissues respond to electrical current in a linear fashion, characterized by Ohm's law $\overrightarrow{j} = \sigma^* \overrightarrow{E}$, where $\overrightarrow{j}$ is the current density, σ^* is the (complex) conductivity, and $\overrightarrow{E}$ is the electrical field. The complex

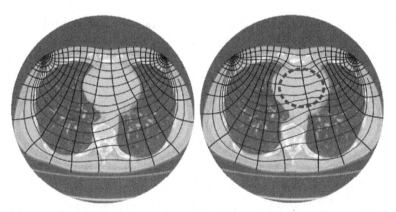

FIGURE 6.1   The propagation of current in a body. In each image, a volumetric model of the thorax is used to simulate the propagation of electrical current from a pair of surface electrodes. Blue lines show current streamlines while the black lines are isopotential surfaces. (Right) An increase of conductivity in the heart (dotted red lines (gray in print)) is simulated, and the resulting change in current flow and body-surface voltage shown. (Note that figures appear in color only in the electronic version of the book.)

$\sigma^* = \sigma + j\omega\epsilon_0\epsilon_r$ characterizes the conductivity $(\sigma[s/m])$ and permittivity $\epsilon_0\epsilon_r$ of tissue, and describes the amplitude and phase of the conductivity response to applied electrical stimulations. However, at the frequencies used in EIT, the phase changes due to permittivity variations are small, and typically less than the phase noise of available systems. For this reason, the complex component of σ^* has often been ignored. Values for different tissues in the chest differ by over an order of magnitude (lung conductivity is small and blood conductivity is large). In addition to static differences, physiological activities result in time changes of the conductivity; in the chest, air movement during breathing increases and decreases lung conductivity, while blood and heart movement changes heart and vascular conductivity.

EIT is considered soft-field imaging modality. Current applied at electrodes propagates in diffusive way: current density and thus sensitivity are largest near the electrodes and lower further away. This is in contrast to "hard-field" imaging techniques such as X-ray CT, where the emitted energy travels in a straight line and the imaging properties vary little with position. Because of the "soft-field" nature, sensitivity varies dramatically with spatial position, leading to a mathematical ill-conditioning. Reconstruction of EIT images is an "inverse problem," and a schematic of image reconstruction is shown in Fig. 6.2, illustrating the role of two parts: the forward model and an inverse model. The goal is to estimate image parameters ($\hat{x}$) which best match the measured data and fit prior constraints such as smoothness (we use the notation of bold x to represent vectors of measured data or image pixel or voxel values).

Forward model

The forward model describes the flow of electrical current and measurement of voltages in EIT. Since analytical solutions of the electrical propagation are not

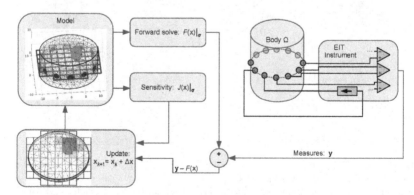

FIGURE 6.2 Illustration of the process of image reconstruction. EIT data, y, are measured with an instrument from body Ω. The model, x_k is improved by updates, Δx, calculated from the mismatch between the current forward estimate and sensitivity, and a prior model. (Note that figures appear in color only in the electronic version of the book.)

possible for arbitrary shapes like the body, numerical approximations are used, the most common being the Finite Element Method (FEM). Using the forward model, we calculate a voltage at each electrode, σ_F, for each applied electrode current. Measuring the electrode voltages for all applied current patterns yields a frame, $v \in \mathbb{C}^{N_m}$, of EIT data. Due to reciprocity [3] the maximum number of independent measurements possible on N_E electrodes is $\frac{1}{2}N_E(N_E - 1)$.

Most EIT systems do not use measurements which are made on stimulation electrodes because such measurements are more sensitive to contact imped-ance and electrode movement, and this removes an additional N measurement. Thus, an EIT frame has $\frac{1}{2}N_E(N_E - 3)$ independent measurements.

The FEM-based forward calculation is represented as

$$y = \mathbf{F}(\mathbf{x})|_{\sigma=\sigma_r}, \tag{6.1}$$

calculated at an assumed reference conductivity, σ_r. The sensitivity (repre-sented by the Jacobian, J) is the change in measurements for a given change in σ^* and characterizes the detection of contrasts of interest of a given EIT configuration (e.g., electrode positions, stimulation and measurement pattern, and body shape). Here, component $[J]_{i,j}$ is the sensitivity of data i to image parameter j.

$$J_{i,j} = \frac{\partial}{\partial \sigma_j} F(x)_i \big|_{\sigma=\sigma_r}. \tag{6.2}$$

Inverse model

Image reconstruction describes the calculation of an image from projection data. Image reconstruction in EIT is challenging because the problem is ill-conditioned (large difference in sensitivity between regions) and ill-posed (greater number of image parameters than independent measurements). A block diagram of image reconstruction is shown in Fig. 6.2, which illustrates the process by which model parameters are iteratively adjusted to fit the measurements (and "prior" image constraints). The reconstructed image is the estimate upon convergence of the iterative scheme. Here we calculate an es-timate, $\widehat{x}$, of the distribution of internal properties, x, which best explains the measurements, y.

Image reconstruction in EIT is divided into absolute and difference EIT. Absolute EIT (aEIT) estimates the conductivity distribution σ at a single point in time from a single set of data, y. Difference EIT calculates the change in conductivity $\Delta\sigma$ between two measurements. The most common subtype is time-difference EIT (tdEIT) where $\Delta\sigma$ represents changes between times t_1 and t_2. Another type is frequency difference EIT (fdEIT) in which measure-ments are made using currents at two stimulation frequencies. In this chapter, we consider exclusively tdEIT because it is better able to reject differences between the forward model and the true geometry of the body and the electrodes. To illustrate this sensitivity, consider that EIT is very sensitive to the placement and contact quality of the electrodes. If an electrode is a few millimeters away from its modeled position, the forward model will not be able to accurately predict measured voltages. With tdEIT, however, the electrode does not move between the two measurements, and these errors can be largely canceled.

Typically difference EIT image reconstruction minimizes the norm

$$\|y - F(\widehat{x})\|_{\mathbf{W}}^2 + \lambda^2 \|\widehat{x} - x_0\|_{\mathbf{Q}}^2. \tag{6.3}$$

The first term, $\|y - F(\widehat{x})\|$ is the "data mismatch" between the measured data and their estimate via the forward model. $\mathbf{W}$ is a data weighting matrix, and represents the inverse covariance of measurements. The second term is the mismatch between the reconstruction estimate, $\widehat{X}$, and an a priori estimate of its value, x_0, where $\mathbf{Q}$ is the "regularization matrix," and the relative weighting between the data and prior mismatch terms is controlled by a hyperparameter, λ.

The norm Eq. (6.3) can be minimized using the iterative Gauss–Newton method. Starting from an estimate x_0, at each step an update, Δx_k, is calculated

$$x_{k+1} = x_k + \Delta x_k \tag{6.4}$$

The update solves the linearized problem around each estimate x_k, using a Jacobian, $\mathbf{J}_k = \frac{\partial}{\partial \mathbf{x}} F(\mathbf{x})|_{\mathbf{x}_k}$.

$$\Delta x_k = \left(\mathbf{J}_k^t \mathbf{W} \mathbf{J}_k + \lambda^2 \mathbf{Q} \right) \left(\mathbf{J}_k^t \mathbf{W} \left(y - F(x_k) + \lambda^2 \mathbf{Q}(x_0 - x_k) \right) \right) \tag{6.5}$$

In most tdEIT systems, a one-step regularized difference EIT solution is used. Here Eq. (6.5) reduces to a matrix multiplication, $\widehat{x} = \mathbf{R}y$, with a reconstruction matrix $\mathbf{R}$, where

$$\mathbf{R} = \left(\mathbf{J}^t \mathbf{W} \mathbf{J} + \lambda^2 \mathbf{Q} \right)^{-1} \mathbf{J}^t \mathbf{W} \tag{6.6}$$

since x_0 and $F(x_0)$ are zero for difference EIT.

The behavior of image reconstruction is controlled by the selection of the regularization matrix $\mathbf{Q}$ and the hyperparameter, λ, which controls the amount of regularization, as illustrated in Fig. 6.3. As λ decreases, the image more accurately reflects the ground truth, but also becomes more sensitive to noise. This "noise-resolution" trade-off is typically the most important parameter choice for an EIT configuration. Its value should be chosen to be the best resolution which is still able to provide relatively noise-free images. Generally, λ is chosen for the EIT system, and its selection is not given to the user.

$\lambda = 0.015$ $\lambda = 0.003$

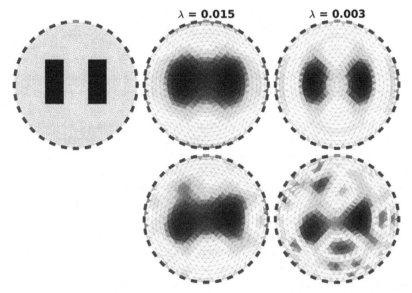

FIGURE 6.3 Simulation phantom (*top left*) and reconstructed images for values of the hyper-parameter, λ. *Top:* Simulated data with no noise. *Bottom:* Simulated data with added Gaussian noise. (Note that figures appear in color only in the electronic version of the book.)

The regularization matrix **Q** imposes a structural penalty onto image elements. In the simplest case, **Q** is diagonal and imposes a penalty on image amplitude, favoring low-amplitude reconstructions; this is the case for a Tikhonov or a diagonal matrix [4]. Next, a matrix with off-diagonal elements can impose a spatial high-pass filter to penalize nonsmooth components in the image, such as for a Laplace spatial filter [5]. One widely used approach to select the image reconstruction parameters is GREIT [6], which choses **Q** and λ through a set of figures of merit of the reconstructed images, which are then implemented via a set of training targets and "desired images." GREIT has lower sensitivity to boundary artifacts, and a more uniform spatial resolution [7].

EIT image reconstruction for wearable sensors

In general, for a wearable system, there is little that needs to change from EIT image reconstruction in static scenarios. The reader who wishes to implement EIT image reconstruction is recommended to take a look at open-source algorithms available in EIDORS [8]. Using a one-step regularized approach, image reconstruction is formulated as matrix multiplication, which can run in real time on commodity hardware. For example, with 16 electrodes and a 32×32 pixel image, the number of floating point operations per reconstructed image frame is $\frac{1}{2} 16 \times (16 - 3) \times 32^2 \approx 100$.

The one aspect of EIT image reconstruction which needs attention in wearable scenarios is the requirement for compensation for poor electrode contact. In a mobile subject, electrode contact quality will vary, and the data from such "failing" electrodes will have large noise. If these data values are used, the resulting EIT images will be unreadable. There are thus two related problems: detection of failing electrodes and correction of the reconstructed images.

Generally, electrode contact impedance can be measured for a pair of electrodes by measuring the voltage across the driven pair (strictly speaking, this gives a measure of the contact impedance of the pair, but we also have measurements across other combinations, allowing the failing electrode to be isolated). Another approach is to compare the reciprocity match between measurements made on complimentary stimulation and measurement pairs [9]. Without electrode errors, reciprocal pairs should give equal values. To compensate, the weighting of poorly matching pairs is reduced.

Correction of failing electrodes can be done heuristically, by setting failing pairs to zero, or through the inverse model [10] and adjusting the weighting matrix, W in Eq. (6.3). The main idea is to model errors as low SNR, and thus high noise, on measurements which use the affected electrodes. We thus create a new reconstruction matrix $\mathbf{R}^* = \left(\mathbf{J}^t \mathbf{W}^* \mathbf{J} + \lambda^2 \mathbf{Q} \right)^{-1} \mathbf{J}^t \mathbf{W}^*$, based on Eq. (6.6) but with each entry in $\mathbf{W}^*$ which uses a failing electrode set to zero.

Many challenges remain in EIT image reconstruction, related to improving the image quality and resolution, and to improving its robustness. The latter is the principle challenge for wearable sensor applications. As the subject moves, many data-quality issues can occur which impact the EIT images. Finally, it is not the EIT images which are of value, but rather the physiological information which can be used for diagnosis, monitoring, and treatment which is useful. To do this, robust functional (fEIT) parameters are required, which we discuss in the next section.

Functional EIT images and measures

EIT is often used to monitor the impedance changes in regions of interest (ROIs). Due to its ability to capture spatial information with high temporal resolution, EIT scans contain valuable functional information that may influence clinical decisions. fEIT images are images that subtract and present amplitude or phase information over a period of time (Fig. 6.4). If the information is summarized as a single number, it may be called an EIT measure or index. The fEIT visualizes regional information, whereas the corresponding EIT measures provide high-level overview so that the trend over time can be easily compared. In a previous consensus paper on thoracic EIT, the fEIT and measures were introduced in a separate chapter and grouped according to their target physiological parameters [1]. In this section, several fEIT images and measures are introduced that are relevant for use in awake subjects/patients using wearable EIT.

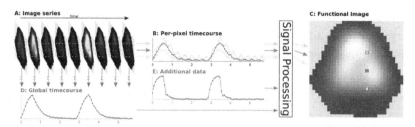

FIGURE 6.4 Conceptual diagram of the process of computing a functional EIT image. (A) A sequence of raw EIT images is obtained; (B) a waveform is obtained for each pixel from the image sequence (A); (C) using various signal processing approaches a functional EIT image pixel value is calculated based on the pixel waveform (B); (D) some approaches require a global waveform (time course) that is obtained by computing the sum of all image pixel values, or other auxiliary data such as pressure or volume recordings from the patient ventilator or spirometer (E). (Note that figures appear in color only in the electronic version of the book.) *Reproduced from I. Frerichs, M.B.P. Amato, A.H. van Kaam, D.G. Tingay, Z. Zhao, B. Grychtol, et al. (2017). Chest electrical impedance tomography examination, data analysis, terminology, clinical use and recommendations: consensus statement of the TRanslational EIT developmeNt stuDy group. Thorax, 72 (1) (2017) 83−93. https://doi.org/10.1136/thoraxjnl-2016-208357.*

Ventilation distribution during tidal breathing − volume changes

Tidal impedance variation (abbreviated as TIV or TV) refers to the ventilation changes within tidal breathing; usually it represents the amplitude changes if not specified otherwise. When the position of the electrodes is adequate [11,12], the change of impedance is proportional to the change of air volume in the lungs. Hence, regional TV reveals regional ventilation distribution of tidal volume. To calculate fEIT-TV, EIT image at end-expiration can be subtracted from the EIT image at end-inspiration. To increase the signal-to-noise level, the following mathematical algorithms may be used:

- Standard deviation [13]

Each pixel in this fEIT image is the standard deviation of the relative impedance change values in each raw EIT images pixel for a certain time period:

$$S_i = \sqrt{\frac{1}{K} \sum_{k=1}^{K} \left(\Delta Z_{i,k} - \text{mean}(\Delta Z_i) \right)^2} \tag{6.7}$$

where S_i is the pixel i in the fEIT image; ΔZ_i represents the relative impedance change of the pixel i in the raw EIT images; K denotes the number of frames for a certain time period; and k is the time point in that period.

- Regression [14]

Each pixel value in this fEIT image is the regression coefficient of the following linear regression formula:

$$\Delta Z_i(t) = \alpha_i \cdot \sum_{m=1}^{M} \Delta Z_m(t) + \beta_i + \varepsilon_i \qquad (6.8)$$

where $\Delta Z_i(t)$ denotes time-dependent relative impedance value of pixel i of the raw EIT images; α_i and β_i are regression coefficients and ε_i is the fitting error of pixel i; M is the total number of pixels in the reconstructed EIT images. As result, α_i will be the value plotted in pixel i in the fEIT image.

- Average TV

After subtracting the end-expiration from the end-inspiration image of single breath to form tidal images, these images within certain time period are averaged, i.e.,

$$V_i = \frac{1}{N} \sum_{n=1}^{N} \left(\Delta Z_{i,Ins,n} - \Delta Z_{i,Exp,n} \right) \qquad (6.9)$$

where V_i is the pixel i in the fEIT image; N is the number of breaths within analyzing period; $\Delta Z_{i,Ins}$ and $\Delta Z_{i,Exp}$ are the pixel values in the raw EIT image at the end-inspiration and end-expiration, respectively.

Ventilation distribution during tidal breathing — time (phase) shift

In presence of heterogeneous airway resistance or nonuniform transmission of pleural pressure generated by diaphragmatic contraction, temporal heterogeneity of the tidal ventilation distribution can be captured by various fEIT.

- Inspiration/expiration time [15,16]

The regional inspiration and/or expiration time can be characterized with absolute time [16] or time constant [15]. Take the calculation of absolute inspiration time, for example, each pixel value of the lung regions in the fEIT image is corresponding to the inspiration time of that pixel region, i.e.,

$$T_i = \frac{1}{N} \sum_{n=1}^{N} \left(k_{i,InsEnd,n} - k_{i,InsBegin,n} \right) / f \qquad (6.10)$$

where T_i is the average inspiration time of the pixel i; N is the number of breaths within the analyzing period; $k_{i,InsEnd}$ and $k_{i,InsBegin}$ are the frame numbers at the inspiration end and inspiration begin identified for pixel i of the raw EIT data, respectively; f is the sampling rate of EIT measurement. This type of fEIT has actually a unit for the pixel values, which is second.

Expiratory time constants proposed by Karagiannidis et al. were determined from the expiratory flow—volume curve using 75% of tidal volume. The first 25% of the local signal amplitude was omitted [15], i.e.,

$$V_i(t) = V_{i\_75\%} \times e^{-t/\tau_i} + C_i \qquad (6.11)$$

where τ_i is the time constant of pixel i *and C_i is the end-expiratory volume.* EIT signals stemming from lung areas with low or no ventilation or non-exponential signals were excluded from τ calculations.

- Regional ventilation delay (RVD) [17]

The RVD index (see Eq. 6.12) characterizes the ventilation delay as pixel impedance rising time compared to the global impedance curve, which may assess tidal recruitment/derecruitment.

$$RVD_i = t_{i,40\%} / T_{inspiration,global} \times 100\% \qquad (6.12)$$

where $t_{i,40\%}$ is the time needed for pixel i to reach 40% of its maximum inspiratory impedance change. $T_{inspiration,global}$ denotes the inspiration time calculated from the global impedance curve.

Ventilation distribution during tidal breathing – regional compliance

Static respiratory system compliance is the ratio of tidal volume and driving pressure (P_{driv}) (see Eq. 6.13). Assuming regional flows are zero at the end-inspiration and end-expiration (no pendelluft), P_{driv} should be similar at all lung regions. Considering that the regional impedance changes are proportional to regional volume changes, regional compliance can be assessed by $\Delta Z/P_{driv}$. Further, ΔZ could be normalized to ΔV in milliliters, so that regional compliance may also have the unit as respiratory system compliance (mL/cmH_2O).

$$C_{reg\_i} = \Delta Z_i / P_{driv} \times \Delta Z_{global} / \Delta V_{global} \qquad (6.13)$$

where C_{reg_i} is the regional compliance for pixel i; ΔZ_{global} and ΔV_{global} are impedance and volume changes for the whole lung, respectively.

Combination of amplitude and time

- Lung perfusion

Lung perfusion represents the blood flow in the pulmonary vascular system. It can be assessed combining EIT with hypertonic saline bolus injection [18,19,20]. Since the saline is conductive, impedance drops after bolus injection and the slope can be used to estimate the regional perfusion (Fig. 6.5).

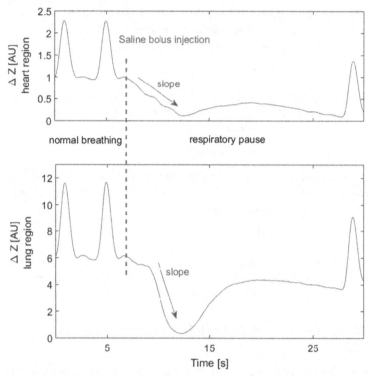

FIGURE 6.5 Ventilation distribution in a subject during spontaneous breathing (left) and under mechanical ventilation (right). GI, the global inhomogeneity index; CoV, ventrodorsal center of ventilation; ROI, region of interest; ΔZ, impedance change; AU, arbitrary units. Highly ventilated regions are marked in light blue (gray in print). (Note that figures appear in color only in the electronic version of the book.)

- Regional flow limitations

Given that airway flow is the derivative of volume, the derivative of relative impedance can be used to assess regional flow limitations in patients with obstructive lung diseases, such as asthma and COPD [21,22,23]. Theoretically the parameters calculated in a spirometry test (e.g., peak expiratory flow, maximum expiratory flow at 25%, 50%, and 75% of vital capacity, and the expiratory volumes) can be deducted and presented at a pixel level. However, in the current EIT devices, only relative impedance values are delivered, which do not represent absolute volumes and flow rates. A direct comparison with predicted normal is therefore not possible [24].

Summarizing fEITs — various measures

To summarize the information in fEIT with single numbers, many EIT-based measures were proposed. A direct way is to calculate the heterogeneities among pixels in the fEITs with, for example, coefficient of variation or standard deviation. Depending on the application and subject's characteristics, the following measures may be considered.

- The global inhomogeneity (GI) index [25]

  The GI index is calculated as follows:

  $$GI = \sum\nolimits_{l \in lung} \left| TV_l - \text{Median}\left(TV_{lung}\right)\right| \Big/ \sum\nolimits_{l \in lung} TV_l \qquad (6.14)$$

  where TV denotes the value of the differential impedance in the tidal images; TV_l is the pixel in the identified lung area. TV_{lung} are all the pixels representing the lung area. The design of GI index is similar to coefficient of variation but substituting mean with median, since the regional lung volume might not be normally distributed in diseased lungs.

- Center of ventilation (CoV) [26]

  CoV depicts ventilation distribution within the studied chest cross section (relative impedance value weighted with location in ventrodorsal or right-to-left coordinate):

  $$CoV = \sum(y_i \times TV_i)\Big/\sum TV_i \times 100\% \qquad (6.15)$$

  where TV_i is the tidal impedance variation in the fEIT image for pixel i, and y_i is the pixel height and of pixel i scaled so the bottom of the image (dorsal) is 100% and the top (ventral) is 0%.

- Impedance distribution in particular ROIs [27]

  For certain type of patients, physicians are mainly interested in particular ROIs, e.g., the gravity-dependent regions (corresponding to dorsal or most dorsal regions when the subject is in supine position). Ventilation distribution (in percentage) in those regions is evaluated to understand the lung status or treatment effect in the patients (Fig. 6.6).

- Size (number of pixels) of certain ROIs [18,28]

  ROIs in fEIT reveal lung status. For example, regions that contain no air at the end of expiration but are ventilated during tidal breathing might be related to tidal recruitment/derecruitment [28]. Regions with good ventilation and low perfusion might represent dead space [19]. The size (number of pixels) of the ROIs can be compared and used to guide treatment strategy (e.g., Ref. [29]).

- Intratidal ventilation distribution (ITVD) [30].

  Ventilation distribution during the courses of inhalation or exhalation may also be heterogeneous. To evaluate the time—impedance trend, inspiration or expiration can be subdivided into several isovolume (or equal time length)

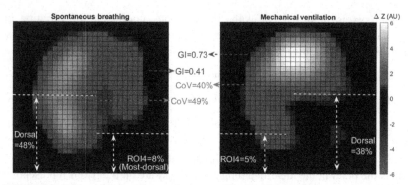

FIGURE 6.6 Time—impedance curves for heart and lung regions during normal breathing and respiratory pause with saline bolus injection. The slopes of impedance drop after bolus injection suggest the blood flow in the corresponding regions. ΔZ, impedance change; AU, arbitrary units. (Note that figures appear in color only in the electronic version of the book.)

periods and corresponding impedance variation images are produced. Taking ITVD, for example, this measure captures the changes of ventilation distribution during inspiration phase to reflect heterogeneous lung mechanics in various regions (Fig. 6.7). The lung is divided into two or four anteroposterior ROIs with equal height. ITVD in each ROI is calculated according to the following equation:

$$\text{ITVD\_ROI} = \Delta Z_{n\_ROI}/\Delta Z_n \qquad (6.16)$$

where ΔZ_{n_ROI} is the sum of impedance changes in the ROIs at n steps of global isovolumetric impedance changes; ΔZ_n is the sum of all ROIs ($n = 1, 2, \dots 8$). Patients with spontaneous breathing may show different ITVD patterns compared to patients under control ventilation [31].

Assumptions and limitations of the fEIT imaging and EIT measures

The fEIT images and EIT measures were developed for specific applications. Therefore, we cannot assume that there is one good fEIT or measure that fits all application needs. To select a suitable fEIT and measures to analyze EIT data, to meet our objective, it is important to understand their assumptions and limitations. For example, change of C_{reg} is widely used for titrating PEEP (see, e.g., [32] [33]). However, depending on the starting and ending PEEP levels, the "optimal" PEEP during the titration would be different [34]. In another word, the optimization with the C_{reg} selects the local not the global "optimum" regarding compliance changes. Another example is RVD, which was validated during a low-flow maneuver (inhalation with an extremely low flow, so the inspiration time is extended) [17]. For normal tidal breathing with brief inspiration time, the number of sampling points is limited, so that the calculation would be too sensitive to noise (especially during spontaneous

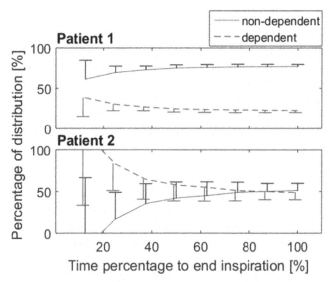

FIGURE 6.7 Intratidal ventilation distribution of two patients during spontaneous breathing trial. Percentages of ventilation distributed in ventral (blue solid lines) and dorsal (red dashed lines) regions were calculated at eight time points during inspiration (X axis, 0%, beginning of inspiration; 100%, end of inspiration). Medians and interquartile ranges of ITVD values during 1-minute period were presented. The distributions may indicate the diaphragm activities of the patients. (Note that figures appear in color only in the electronic version of the book.)

breathing). Selecting fEIT and EIT measures wisely according to the study objective or monitoring and therapy aims may help to utilize the EIT data in a productive way. The most suitable fEIT images and measures, as well as their meaningful combinations, for use in spontaneously breathing patients under wearable monitoring still need to be defined.

Challenges of data acquisition and analysis using wearable EIT

The development of wearable EIT comes along with new challenges regarding both data acquisition and analysis. In conventional chest EIT examinations, a set of electrodes is placed on the chest circumference by the examiner. The examiner selects a position regarded to be sensitive to the phenomena of interest, i.e., respiration and/or hemodynamics. In wearable devices, electrodes are integrated into belts or vests, thus minimizing user intervention concerning placement issues like electrode spacing and adjustment. In most EIT device configurations 16 or 32 electrodes are used typically placed in one transverse (or slightly oblique) plane. In the following, we describe the main factors influencing chest EIT measurements and present current approaches to circumvent their negative effects in wearable EIT:

Electrode plane location

A number of studies have shown that EIT-based findings are affected by the location of the transverse plane defined by the electrodes [11,12,35,36]. As already mentioned, in wearable EIT devices, this location is fixed through the integration of the electrodes in the vest. Also, it is carefully selected by the vest designer so that possible undesirable phenomena such as diaphragm entry during respiration into the measuring plane do not occur. Therefore, in wearable EIT devices, no special adjustment is required.

Electrode contact

Low electrical impedance between electrodes and skin ensures good electrode contact, which is a sufficient condition for recording high-quality data. There is experimental evidence that EIT measurements can be performed even when the impedance at the body—electrode interface is high (e.g., when the skin is dry) or nonuniform; however, their quality may be low resulting in numerical instabilities during subsequent processing such as reconstruction [9,10]. Similarly, low-quality EIT measurements may result in cases where there is excessive sweat (or other conductive fluid) creating conductivity between electrodes. It is agreed that in optimal EIT recording conditions the body—electrode impedance is low, stable, and similar among the electrodes. Modern electrodes can sense and continuously monitor the electrode—skin contact quality. Therefore, EIT data acquisition periods corresponding to low-quality electrode contact are detected and either rejected or processed with special methods.

Body movement

Movement of patients during EIT data acquisition causes significant effects on the recorded measurements [37]. In addition, movement may modify the quality of electrode contact which also affects measurements. Wearable devices make extensive use of accelerometers, which can accurately sense and record all kinds of movement in three-dimensional space in real time. Hence, EIT measurements recorded during body movement can be identified and excluded from the subsequent EIT processing steps.

Missing/faulty electrodes

EIT images of the best quality are achieved when the EIT signal quality is good at all electrodes. However, even with missing or faulty electrodes EIT images can be generated, albeit of slightly lower quality. With this respect, it is important that the signal quality at individual electrodes is monitored and that low-quality and/or missing data are taken into account during EIT image reconstruction (see also Section "EIT image reconstruction for wearable sensors"). As mentioned above, modern EIT wearable vests automatically record electrode contact information during acquisition enabling the detection of time periods of missing or low-quality data [9,10]. The data of the remaining high-quality recording periods can be safely used for subsequent analysis such as reconstruction and processing.

Posture

Optimal EIT data acquisition protocols require specific body posture because body position affects the EIT findings [38,39,36]. However, during various pulmonary maneuvers, but even during quiet tidal breathing, subjects move involuntarily to facilitate and improve their ventilation. This can be observed particularly in patients suffering from respiratory disorders with airway obstruction. This body movement often results in the leaning forward position when the subjects bend their torsos from the waist and may be accompanied by the elevation of arms. The influence of the torso and the arm position and movements on EIT chest examinations (Fig. 6.8) was recently studied in healthy subjects and in patients with obstructive lung diseases [40].

The main findings of the study were that the forward movement of the torso and the elevation of the arms cause a significant effect on the EIT waveforms registered during otherwise undisturbed tidal breathing. The observed changes in the end-expiratory impedance, regional tidal impedance variation, and ventilation heterogeneity may influence the findings during pulmonary function testing. Specifically, the effects on the end-expiratory impedance level are intense. Fortunately, torso position and arm movement can be detected and recorded by accelerometers. It is noted that arm movement determination may require additional sensors external to the EIT vest.

Type of ventilation

Clinical studies have shown that the type of ventilation affects the EIT findings [26,41,42,43,44] and, therefore, ventilation type should be recorded to ease their interpretation. The factors affecting EIT acquisition presented so far (contact, posture, and movement) can be identified automatically via advanced technological solutions such as smart electrodes/sensors and accelerometers. This is not possible for the identification of the ventilation type (tidal breathing, deep breathing, or forced maneuver) and, therefore, data-driven approaches are used: the ventilation type is inferred from the analysis of the EIT measurements as described in the following section.

Ventilation type detection

The main breathing patterns encountered in a chest EIT examination carried out in awake subjects, namely, tidal breathing, deep breathing, and forced

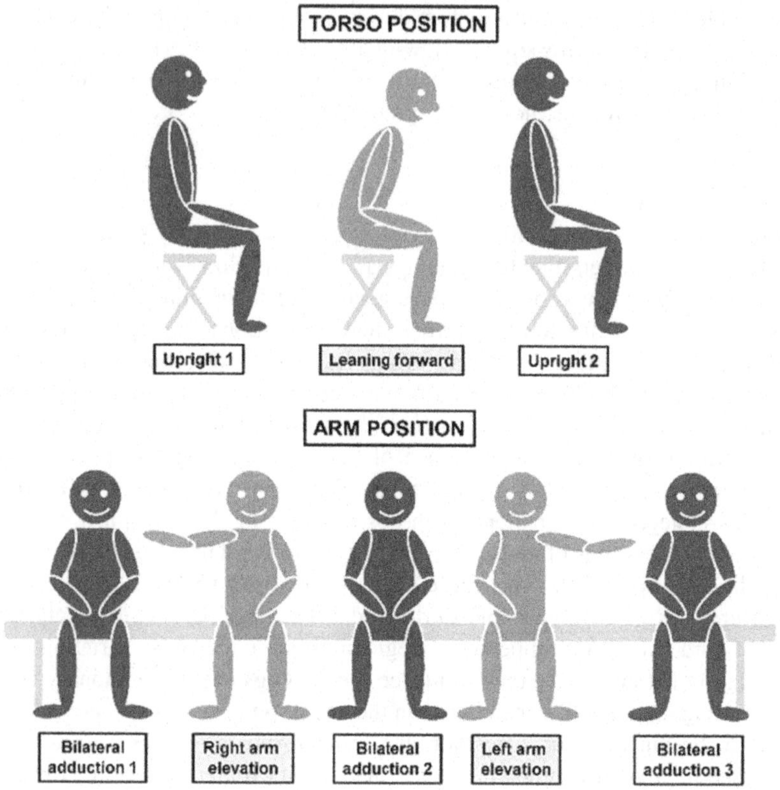

FIGURE 6.8 Examination of the effects of torso and arm positions on EIT recordings. Various torso (*top*) and arm positions (*bottom*) were studied during quiet tidal breathing in the sitting position. *Reproduced by permission of IOP Publishing from B. Vogt, L. Mendes, I. Chouvarda, E. Perantoni, E. Kaimakamis, T. Becher, et al. (2016). Influence of torso and arm positions on chest examinations by electrical impedance tomography. Physiol. Meas. 37 (6) (2016) 904–921. https:// doi.org/10.1088/0967-3334/37/6/904.* © *Institute of Physics and Engineering in Medicine. All rights reserved.*

maneuver, are detected by computing and analyzing the global (impedance) waveform as follows: (i) Initially, the raw EIT images are reconstructed and then, (ii) the raw global waveform is computed by taking the mean value of each raw image (Fig. 6.9, blue waveform in the electronic version of the book). The respiratory component of the global waveform is isolated by appropriate low-pass filtering (Fig. 6.9, red waveform in the electronic version of the book). In the resulting respiratory global waveform, the identification of the end-inspiratory and end-expiratory points is reduced to local maxima and minima detection.

Having identified the respiration cycles allows the detection of the so-called *Tidal Breathing Periods* (*TBPs*) which are defined as time periods

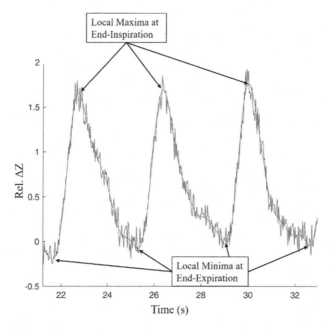

FIGURE 6.9   Respiratory cycle identification. The raw Global Impedance Waveform (blue line (light gray in print)) contains respiratory and cardiac components. The respiratory component of the waveform (red line (dark gray in print)) is isolated by low-pass filtering. In the resulting waveform, the respiratory cycle identification is reduced to simple local minima and maxima detection. Rel. ΔZ, relative impedance change. (Note that figures appear in color only in the electronic version of the book.)

consisting of consecutive respiratory cycles of similar amplitude and duration (Fig. 6.10) [45]. TBPs represent periods of quiet spontaneous tidal breathing (first ventilation type).

For the other two ventilation maneuvers (*Forced* and *Deep*), specific three-step protocols are applied: the patient is instructed to breathe quietly for a number of respiratory cycles (at least 4), execute the maneuver, and then continue breathing quietly. As expected, the above sequence of respiratory cycles is projected to the respiratory global waveform as shown in Fig. 6.10 for a forced maneuver. Specifically, the pattern is: (i) there is a TBP corresponding to the initial quiet breathing, (ii) it follows the forced maneuver respiratory cycle with amplitude at least three times the one of the preceding TBP, and (iii) there is another TBP for the last phase of the maneuver. Similarly, a sequence of respiratory cycles is identified as *Deep Breathing Period* (DBP) if there is a preceding and a succeeding TBP with a mean amplitude less than half of DBP.

It should be noted that the above predetermined protocols are applied in order to exclude other ventilation types (like coughing) that may happen to produce similar breathing patterns to the forced maneuver or deep breathing.

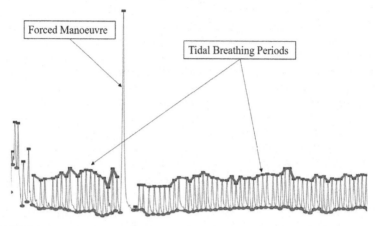

FIGURE 6.10 Ventilation type detection. Tidal Breathing Periods are detected as consecutive respiratory cycles with similar amplitude and duration (two such periods are shown). By convention, a Forced Maneuver has an amplitude at least three times larger and is located between tidal breathing periods. (Note that figures appear in color only in the electronic version of the book.)

These protocols combined with the synchronized data acquired by the sensors and the accelerometers integrated into the wearable vest make the unambiguous ventilation type identification possible.

Current experience with wearable EIT for ventilation monitoring

By virtue of its measuring principle EIT can already be regarded as a partially wearable medical technology because the EIT electrodes are placed ("worn") on the chest of the examined person. The commercially available EIT devices use either an array of single electrodes or wearable fabric or silicone belts into which the electrodes are integrated. Albeit the current EIT devices are portable, the examined subject still cannot "leave" the vicinity of the device because the electrodes/sensors are connected with the device via leads. Thus, the essential innovation needed to make chest EIT truly wearable is the omission of the leads with either a physical separation of the device HW from the sensors (and wireless communication between them) or the integration of all EIT system components into just one wearable. To enable the option of free movement of the examined patient even outside the home environment and, thus, the full capacity of remote monitoring, the latter approach is clearly the preferred one.

Thanks to ongoing miniaturization of HW components and progress in modern communication technologies the realization of fully wearable chest EIT instruments seems to be feasible. First attempts addressing partial aspects of the necessary development have already been realized. Low-power consumption is an essential prerequisite for the development of wearable EIT. Menolotto and colleagues built such a low-power architecture for a wearable electrical bioimpedance sensor and tested its performance through simulations in anatomical chest models and in phantoms [46]. The low-power design was associated with a low signal-to-noise ratio; nonetheless, its effect on the EIT image quality was found to be acceptable and not precluding further utilization in a wearable EIT imaging system.

The wearable design creates new requirements regarding the EIT system's accuracy and adaptability because the measuring circumstances are more variable than in the relatively stable scenario encountered during examinations with stationary devices. In the wearable setting, the current source still needs to drive accurately large complex loads. Klum and colleagues created a new current source architecture suitable for adaptive electrical bioimpedance measurements [47]. The proposed current source allowed for adaptive frontend calibration, output current monitoring, and increased common mode rejection with balancing while utilizing the full available system voltage range. Its performance was tested under in vitro conditions and the current source was regarded as suitable for bioimpedance measurements in a wearable setup.

Rapin and colleagues developed an electronic sensing architecture allowing flexible current stimulation and voltage measurement patterns using frequency-division multiplexing. The system was not only fully parallel, but it also exhibited reduced cabling complexity suitable for chest EIT use [48]. Minimal wiring was achieved by connecting the sensors in a bus arrangement via two unshielded wires. Using this novel cooperative sensor architecture not only multichannel bioimpedance measurement used for EIT imaging was accomplished, but also multi-lead ECG data were acquired. The cooperative sensors were tested under in vivo conditions in a healthy human subject and the findings demonstrated the feasibility of the proposed system. The tested system was based on discrete components which impacted the power consumption and limited the possible miniaturization of the sensor size. The authors concluded that development of application-specific integrated circuits (ASICs) would allow future optimization of these features.

Such implementation of ASICs in an electronic architecture for wearable EIT was recently reported [49]. The authors aimed at developing a wearable EIT belt with active electrodes and with flexible current drive and voltage measurement patterns for neonatal chest monitoring. The system exhibited a fast sampling rate of 122 frames/s, a wide bandwidth of 1 MHz, and multi-frequency operation. The described design, however, in a larger size than the one intended for neonates, was tested in one adult subject and the findings demonstrated the capability of the system to capture ventilation-related impedance changes in the chest based on analysis of EIT images reconstructed offline.

To the best of our knowledge, the only wearable EIT system with an implemented feature of real-time EIT image generation was described by Hong and colleagues [50]. It was based on a simple beltlike structure, which contained 32 electrodes and the EIT integrated circuit. The external imaging application ran on a mobile device. The images were reconstructed at a rate of 20 images/s and continuously displayed (Fig. 6.11). In addition, the relative tidal impedance variation was calculated during each respiratory cycle. The system only acquired EIT data and no other biosignals. The performance of the proposed instrument was tested in one healthy adult male subject.

The only study testing the performance of a wearable EIT system in a large group of subjects under conditions comparable with the intended use was conducted by Frerichs and colleagues [51]. Fifty healthy adult volunteers were examined during quiet tidal breathing, deep breathing, and while performing forced full inspiration and expiration maneuvers. Such maneuvers are well established in conventional pulmonary function testing and used to determine different pulmonary function measures characterizing the lung status. The study participants were wearing newly designed washable vests into which a total of 21 detachable sensors were integrated. The sensors followed the

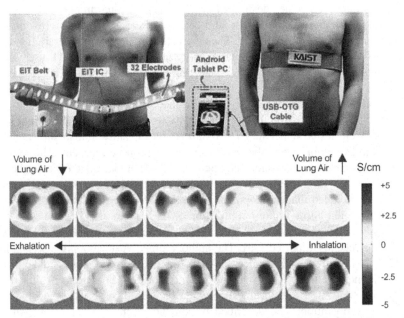

FIGURE 6.11 Application of a belt-type EIT system in a healthy male adult subject. Proposed belt-type EIT system with 32 measuring electrodes (top) and a series of 10 reconstructed EIT images (bottom). (Note that figures appear in color only in the electronic version of the book.) *Reproduced with permission from S. Hong, J. Lee, H.-J. Yoo. Wearable lung-health monitoring system with electrical impedance tomography. Annu. Int. Conf. IEEE Eng. Med. Biol. Soc. 2015 (2015) 1707–1710. https://doi.org/10.1109/EMBC.2015.7318706*

cooperative sensor design as described above [48]. They enabled the continuous acquisition of multiple biosignals: electrical bioimpedance of the chest for EIT and respiratory rate assessment, peripheral oxygen saturation, chest sounds, electrocardiography for heart rate measurement, body activity, and posture. Sixteen of the 21 sensors were directly involved in the generation of EIT data at a sampling rate of 80 frames/s. Selected signals (respiratory rate, heart rate, one-lead ECG, body posture and activity, peripheral oxygen saturation, and the quality of the skin-to-sensor contact at preselected sensors) were streamed onto a mobile phone via Bluetooth in real time using a dedicated application.

An important feature of the proposed vest was that it took the differences in chest anatomy between the genders into account. Specific male and female designs were created; the female vest had the 16 sensors needed for EIT placed in a fabric stripe with a bralike closure at the front of the chest (Fig. 6.12). The performance of the wearable could therefore be reliably tested not only in men but also in women, who comprised 46% of the study cohort. Since both male and female vests were produced in various sizes (M, L, XL, and XXL) the effect of different body shapes could be tested.

 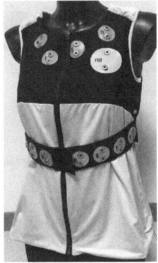

FIGURE 6.12 Female model of the multisensor vest for remote chest monitoring. The vest is shown inverted inside out. The color circles indicate the locations at which individual sensors are inserted using snap buttons. The large gray circle shows the position of the master (reference) sensor, the small gray circles of the current-injecting sensors, and the orange circles of the voltage-measuring sensors. (Note that figures appear in color only in the electronic version of the book.) *Reproduced by permission of IOP Publishing from I. Frerichs, B. Vogt, J. Wacker, R. Paradiso, F. Braun, M. Rapin, et al. Multimodal remote chest monitoring system with wearable sensors: a validation study in healthy subjects. Physiol. Meas. 41 (1) (2020) 015006. https://doi.org/10.1088/ 1361-6579/ab668f. © Institute of Physics and Engineering in Medicine. All rights reserved.*

The study participants were examined during two sessions between which the vests were taken off and then put on again. During the first session, the subjects were studied in a sitting posture during quiet tidal breathing, deep breathing, and two repeated forced ventilation maneuvers followed by quiet breathing (Fig. 6.13). During the second session, the same phases were examined but data were additionally acquired also after standing up, walking, sitting down, lying in a supine position, and even during writing while sitting at a table (Fig. 6.14). By using this complex study protocol, situations of vest applications typical for possible future home use were simulated. Thus, the performance of the vest could be tested in various postures, during body movement and activity as well as during different types of ventilation.

In total, about 125,000 EIT images were acquired in the examined subjects during the study. The raw EIT images were reconstructed offline and in the next step functional images were generated showing the spatial distribution of selected relevant ventilation measures, like regional tidal volumes during quiet and deep breathing and forced vital capacity (Figs. 6.13 and 6.14). Generally, the ventilation-related changes in regional electrical bioimpedance were reliably identified in the majority of subjects (34 out of 50 subjects). However,

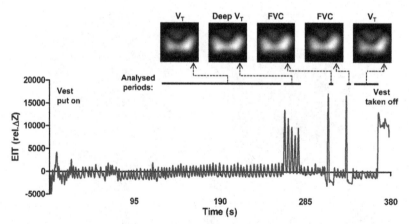

FIGURE 6.13   EIT recording obtained with a wearable electrical impedance tomography (EIT) instrument. Original EIT waveform of relative impedance change (rel. ΔZ) recorded in a 45-year-old woman (body height 172 cm and weight 67 kg) using a female multisensor vest during a single examination session. The analyzed periods of the recording representing quiet tidal breathing, deep breathing, two forced full expiration maneuvers, and repeated tidal breathing in sitting posture are highlighted by thick black lines above the waveform. Autoscaled functional EIT images generated from the analyzed periods of the EIT recording show the distribution of tidal volume (V_T) and forced vital capacity (FVC) in the chest cross section in the respective periods. (Note that figures appear in color only in the electronic version of the book.) *Reproduced by permission of IOP Publishing from I. Frerichs, B. Vogt, J. Wacker, R. Paradiso, F. Braun, M. Rapin, et al. Multimodal remote chest monitoring system with wearable sensors: a validation study in healthy subjects. Physiol. Meas. 41 (1) (2020) 015006. https://doi.org/10.1088/1361-6579/ab668f.* © *Institute of Physics and Engineering in Medicine. All rights reserved.*

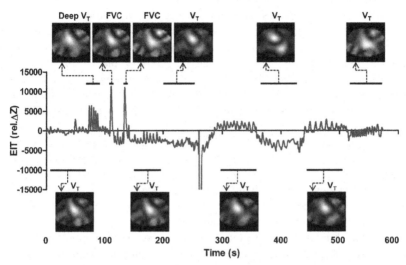

FIGURE 6.14 EIT recording obtained with a wearable electrical impedance tomography (EIT) instrument. Original EIT waveform recorded in a 22-year-old woman (body height 178 cm and weight 70 kg) using a wearable female multisensor vest. The analyzed periods of the recording representing quiet tidal breathing, deep breathing, two forced full expiration maneuvers, repeated tidal breathing in sitting posture, followed by periods of walking, sitting, supine horizontal position, and final sitting phase (passive and while writing) are highlighted by thick black lines above and below the waveform. Functional EIT images generated from the analyzed periods of the EIT recording should have shown the distribution of tidal volume (V_T) and forced vital capacity (FVC) within the lung regions in the chest cross section in the respective periods but are dominated by artifacts. Note that to better visualize the ventilation-related changes in the EIT signal during the analyzed measurement periods, the first 3 minutes of the recording are not shown and the large singular disturbance during walking between 259 and 265 s is not plotted with its full amplitude. (Note that figures appear in color only in the electronic version of the book.) *Reproduced by permission of IOP Publishing from I. Frerichs, B. Vogt, J. Wacker, R. Paradiso, F. Braun, M. Rapin, et al. Multimodal remote chest monitoring system with wearable sensors: a validation study in healthy subjects. Physiol. Meas. 41 (1) (2020) 015006. https://doi.org/10.1088/1361-6579/ab668f.*

good-quality EIT images were obtained in only about half of the studied women and men, without gender-related differences. As shown in the example recording in Fig. 6.14, fEIT images exhibiting multiple artifacts were observed even when the EIT waveform clearly reflected the different ventilation during the individual session phases. The presence of less satisfactory EIT image quality was attributed to the following factors. The sensors were fixed in their positions in the vests using snap buttons. This solution was not optimal because the contact at snap buttons deteriorated in the course of the study. Moreover, it was cumbersome to use. The tested vests were not tailored to the individual body sizes of the study participants. The sensor—skin contact was monitored in real time only at 12 voltage-measuring sensors (8 of which were needed for EIT), but not at the 8 current-injecting ones. Thus, insufficient skin contact at the latter sensors could not be identified and corrected during the examination. The spatial image resolution was relatively low because the chosen design reduced the number of independent measurements of transthoracic bioimpedance. Preferably all 16 sensors placed on the chest circumference should be utilized for both voltage measurement and current application, and skin contact must be monitored at all sensors involved in EIT. These improvements are intended in a planned follow-up study using an improved vest design (www.welmo-project.eu).

The development of wearable chest EIT represents a very recent research activity. Therefore, most of the studies conducted so far and presented in this chapter were proof-of-concept or pilot studies. The increased interest in wearable EIT beyond the existing bedside EIT systems approved for clinical use has been triggered, on one hand, by the general technological advancements in e-health solutions and, on the other hand, by the clinical need for remote health-related information that could be meaningfully utilized for monitoring of patient's pulmonary status and treatment effects. Medical decisions are typically based on complex interrelated information. Therefore, remote pulmonary monitoring should ideally provide an imaging feature that could be covered by EIT but it should implement also additional biosignals as will be addressed in the following section.

Use of wearable EIT beyond ventilation monitoring

Even though wearable EIT can be used for a variety of other applications as diverse as bladder volume estimation [52], hand gesture recognition [53], and upper airway monitoring in sleep apnea [54], this section focuses on the use of thoracic EIT for cardiovascular monitoring which was intensively investigated in the last two decades.

The motivation for EIT-based cardiovascular monitoring is that it allows for the safe, noninvasive, and continuous measurement of central hemodynamic parameters — including stroke volume (SV) and pulmonary artery pressure (PAP) — which are usually difficult to measure as they often require invasive catheterization. In addition, EIT has the big advantage that it can provide simultaneous, breath-by-breath ventilation information [1].

The typical processing steps for thoracic EIT are illustrated in Fig. 6.15. A crucial preprocessing step is the separation of cardiovascular information from ventilation since the former is about 10-fold weaker than the latter. A variety of techniques exist (ECG gating, blind source separation, and frequency domain filtering) which are detailed in Ref. [55]. The resulting cardiovascular signals are related to ventricular activity in the heart region [56,57] (bottom red signal in Fig. 6.15C) and the pulsatile activity of the pulmonary arteries in the lung region [58] (middle blue signal in Fig. 6.15C). However, the exact origin of these signals is still not fully understood [59,60,61] and subject to investigations. First, it is worth noting that the signals in the lung region are

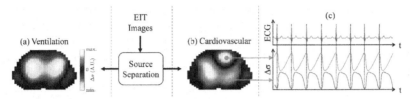

FIGURE 6.15   Typical processing steps for thoracic EIT. Source separation is applied to separate (A) ventilation-related and (B) cardiovascular activity. The images in (A) and (B) show the conductivity change $\Delta\sigma$ of (A) one breath (inspiration vs. full expiration) and (B) one heartbeat (end systole vs. end diastole). Note that (A) and (B) share the same color bar but are scaled individually to their corresponding minimum and maximum because cardiovascular activity in (B) is about factor 10 weaker than ventilation shown in (A). The image in (C) shows an illustrative example of cardiovascular EIT signals in the heart region (blue (light gray in print)) and the lung region (red (dark gray in print)) synchronously with ECG (green (gray in print)). The vertical black dotted lines represent the timing of the ventricular contraction (ECG R-wave peak). (Note that figures appear in color only in the electronic version of the book.)

often mistakenly called *perfusion* signals which is not entirely correct as they do not reflect pulmonary perfusion alone [61,62]. We therefore suggest using the term *pulsatility* signals [61]. Second, the EIT signal in the heart region does not solely originate from changes in ventricular blood volume but is also affected by cardioballistic effects [56].

The EIT-derived signals illustrated in Fig. 6.15C are used for extracting cardiovascular parameters with techniques presented in the following sections for the four main applications (SV, PAP, aortic blood pressure (ABP), and pulmonary perfusion monitoring). This section is concluded by addressing practical challenges of wearable cardiovascular EIT and by giving an outlook on the future of this technology.

Cardiovascular parameter estimation

Pulmonary artery pressure

The assessment of PAP at regular intervals is of importance in patients with chronic pulmonary artery hypertension as it has shown to improve patient outcomes [63]. Currently, right heart catheterization is the gold-standard technique for PAP monitoring, but its use has been sharply reduced over the last years due to the risks associated with this highly invasive procedure [64]. Therefore, there is a strong, but unmet, clinical need for a noninvasive alternative to right heart catheterization. Even though, transthoracic echocardiography (TTE) provides a noninvasive alternative for PAP monitoring, it is ill-suited for frequent or continuous monitoring because TTE measurements are operator dependent and not applicable in 20%–50% of patients. In contrast, EIT provides a promising noninvasive alternative for wearable, continuous, and unsupervised monitoring of PAP [58,65].

The method for EIT-based PAP monitoring makes use of the physiological link between the PAP and the velocity at which the blood pressure waves travel along the walls of the pulmonary arteries. This pulse wave velocity (PWV) and PAP are intrinsically linked, i.e., the PWV increases as PAP increases [66]. Estimating PAP via PWV is done by measuring the arrival of the blood pressure pulse at two locations separated by a distance D along the pulmonary arterial tree, resulting in the pulse transit time (PTT), as illustrated in Fig. 6.16. For PTT-based PAP estimation, the first location is the pulmonary valve, and the second a distal location in the lungs. As the former is difficult to measure in practice, PTT is approximated by the pulse arrival time (PAT), replacing the timing of the pulmonary valve by the timing of the ECG's R-wave peak [67,60]. Cardiovascular EIT is used to detect the arrival of the pressure pulse in the lungs to obtain the second timing. PAP is then estimated from the EIT-derived PAT: PAP $= f(\text{PAT})$. Although the function $f(\cdot)$ is typically an inverse exponential [68,69], monitoring relatively small PAP changes around a certain baseline value allows to locally approximate $f(\cdot)$ by a simple affine function: PAP $= \alpha \cdot \text{PAT} + \beta$ [65], where only the coefficient β is adjusted for each subject, while the coefficient α is identical for all subjects. The pulmonary arteries being not subjected to vasomotion [70], the parameters of this calibration function are expected to remain valid over a sustained period of time.

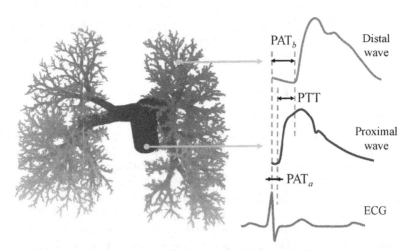

FIGURE 6.16 Measurement of the pulmonary pulse transit time (PTT) as the difference between the pulse arrival times (PATs) at location a and b in the arterial tree. (Note that figures appear in color only in the electronic version of the book.) *Adapted from M. Proença. Non-invasive Hemodynamic Monitoring by Electrical Impedance Tomography, 2017. https://doi.org/10.5075/epfl-thesis-7444.*

This approach was first proposed by Proença et al. [67] and its feasibility was validated in model-based simulations [60,60]. In a controlled hypoxemia study on 24 healthy volunteers [65], EIT-based PAP has shown good agreement with TTE with an error of -0.1 ± 4.5 mmHg (see Fig. 6.17). Validation experiments in animals with comparison to gold-standard right heart catheterization are ongoing [71]. One limitation of the approach is that — at higher pressures — the PAT becomes decreasingly smaller, and the resulting measurement errors larger. Future work requires the validation of this approach in critically ill patients with comparison to invasive PAP reference measurements.

Aortic blood pressure

Similar to PAP, one can apply the same approach for EIT-based monitoring of central ABP as initially proposed by Solà and colleagues [72]. Nevertheless, the aortic EIT signal is much weaker than the surrounding cardiac and pulmonary signals which makes it difficult to detect the aorta in EIT [73,74,75]. In addition — in contrast to PAP — there exist various noninvasive alternatives

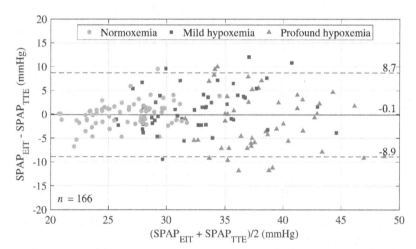

FIGURE 6.17 Bland–Altman plot showing the agreement of systolic pulmonary artery pressure (SPAP) estimated via EIT and compared to transthoracic echocardiography (TTE). Measurements were performed in 24 healthy subjects exposed to normobaric hypoxemia which induces increases in SPAP. The solid line shows the cohort-wise bias of -0.1 mmHg, and the two dashed lines encompass the 95% limits of agreement (bias ± 1.96 SD), thus depicting a cohort-wise SD of 4.5 mmHg. (Note that figures appear in color only in the electronic version of the book.) *Adapted from M. Proença, F. Braun, M. Lemay, J. Solà, A. Adler, T. Riedel, et al. Non-invasive pulmonary artery pressure estimation by electrical impedance tomography in a controlled hypoxemia study in healthy subjects. Sci. Rep. 10 (2020) 21462. https://doi.org/10.1038/s41598-020-78535-4.*

to assess ABP in a continuous and unsupervised manner. This might explain why the approach of EIT-based ABP estimation has been less pursued in the last years.

Stroke volume and cardiac output

The measurement of SV and the related cardiac output (CO) are of high clinical interest as they are closely linked with oxygen delivery and the health of the heart. However, none of the existing noninvasive techniques have proven reliable enough to continuously measure CO and SV in clinical settings.

Since the first investigations 20 years ago by Vonk-Noordegraaf et al. [76], EIT has been investigated as potential alternative for CO and SV monitoring. However, most of the studies with promising outcomes were performed in animals [77,78,79] under controlled laboratory conditions. In addition, these animal studies have partly contradictory findings as further discussed in Refs. [59,80]. In short, all approaches are based on the hypothesis that the amplitude of the EIT-derived impedance changes in either (a) the heart [78,79] or (b) the lung region [77] is proportional to SV. While some studies could confirm hypothesis (a) but not (b) it is the opposite for other studies. However, when tested in healthy volunteers [59] none of the two hypotheses could be confirmed. This was different for a study in critically ill patients on mechanical ventilation [80] where (b), the impedance changes in the lung region could be used to monitor changes in SV (see Fig. 6.18).

The sensitivity of impedance signals in cardiovascular EIT to body position [59,81,82], lung air volume [83,84,82], or electrode–skin contact [59] might explain the mixed outcomes of the abovementioned studies. This might limit the EIT-based SV monitoring to scenarios where minimal changes in the EIT

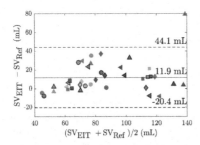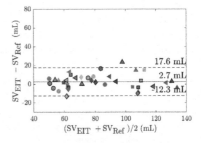

FIGURE 6.18 Bland–Altman analysis comparing EIT-based SV estimates (SV_{EIT}) to transpulmonary thermodilution measurements (SV_{REF}) for 39 measurement points from 20 patients during a fluid challenge. SV_{EIT} was estimated from the impedance changes in the heart region (left) or the lung region (right). (Note that figures appear in color only in the electronic version of the book.) *Adapted from F. Braun, M. Proença, A. Wendler, J. Solà, M. Lemay, J.-P. Thiran, et al. Noninvasive measurement of stroke volume changes in critically ill patients by means of electrical impedance tomography. J. Clin. Monit. Comput. 34 (5) (2020) 903–911. https://doi.org/10.1007/s10877-019-00402-z.*

measurement setup occur (i.e., unchanged body position, stable electrode contact, and stable ventilation). Future studies should systematically investigate potential limitations in real clinical scenarios on patients.

Lung perfusion

The assessment of lung perfusion is of high importance in mechanically ventilated patients since — together with the regional ventilation distribution — it allows for the assessment of the ventilation/perfusion distribution mismatch and the subsequent optimization of mechanical ventilation.

Some studies have addressed the estimation of lung perfusion directly via the amplitude of the cardiovascular EIT signal [55]. This approach remains questionable since the pulmonary signal does not purely reflect perfusion but rather pulsatility [61,62]. The remaining and validated approaches of EIT-based lung perfusion measurement are invasive since they rely on the infusion of a conductivity-contrasting bolus (e.g., hypertonic saline bolus) [85,20,86]. Therefore, we consider these approaches out of scope for wearable EIT monitoring and refer the interested reader to Ref. [1] for an overview of different studies.

Challenges of cardiovascular monitoring using wearable EIT devices

As alluded to in the previous sections, the cardiovascular EIT signals can be perturbated by various factors including changes in body position [59,81,82], lung air volume [84,82], or electrode—skin contact [59,57]. While such perturbations might be reduced to a minimum during controlled laboratory experiments (e.g., in animals) or limited in critically ill patients, they will be a great challenge for ambulatory monitoring using wearable EIT devices.

Future efforts should therefore concentrate on studies to quantify the impact of each potential limiting factor and on strategies to mitigate these perturbations. Such strategies include the fully automatic rejection of erroneous measurements from faulty electrodes or the restriction of measurements to a specific range of body positions and lung air volumes with known outcomes.

Finally, independently of whether EIT is used with a wearable or stationary device, the limited understanding of the origin of cardiovascular EIT signals [59,60,61] poses a problem requiring in-depth investigations. The knowledge gained from such investigations will undoubtedly help to further improve the reliability and accuracy of wearable EIT for cardiovascular monitoring.

The future of EIT-based cardiovascular monitoring

Despite the limitations and challenges related to cardiovascular EIT, there is promise for using a wearable EIT device for unobtrusive and noninvasive cardiovascular monitoring. Embedded in a T-shirt [48,51], a wearable EIT system could be used for the unsupervised and long-term monitoring of PAP in patients with pulmonary hypertension. Even though the continuous monitoring during 24 h is probably difficult to realize, the nocturnal monitoring of PAP and SV — together with ventilation-related parameters — is easy to imagine and would enable the long-term, ambulatory monitoring in a variety of patients.

Integration of EIT findings with other biosignals

Recent technological advances in data acquisition and biosignal processing are paving the way for the optimal integration and fusion of complementary data modalities in a wide variety of clinical settings and wearable devices. The diversified nature of data sources coming from different clinical analyses and acquisition modalities presents big challenges at the integration level. Therefore, the main objective of data fusion is to exploit complementary properties of several single-modality methods to improve each of them considered separately, promoting a synergy effect.

In general, all tasks that demand any type of parameter estimation from multiple sources can benefit from the use of data/information fusion methods [87]. This integration of information can be divided into three main categories, namely, complementary, redundant, and cooperative [87]. Moreover, it can also happen at different levels of abstraction, such as the raw measurements, signals and characteristics, and decisions [87,88]. Multimodal approaches could bring complementary information and reduce the level of uncertainty. Furthermore, this integration can also increase the confidence and enhance the robustness and reliability of the developed methods [89].

One of the major areas of development and research in the health-care area is the development of wearable body sensor networks (BSNs). Multisensor data fusion using BSNs is a technology that enables the combination of information retrieved from several sources in order to form a unified picture [90]. This type of technology is particularly suitable to be deployed in wearable devices as an integrated system that offers the possibility to remotely monitor patients and closely monitor their physiological parameters in an everyday environment [48]. Therefore, significant research efforts have been made to develop this kind of sensors and implement EIT in wearable devices [48,91,51,50]. Notwithstanding, these studies are mostly limited to proofs-of-concept.

Essentially, two different approaches have been used to collect and integrate simultaneous data from EIT and other data source: (i) using cooperative sensors to capture both sources; (ii) capture each source individually. From a conceptual standpoint, using the same sensor is the best solution to avoid the use of separate equipment, which can introduce additional steps in the integration process, such as the synchronism/alignment of both sources in postacquisition. However, this approach can introduce other technical challenges, such as the contamination of the signals with artifacts. In Refs. [48] and [92], two architectures based on frequency-multiplexed sensors were proposed to record EIT in simultaneous with ECG and EEG, respectively. In both studies, it was possible to successfully record EIT alongside with other biosignal without contamination or interference, and the cooperative sensors were tested in in vivo conditions.

In the past, EIT has been combined with other biosignals. These include mostly ECG signals. Frequency filtering and ECG-triggered data acquisition have been used to access the cardiac and perfusion-related components of the EIT signal without interference from ventilation and noise [83]. The ECG-gated method is the first approach to accomplish the separation of cardiac and respiration changes in EIT. McArdle et al. performed one of the first studies where EIT has been used on pair with ECG [93].

Despite the many use cases of EIT as successful impedance imaging method, it is still difficult to extract cardiac-related conductivity changes and respiratory-related conductivity changes in spontaneous breathing subjects (see also Section "Use of wearable EIT beyond ventilation monitoring"). The impedance change of the cardiac signal blends with the impedance change of respiration signal and impedance change of blood flow. The EIT signal gets demodulated with the respiration signal of the human body and the cardiac signal gets demodulated too (see Fig. 6.19). In Ref. [94], the authors successfully applied independent component analysis (ICA) to perform dynamic

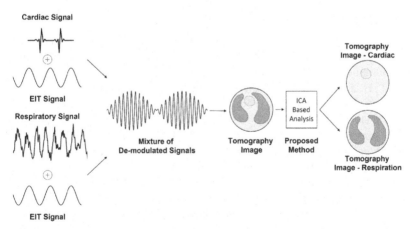

FIGURE 6.19 Separation of cardiac and respiration signals. *Adapted from T. Rahman, M. Hasan, A. Farooq, M.Z. Uddin, Extraction of cardiac and respiration signals in electrical impedance tomography based on independent component analysis, J. Electr. Bioimpedance 4 (1) (2013) 38–44, https://doi.org/10.5617/ jeb.553.*

extraction of cardiac and respiration changes in EIT image. Their method does not rely on averaging over several heart cycles or on frequency domain filtering.

One of the main disadvantages of EIT is its spatial resolution. While conventional imaging methods, such as CT-based techniques and MRI, can provide high anatomical detailed resolution images, EIT images have a low spatial resolution, even with more advanced EIT devices. Plus, with EIT it is only possible to obtain impedance images of an axial cross-sectional slice of 5−10 cm of thorax, leaving out the rest of the lung parenchyma and assuming that other lung regions behave similarly [95]. For those reasons, EIT is unlikely to ever compete, at an anatomical level, with established medical imaging techniques such as CT and MRI, as the necessary spatial resolution cannot be achieved [96]. In Ref. [97], a novel and effective methodology of combining images obtained by EIT and MRI data was proposed. The combination of both imaging modalities enables an accurate and quantitative comparison of the structural and functional information provided by EIT with the complementary structural and functional information provided by MRI. To develop and test their system, the authors have used a tissue-realistic material, a TX151 gel phantom, which has previously been used in the context both of EIT and of MRI [97]. After EIT and MRI measures have been acquired, the open-source software package 3D Slicer was used to manipulate and visualize the resulting data. The visual inspection of the resulting image suggests good correspondence between the two modalities. Even though this work was not conducted on human subjects, it successfully demonstrates the concept of data integration between EIT and MRI. The fusion of both modalities is an essential step in the development of combined dynamic lung function imaging, allowing to obtain high spatial and temporal resolution images.

Another of the signals that could potentially be integrated with EIT is lung sound. Currently, to the best of our knowledge, there are no studies available in the literature combining successfully the EIT and lung sounds findings. In Ref. [51], a study was conducted on healthy volunteers using a wearable device capable of recording several biosignals (see also Section "Current experience with wearable EIT for ventilation monitoring"). The wearable vest used cooperative sensors, described in Ref. [48], and allowed the recording of EIT along with peripheral oxygen saturation, chest sounds, electrocardiography, body activity, and posture. Despite the device's capability to record sound, the chest sound recordings did not show plausible waveforms. In fact, the phonograms of almost all subjects did not exhibit any discernible modulation of the signal, even under the conditions of deep breathing and forced expiration [51]. Since EIT and the analysis of lung sounds (specifically adventitious sounds) address two of the most relevant aspects related to the respiratory health, the integration of these two sources can provide relevant complementary information, which still needs to be demonstrated.

The integration of lung sound and EIT can be performed at different levels, ranging from lower levels of integration to higher ones. At a lower level, one of the most obvious use cases for the fusion of the two data sources is the identification of the respiratory cycle/phase using EIT. Through the identification of end-inspiration and end-expiration moments with EIT, it is possible to detect in which phase a certain adventitious sound was produced. For instance, in Fig. 6.20, with the use of EIT and lung sound it is possible to determine that the wheeze event in the sound recording (left red box) has occurred in the early expiratory phase. In addition to monitoring the presence of adventitious sounds, it is relevant for clinicians to know their timing in the respiratory cycles, i.e., early/mid/late inspiratory or expiratory. This information might be of clinical relevance for the assessment of the respiratory function of the patients and the corresponding differential diagnosis of cardiovascular diseases [98]. Moreover, in this figure it is also possible to detect a period of apnea after deep inspiration with EIT that was interrupted by a cough event (right green box). With the use of complementary information, it is conceivable to provide a better and more accurate characterization of the

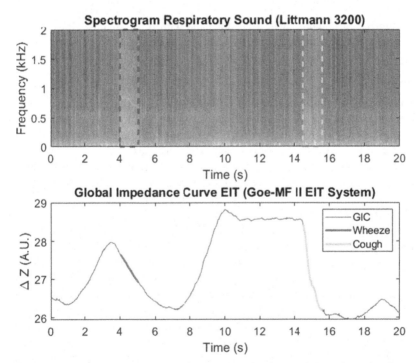

FIGURE 6.20 Synchronous collection of EIT and respiratory sound data. (Note that figures appear in color only in the electronic version of the book.)

respiratory sound. Furthermore, other options to achieve the integration of lung sound with EIT might also include the extraction of global parameters that can be used in a higher-level domain to better diagnose and monitor respiratory conditions.

Besides the mentioned signals, other types of less complex signals are also suitable to be integrated alongside EIT. For instance, biosignals such as transcutaneous CO_2 and oxygen saturation may help to provide a better overall snapshot of the respiratory function. EIT does not offer direct measurement of oxygenation and ventilation. Rather, it provides information regarding ventilation and perfusion matching. Using O_2 and CO_2, real-time quantification along with regional perfusion and ventilation may offer important insights into certain diseases and offer a means of assessing interventions aimed at improving gas exchange in critically ill patients [99]. Another advantage of considering such signals is the fact that they can also be deployed in wearable devices, allowing for continuous monitoring of the patients' respiratory function. Besides, since EIT might be affected by the body posture of the subjects, as demonstrated in Ref. [40], sensors such as accelerometers and gyroscopes can be used with EIT to determine the patient's positioning.

We may summarize that significant research efforts have been targeting the development of wearable devices that would allow continuous and remote monitoring of patients using EIT alongside with other signals. However, at a commercial level, those are yet to become a reality.

Conclusions

Lung imaging is an essential and established part of the management of patients suffering from respiratory diseases. At the moment, it has mostly been neither considered nor implemented in the newly developed wearable medical devices for monitoring of patients with such diseases.

We believe that research efforts should be undertaken to make EIT part of the future wearable devices designed for this patient group. This belief is based not just on the fact that, thanks to its noninvasive radiation-free measuring principle, such wearable EIT design is realistic and feasible, in contrast to other medical imaging methods. The main argument in favor of fully wearable EIT is the capacity of this method to provide information on regional lung function and its deterioration induced by pulmonary diseases (like COPD, asthma, or cystic fibrosis) in the course of the natural disease progression or as a consequence of inadequate therapy or disease exacerbation. EIT can also detect improvements in regional lung function during successful therapy. The timely knowledge of this type of information derived from EIT data may be used to optimize the overall management of the patients, including pharmacological and physical therapy, and potentially to identify exacerbations early and prevent hospitalization.

The current established clinical use of EIT is limited to the detection of pulmonary aeration and ventilation changes. However, as described in detail in this chapter, chest EIT allows the assessment of cardiovascular function as well. This research field is new and the possibilities and limitations of EIT in this application still need to be explored. However, the potential for determining pulmonary perfusion, cardiac output (CO), and PAP simultaneously with regional lung ventilation, respiratory mechanics, and aeration shifts using EIT is intriguing.

Another important and also relatively unexplored topic is the integration of EIT with other biosignals and data sources. Even though previous work has been published in this field, the full potential of EIT integration is still largely unknown. Ultimately, the integration of EIT with other sources of information can potentially enhance the monitoring and diagnosing capability of this imaging modality to assess the respiratory system. This integration can provide a complex and thorough solution capable of generating a complete picture of patients' condition and respiratory function.

First successful attempts at developing fully wearable EIT for remote patient monitoring have been conducted and the feasibility of this solution with integrated acquisition of other biosignals demonstrated. However, further technological development, miniaturization of the sensors, improvements in the ergonomics, and design of such wearable medical devices along with enhanced data exploitation and analysis are still needed. The clinical relevance and impact need to be proved in large-scale validation studies with the end-users properly involved.

References

[1] I. Frerichs, M.B.P. Amato, A.H. van Kaam, D.G. Tingay, Z. Zhao, B. Grychtol, M. Bodenstein, H. Gagnon, S.H. Böhm, E. Teschner, O. Stenqvist, T. Mauri, V. Torsani, L. Camporota, A. Schibler, G.K. Wolf, D. Gommers, S. Leonhardt, A. Adler, Chest electrical impedance tomography examination, data analysis, terminology, clinical use and recommendations: consensus statement of the TRanslational EIT developmeNt stuDy group, Thorax 72 (1) (2017) 83–93, https://doi.org/10.1136/thoraxjnl-2016-208357.

[2] L. Sang, Z. Zhao, Z. Lin, X. Liu, N. Zhong, Y. Li, A narrative review of electrical impedance tomography in lung diseases with flow limitation and hyperinflation: methodologies and applications, Ann. Transl. Med. 8 (2020) 1688, https://doi.org/10.21037/atm-20-4984.

[3] H.A. Lorentz, The theorem of Poynting concerning the energy in the electro-magnetic field and two general propositions concerning the propagation of light, Amsterdammer Akademie der Wetenschappen 4 (1896) 176–187.

[4] M. Cheney, D. Isaacson, J.C. Newell, S. Simske, J. Goble, NOSER: an algorithm for solving the inverse conductivity problem, Int. J. Imag. Syst. Technol. 2 (2) (1990) 66–75, https://doi.org/10.1002/ima.1850020203.

[5] N. Polydorides, W.R.B. Lionheart, A Matlab toolkit for three-dimensional electrical impedance tomography: a contribution to the Electrical Impedance and Diffuse Optical Reconstruction Software project, Meas. Sci. Technol. 13 (2002) 1871−1883, https://doi.org/10.1088/0957-0233/13/12/310. Pii S0957-0233(02)35899-5.

[6] A. Adler, J.H. Arnold, R. Bayford, A. Borsic, B. Brown, P. Dixon, T.J. Faes, I. Frerichs, H. Gagnon, Y. Garber, B. Grychtol, G. Hahn, W.R. Lionheart, A. Malik, R.P. Patterson, J. Stocks, A. Tizzard, N. Weiler, G.K. Wolf, GREIT: a unified approach to 2D linear EIT reconstruction of lung images, Physiol. Meas. 30 (2009) S35−S55, https://doi.org/10.1088/0967-3334/30/6/S03.

[7] B. Grychtol, B. Muller, A. Adler, 3D EIT image reconstruction with GREIT, Physiol. Meas. 37 (2016) 785−800, https://doi.org/10.1088/0967-3334/37/6/785.

[8] A. Adler, W.R. Lionheart, Uses and abuses of EIDORS: an extensible software base for EIT, Physiol. Meas. 27 (2006) S25−S42, https://doi.org/10.1088/0967-3334/27/5/S03.

[9] A.E. Hartinger, R. Guardo, A. Adler, H. Gagnon, Real-time management of faulty electrodes in electrical impedance tomography, IEEE Trans. Biomed. Eng. 56 (2) (2009) 369−377, https://doi.org/10.1109/TBME.2008.2003103.

[10] A. Adler, Accounting for erroneous electrode data in electrical impedance tomography, in: Physiological Measurement, vol. 25, 2004, pp. 227−238, https://doi.org/10.1088/0967-3334/25/1/028, 1.

[11] J. Karsten, T. Stueber, N. Voigt, E. Teschner, H. Heinze, Influence of different electrode belt positions on electrical impedance tomography imaging of regional ventilation: a prospective observational study, Crit. Care 20 (2016) 1, https://doi.org/10.1186/s13054-015-1161-9.

[12] S. Krueger-Ziolek, B. Schullcke, J. Kretschmer, U. Müller-Lisse, K. Möller, Z. Zhao, Positioning of electrode plane systematically influences EIT imaging, in: Physiological Measurement, vol. 36, Institute of Physics Publishing, 2015, pp. 1109−1118, https://doi.org/10.1088/0967-3334/36/6/1109, 6.

[13] G. Hahn, I. Šipinková, F. Baisch, G. Hellige, Changes in the thoracic impedance distribution under different ventilatory conditions, Physiol. Meas. 16 (1995) A161−A173, https://doi.org/10.1088/0967-3334/16/3A/016.

[14] G. Kühnel, G. Hahn, I. Frerichs, T. Schroder, G. Hellige, New methods for improving the image quality of functional electric impedance tomography, Biomed. Tech. 42 (1997) 470−471. Suppl. 1.

[15] C. Karagiannidis, A.D. Waldmann, P.L. Roka, T. Schreiber, S. Strassmann, W. Windisch, S.H. Bohm, Regional expiratory time constants in severe respiratory failure estimated by electrical impedance tomography: a feasibility study, Crit. Care 22 (2018) 221, https://doi.org/10.1186/s13054-018-2137-3.

[16] Z. Zhao, P.J. Yun, Y.L. Kuo, F. Fu, M. Dai, I. Frerichs, K. Moller, Comparison of different functional EIT approaches to quantify tidal ventilation distribution, Physiol. Meas. 39 (2018) 01NT01, https://doi.org/10.1088/1361-6579/aa9eb4.

[17] T. Muders, H. Luepschen, J. Zinserling, S. Greschus, R. Fimmers, U. Guenther, M. Buchwald, D. Grigutsch, S. Leonhardt, C. Putensen, H. Wrigge, Tidal recruitment assessed by electrical impedance tomography and computed tomography in a porcine model of lung injury, Crit. Care Med. 40 (2012) 903−911, https://doi.org/10.1097/CCM.0b013e318236f452.

[18] H. He, Y. Chi, Y. Long, S. Yuan, I. Frerichs, K. Moller, F. Fu, Z. Zhao, Influence of overdistension/recruitment induced by high positive end-expiratory pressure on ventilation-perfusion matching assessed by electrical impedance tomography with saline bolus, Crit. Care 24 (2020) 586, https://doi.org/10.1186/s13054-020-03301-x.

[19] H. He, Y. Chi, Y. Long, S. Yuan, R. Zhang, I. Frerichs, K. Moller, F. Fu, Z. Zhao, Bedside evaluation of pulmonary embolism by saline contrast electrical impedance tomography method: a prospective observational study, Am. J. Respir. Crit. Care Med. 202 (2020) 1464−1468, https://doi.org/10.1164/rccm.202005-1780LE.

[20] M. Kircher, G. Elke, B. Stender, M. Hernandez Mesa, F. Schuderer, O. Dossel, M.K. Fuld, A.F. Halaweish, E.A. Hoffman, N. Weiler, I. Frerichs, Regional lung perfusion analysis in experimental ARDS by electrical impedance and computed tomography, IEEE Trans. Med. Imag. 40 (2021) 251−261, https://doi.org/10.1109/TMI.2020.3025080.

[21] I. Frerichs, Z. Zhao, T. Becher, P. Zabel, N. Weiler, B. Vogt, Regional lung function determined by electrical impedance tomography during bronchodilator reversibility testing in patients with asthma, Physiol. Meas. 37 (2016) 698−712, https://doi.org/10.1088/0967-3334/37/6/698.

[22] L. Lasarow, B. Vogt, Z. Zhao, L. Balke, N. Weiler, I. Frerichs, Regional lung function measures determined by electrical impedance tomography during repetitive ventilation manoeuvres in patients with COPD, Physiol. Meas. 42 (2021) 015008, https://doi.org/10.1088/1361-6579/abdad6.

[23] B. Vogt, Z. Zhao, P. Zabel, N. Weiler, I. Frerichs, Regional lung response to bronchodilator reversibility testing determined by electrical impedance tomography in chronic obstructive pulmonary disease, Am. J. Physiol. Lung Cell Mol. Physiol. 311 (1) (2016) L8−L19, https://doi.org/10.1152/ajplung.00463.2015.

[24] Z. Zhao, R. Fischer, I. Frerichs, U. Muller-Lisse, K. Moller, Regional ventilation in cystic fibrosis measured by electrical impedance tomography, J. Cyst. Fibros. 11 (2012) 412−418, https://doi.org/10.1016/j.jcf.2012.03.011.

[25] Z. Zhao, K. Moller, D. Steinmann, I. Frerichs, J. Guttmann, Evaluation of an electrical impedance tomography-based global inhomogeneity index for pulmonary ventilation distribution, Intensive Care Med. 35 (2009) 1900−1906, https://doi.org/10.1007/s00134-009-1589-y.

[26] I. Frerichs, G. Hahn, W. Golisch, M. Kurpitz, H. Burchardi, G. Hellige, Monitoring perioperative changes in distribution of pulmonary ventilation by functional electrical impedance tomography, Acta Anaesthesiol. Scand. 42 (1998) 721−726, https://doi.org/10.1111/j.1399-6576.1998.tb05308.x.

[27] S. Pulletz, H.R. van Genderingen, G. Schmitz, G. Zick, D. Schadler, J. Scholz, N. Weiler, I. Frerichs, Comparison of different methods to define regions of interest for evaluation of regional lung ventilation by EIT, Physiol. Meas. 27 (2006) S115−S127, https://doi.org/10.1088/0967-3334/27/5/S10.

[28] S. Liu, L. Tan, K. Moller, I. Frerichs, T. Yu, L. Liu, Y. Huang, F. Guo, J. Xu, Y. Yang, H. Qiu, Z. Zhao, Identification of regional overdistension, recruitment and cyclic alveolar collapse with electrical impedance tomography in an experimental ARDS model, Crit. Care 20 (119) (2016), https://doi.org/10.1186/s13054-016-1300-y.

[29] R. Zhang, H. He, L. Yun, X. Zhou, X. Wang, Y. Chi, S. Yuan, Z. Zhao, Effect of postextubation high-flow nasal cannula therapy on lung recruitment and overdistension in high-risk patient, Crit. Care 24 (82) (2020), https://doi.org/10.1186/s13054-020-2809-7.

[30] K. Lowhagen, S. Lundin, O. Stenqvist, Regional intratidal gas distribution in acute lung injury and acute respiratory distress syndrome-assessed by electric impedance tomography, Minerva Anestesiol. 76 (2010) 1024−1035. R02106251 [pii].

[31] Z. Zhao, S.Y. Peng, M.Y. Chang, Y.L. Hsu, I. Frerichs, H.T. Chang, K. Moller, Spontaneous breathing trials after prolonged mechanical ventilation monitored by electrical impedance tomography: an observational study, Acta Anaesthesiol. Scand. 61 (2017) 1166–1175, https://doi.org/10.1111/aas.12959.

[32] S.J.H. Heines, U. Strauch, M.C.G. van de Poll, P. Roekaerts, D. Bergmans, Clinical implementation of electric impedance tomography in the treatment of ARDS: a single centre experience, J. Clin. Monit. Comput. 33 (2019) 291–300, https://doi.org/10.1007/s10877-018-0164-x.

[33] Z. Zhao, M.Y. Chang, C.H. Gow, J.H. Zhang, Y.L. Hsu, I. Frerichs, H.T. Chang, K. Moller, Positive end-expiratory pressure titration with electrical impedance tomography and pressure-volume curve in severe acute respiratory distress syndrome, Ann. Intensive Care 9 (2019) 7, https://doi.org/10.1186/s13613-019-0484-0.

[34] Z. Zhao, L.C. Lee, M.Y. Chang, I. Frerichs, H.T. Chang, C.H. Gow, Y.L. Hsu, K. Moller, The incidence and interpretation of large differences in EIT-based measures for PEEP titration in ARDS patients, J. Clin. Monit. Comput. 34 (2020) 1005–1013, https://doi.org/10.1007/s10877-019-00396-8.

[35] I. Frerichs, G. Hahn, G. Hellige, Thoracic electrical impedance tomographic measurements during volume controlled ventilation-effects of tidal volume and positive end-expiratory pressure, IEEE Trans. Med. Imag. 18 (9) (1999) 764–773, https://doi.org/10.1109/42.802754.

[36] F. Reifferscheid, G. Elke, S. Pulletz, B. Gawelczyk, I. Lautenschläger, M. Steinfath, N. Weiler, I. Frerichs, Regional ventilation distribution determined by electrical impedance tomography: Reproducibility and effects of posture and chest plane, Respirology 16 (3) (2011) 523–531, https://doi.org/10.1111/j.1440-1843.2011.01929.x.

[37] S. Pulletz, G. Elke, G. Zick, D. Schädler, F. Reifferscheid, N. Weiler, I. Frerichs, Effects of restricted thoracic movement on the regional distribution of ventilation, Acta Anaesthesiol. Scand. 54 (6) (2010) 751–760, https://doi.org/10.1111/j.1399-6576.2010.02233.x.

[38] N. Coulombe, H. Gagnon, F. Marquis, Y. Skrobik, R. Guardo, A parametric model of the relationship between EIT and total lung volume, Physiol. Meas. 26 (4) (2005) 401–411, https://doi.org/10.1088/0967-3334/26/4/006.

[39] I. Frerichs, T. Dudykevych, J. Hinz, M. Bodenstein, G. Hahn, G. Hellige, Gravity effects on regional lung ventilation determined by functional EIT during parabolic flights, J. Appl. Physiol. 91 (1) (2001) 39–50, https://doi.org/10.1152/jappl.2001.91.1.39.

[40] B. Vogt, L. Mendes, I. Chouvarda, E. Perantoni, E. Kaimakamis, T. Becher, N. Weiler, V. Tsara, R.P. Paiva, N. Maglaveras, I. Frerichs, Influence of torso and arm positions on chest examinations by electrical impedance tomography, Physiol. Meas. 37 (6) (2016) 904–921, https://doi.org/10.1088/0967-3334/37/6/904.

[41] I. Frerichs, P. Braun, T. Dudykevych, G. Hahn, D. Genee, G. Hellige, Distribution of ventilation in young and elderly adults determined by electrical impedance tomography, Respir. Physiol. Neurobiol. 143 (2004) 63–75, https://doi.org/10.1016/j.resp.2004.07.014.

[42] T. Yoshida, V. Torsani, S. Gomes, R.R.D. Santis, M.A. Beraldo, E.L.V. Costa, M.R. Tucci, W.A. Zin, B.P. Kavanagh, M.B.P. Amato, Spontaneous effort causes occult pendelluft during mechanical ventilation, Am. J. Respir. Crit. Care Med. 188 (12) (2013) 1420–1427, https://doi.org/10.1164/rccm.201303-0539OC.

[43] P. Blankman, D. Hasan, M.S. Van Mourik, D. Gommers, Ventilation distribution measured with EIT at varying levels of pressure support and neurally adjusted ventilatory assist in patients with ALI, Intensive Care Med. 39 (6) (2013) 1057–1062, https://doi.org/10.1007/s00134-013-2898-8.

[44] T. Mauri, G. Bellani, A. Confalonieri, P. Tagliabue, M. Turella, A. Coppadoro, G. Citerio, N. Patroniti, A. Pesenti, Topographic distribution of tidal ventilation in acute respiratory distress syndrome: effects of positive end-expiratory pressure and pressure support, Crit. Care Med. 41 (7) (2013) 1664−1673, https://doi.org/10.1097/CCM.0b013e318287f6e7.

[45] K. Haris, B. Vogt, C. Strodthoff, D. Pessoa, G.A. Cheimariotis, B. Rocha, G. Petmezas, N. Weiler, R.P. Paiva, P. de Carvalho, N. Maglaveras, I. Frerichs, Identification and analysis of stable breathing periods in electrical impedance tomography recordings, Physiol. Meas. 42 (2021) 064003, https://doi.org/10.1088/1361-6579/ac08e5.

[46] M. Menolotto, S. Rossi, P. Dario, L. Della Torre, Towards the development of a wearable electrical impedance tomography system: a study about the suitability of a low power bioimpedance front-end, Annu. Int. Conf. IEEE Eng. Med. Biol. Soc 2015 (2015) 3133−3136, https://doi.org/10.1109/EMBC.2015.7319056.

[47] M. Klum, M. Schmidt, J. Klaproth, A.-G. Pielmus, T. Tigges, R. Orglmeister, Balanced adjustable mirrored current source with common mode Feedback and output measurement for bioimpedance applications, Annu. Int. Conf. IEEE Eng. Med. Biol. Soc. 2019 (2019) 1278−1281, https://doi.org/10.1109/EMBC.2019.8856325.

[48] M. Rapin, F. Braun, A. Adler, J. Wacker, I. Frerichs, B. Vogt, O. Chetelat, Wearable sensors for frequency-multiplexed EIT and Multilead ECG data acquisition, IEEE Trans. Biomed. Eng. 66 (3) (2019) 810−820, https://doi.org/10.1109/TBME.2018.2857199.

[49] Y. Wu, D. Jiang, A. Bardill, R. Bayford, A. Demosthenous, A 122 fps, 1 MHz bandwidth multi-frequency wearable EIT belt featuring novel active electrode architecture for neonatal thorax vital sign monitoring, IEEE Trans Biomed Circuits Syst 13 (5) (2019) 927−937, https://doi.org/10.1109/TBCAS.2019.2925713.

[50] S. Hong, J. Lee, H.-J. Yoo, Wearable lung-health monitoring system with electrical impedance tomography, Annu. Int. Conf. IEEE Eng. Med. Biol. Soc. 2015 (2015) 1707−1710, https://doi.org/10.1109/EMBC.2015.7318706.

[51] I. Frerichs, B. Vogt, J. Wacker, R. Paradiso, F. Braun, M. Rapin, L. Caldani, O. Chételat, N. Weiler, Multimodal remote chest monitoring system with wearable sensors: a validation study in healthy subjects, Physiol. Meas. 41 (1) (2020) 015006, https://doi.org/10.1088/1361-6579/ab668f.

[52] D. Leonhäuser, C. Castelar, T. Schlebusch, M. Rohm, R. Rupp, S. Leonhardt, M. Walter, J.O. Grosse, Evaluation of electrical impedance tomography for determination of urinary bladder volume: comparison with standard ultrasound methods in healthy volunteers, Biomed. Eng. Online 17 (1) (2018) 95, https://doi.org/10.1186/s12938-018-0526-0.

[53] D. Jiang, Y. Wu, A. Demosthenous, Hand gesture recognition using three-dimensional electrical impedance tomography, IEEE Trans. Circuits Syst. II Express Briefs 67 (2020) 1554−1558, https://doi.org/10.1109/TCSII.2020.3006430.

[54] G. Ayoub, T.H. Dang, T.I. Oh, S.-W. Kim, E.J. Woo, Feature extraction of upper airway dynamics during sleep apnea using electrical impedance tomography, Sci. Rep. 10 (1) (2020) 1637, https://doi.org/10.1038/s41598-020-58450-4.

[55] D.T. Nguyen, C. Jin, A. Thiagalingam, A.L. McEwan, A review on electrical impedance tomography for pulmonary perfusion imaging, Physiol. Meas. 33 (5) (2012) 695−706, https://doi.org/10.1088/0967-3334/33/5/695.

[56] M. Proença, F. Braun, M. Rapin, J. Solà, A. Adler, B. Grychtol, S.H. Bohm, M. Lemay, J.-P. Thiran, Influence of heart motion on cardiac output estimation by means of electrical impedance tomography: a case study, Physiol. Meas. 36 (6) (2015) 1075−1091, https://doi.org/10.1088/0967-3334/36/6/1075.

[57] F. Braun, Noninvasive Stroke Volume Monitoring by Electrical Impedance Tomography, 2018, https://doi.org/10.5075/epfl-thesis-8343.

[58] M. Proenca, Non-invasive Hemodynamic Monitoring by Electrical Impedance Tomography, 2017, https://doi.org/10.5075/epfl-thesis-7444.

[59] F. Braun, M. Proença, A. Adler, T. Riedel, J.-P. Thiran, J. Solà, Accuracy and reliability of noninvasive stroke volume monitoring via ECG-gated 3D electrical impedance tomography in healthy volunteers, PLoS One 13 (1) (2018) e0191870, https://doi.org/10.1371/journal.pone.0191870.

[60] M. Proença, F. Braun, J. Solà, J.-P. Thiran, M. Lemay, Noninvasive pulmonary artery pressure monitoring by EIT: a model-based feasibility study, Med. Biol. Eng. Comput. 55 (6) (2017) 949–963, https://doi.org/10.1007/s11517-016-1570-1.

[61] A. Adler, M. Proença, F. Braun, J. Brunner, J. Solà, Origins of cardiosynchronous signals in EIT. In: The Proceedings of the18th International Conference on Biomedical Applications of Electrical Impedance Tomography. Edited by Alistair Boyle, Ryan Halter, Ethan Murphy, Andy Adler. June 21-24, Electrical Impedance Tomography 73 (2017), https://doi.org/10.5281/zenodo.557093.

[62] G. Hellige, G. Hahn, Cardiac-related impedance changes obtained by electrical impedance tomography: an acceptable parameter for assessment of pulmonary perfusion? Crit. Care 15 (3) (2011) 430, https://doi.org/10.1186/cc10231.

[63] W.T. Abraham, P.B. Adamson, R.C. Bourge, M.F. Aaron, M.R. Costanzo, L.W. Stevenson, W. Strickland, S. Neelagaru, N. Raval, S. Krueger, S. Weiner, D. Shavelle, B. Jeffries, J.S. Yadav, C.T.S. Group, Wireless pulmonary artery haemodynamic monitoring in chronic heart failure: a randomised controlled trial, Lancet 377 (2011) 658–666, https://doi.org/10.1016/S0140-6736(11)60101-3.

[64] S. Harvey, D.A. Harrison, M. Singer, J. Ashcroft, C.M. Jones, D. Elbourne, W. Brampton, D. Williams, D. Young, K. Rowan, P. A.-M. s collaboration, Assessment of the clinical effectiveness of pulmonary artery catheters in management of patients in intensive care (PAC-Man): a randomised controlled trial, Lancet 366 (2005) 472–477, https://doi.org/10.1016/S0140-6736(05)67061-4.

[65] M. Proença, F. Braun, M. Lemay, J. Solà, A. Adler, T. Riedel, F.H. Messerli, J.-P. Thiran, S.F. Rimoldi, E. Rexhaj, Non-invasive pulmonary artery pressure estimation by electrical impedance tomography in a controlled hypoxemia study in healthy subjects, Sci. Rep. 10 (2020) 21462, https://doi.org/10.1038/s41598-020-78535-4.

[66] W.W. Nichols, M.F. O'Rourke, C. Vlachopoulos, McDonald's Blood Flow in Arteries: Theoretic, Experimental, and Clinical Principles, sixth ed., CRC Press, 2011 https://doi.org/10.1201/b13568.

[67] M. Proença, F. Braun, J. Solà, A. Adler, M. Lemay, J.-P. Thiran, S.F. Rimoldi, Non-invasive monitoring of pulmonary artery pressure from timing information by EIT: experimental evaluation during induced hypoxia, Physiol. Meas. 37 (6) (2016) 713–726, https://doi.org/10.1088/0967-3334/37/6/713.

[68] E.M. Lau, N. Iyer, R. Ilsar, B.P. Bailey, M.R. Adams, D.S. Celermajer, Abnormal pulmonary artery stiffness in pulmonary arterial hypertension: in vivo study with intravascular ultrasound, PLoS One 7 (2012) e33331, https://doi.org/10.1371/journal.pone.0033331.

[69] J. Sanz, M. Kariisa, S. Dellegrottaglie, S. Prat-Gonzalez, M.J. Garcia, V. Fuster, S. Rajagopalan, Evaluation of pulmonary artery stiffness in pulmonary hypertension with cardiac magnetic resonance, JACC Cardiovasc Imaging 2 (2009) 286–295, https://doi.org/10.1016/j.jcmg.2008.08.007.

[70] J.R. Levick, An Introduction to Cardiovascular Physiology, fifth ed., Hodder Arnold, London, UK, 2010.

[71] F. Braun, M. Proença, M. Sage, J.-P. Praud, M. Lemay, A. Adler, EIT Measurement of Pulmonary Artery Pressure in Neonatal Lambs, 2019, p. 33, https://doi.org/10.5281/zenodo.269170.

[72] J. Solà, A. Adler, A. Santos, G. Tusman, F.S. Sipmann, S.H. Bohm, Non-invasive monitoring of central blood pressure by electrical impedance tomography: first experimental evidence, Med. Biol. Eng. Comput. 49 (4) (2011) 409−415, https://doi.org/10.1007/s11517-011-0753-z.

[73] F. Thürk, A. Waldmann, K.H. Wodack, C.J. Trepte, D. Reuter, S. Kampusch, E. Kaniusas, Evaluation of reconstruction parameters of electrical impedance tomography on aorta detection during saline bolus injection, Curr. Dir. Biomed. Eng. 2 (1) (2016), https://doi.org/10.1515/cdbme-2016-0113.

[74] K.H. Wodack, S. Buehler, S.A. Nishimoto, M.F. Graessler, C.R. Behem, A.D. Waldmann, B. Mueller, S.H. Böhm, E. Kaniusas, F. Thürk, A. Maerz, C.J.C. Trepte, D.A. Reuter, Detection of thoracic vascular structures by electrical impedance tomography: a systematic assessment of prominence peak analysis of impedance changes, Physiol. Meas. 39 (2) (2018) 024002, https://doi.org/10.1088/1361-6579/aaa924.

[75] F. Braun, M. Proença, M. Rapin, M. Lemay, A. Adler, B. Grychtol, J. Solà, J.-P. Thiran, Aortic blood pressure measured via EIT: investigation of different measurement settings, Physiol. Meas. 36 (6) (2015) 1147−1159, https://doi.org/10.1088/0967-3334/36/6/1147.

[76] A. Vonk-Noordegraaf, A. Janse, J.T. Marcus, J.G. Bronzwaer, P.E. Postmust, T.J. Faes, P.M. De Vries, Determination of stroke volume by means of electrical impedance tomography, Physiol. Meas. 21 (2) (2000) 285−293.

[77] F.J. da Silva Ramos, A. Hovnanian, R. Souza, L.C.P. Azevedo, M.B.P. Amato, E.L.V. Costa, Estimation of stroke volume and stroke volume changes by electrical impedance tomography, Anesth. Analg. 126 (1) (2018) 102−110, https://doi.org/10.1213/ANE.0000000000002271.

[78] G.Y. Jang, Y.J. Jeong, T. Zhang, T.I. Oh, R.-E. Ko, C.R. Chung, G.Y. Suh, E.J. Woo, Noninvasive, simultaneous, and continuous measurements of stroke volume and tidal volume using EIT: feasibility study of animal experiments, Sci. Rep. 10 (1) (2020) 11242, https://doi.org/10.1038/s41598-020-68139-3.

[79] R. Pikkemaat, S. Lundin, O. Stenqvist, R.-D. Hilgers, S. Leonhardt, Recent advances in and limitations of cardiac output monitoring by means of electrical impedance tomography, Anesth. Analg. 119 (1) (2014) 76−83, https://doi.org/10.1213/ANE.0000000000000241.

[80] F. Braun, M. Proença, A. Wendler, J. Solà, M. Lemay, J.-P. Thiran, N. Weiler, I. Frerichs, T. Becher, Noninvasive measurement of stroke volume changes in critically ill patients by means of electrical impedance tomography, J. Clin. Monit. Comput. 34 (5) (2020) 903−911, https://doi.org/10.1007/s10877-019-00402-z.

[81] M. Graf, T. Riedel, Electrical impedance tomography: Amplitudes of cardiac related impedance changes in the lung are highly position dependent, PLoS One 12 (11) (2017) e0188313, https://doi.org/10.1371/journal.pone.0188313.

[82] R.P. Patterson, J. Zhang, L.I. Mason, M. Jerosch-Herold, Variability in the cardiac EIT image as a function of electrode position, lung volume and body position, Physiol. Meas. 22 (1) (2001) 159−166.

[83] I. Frerichs, S. Pulletz, G. Elke, F. Reifferscheid, D. Schadler, J. Scholz, N. Weiler, Assessment of changes in distribution of lung perfusion by electrical impedance tomography, Respiration 77 (2009) 282−291, https://doi.org/10.1159/000193994.

[84] S. Krueger-Ziolek, B. Gong, B. Laufer, K. Möller, Impact of lung volume changes on perfusion estimates derived by Electrical Impedance Tomography, Curr. Dir. Biomed. Eng. 5 (2019) 199–202, https://doi.org/10.1515/cdbme-2019-0051.

[85] I. Frerichs, J. Hinz, P. Herrmann, G. Weisser, G. Hahn, M. Quintel, G. Hellige, Regional lung perfusion as determined by electrical impedance tomography in comparison with electron beam CT imaging, IEEE Trans. Med. Imag. 21 (2002) 646–652, https://doi.org/10.1109/TMI.2002.800585.

[86] J.B. Borges, F. Suarez-Sipmann, S.H. Bohm, G. Tusman, A. Melo, E. Maripuu, M. Sandström, M. Park, E.L.V. Costa, G. Hedenstierna, M. Amato, Regional lung perfusion estimated by electrical impedance tomography in a piglet model of lung collapse, J. Appl. Physiol. 112 (1) (2012) 225–236, https://doi.org/10.1152/japplphysiol.01090.2010.

[87] F. Castanedo, A review of data fusion techniques, Sci. World J. 2013 (2013) 704504, https://doi.org/10.1155/2013/704504.

[88] D.P. Mandic, D. Obradovic, A. Kuh, T. Adali, U. Trutschel, M. Golz, P. De Wilde, J. Barria, A. Constantinides, J. Chambers, Data fusion for modern engineering applications: an overview. Artificial neural networks: formal models and their applications—Icann 2005, Pt 2, Proceedings 3697 (2005) 715–721.

[89] R. Gravina, P. Alinia, H. Ghasemzadeh, G. Fortino, Multi-sensor fusion in body sensor networks: state-of-the-art and research challenges, Inf. Fusion 35 (2017) 68–80, https://doi.org/10.1016/j.inffus.2016.09.005.

[90] B. Khaleghi, A. Khamis, F.O. Karray, S.N. Razavi, Multisensor data fusion: a review of the state-of-the-art, Inf. Fusion 14 (2013), 562–562.10.1016/j.inffus.2012.10.004.

[91] Y. Wu, D. Jiang, A. Bardill, S. de Gelidi, R. Bayford, A. Demosthenous, A high frame rate wearable EIT system using active electrode ASICs for lung respiration and heart rate monitoring, IEEE Trans.Biomed. Circuits Syst. I Regul. Pap. 65 (2018) 3810–3820, https://doi.org/10.1109/Tcsi.2018.2858148.

[92] J. Avery, T. Dowrick, A. Witkowska-Wrobel, M. Faulkner, K. Aristovich, D. Holder, Simultaneous EIT and EEG using frequency division multiplexing, Physiol. Meas. 40 (2019) 034007, https://doi.org/10.1088/1361-6579/ab0bbc.

[93] F.J. McArdle, A.J. Suggett, B.H. Brown, D.C. Barber, An assessment of dynamic images by applied potential tomography for monitoring pulmonary perfusion, Clin. Phys. Physiol. Meas. 9 (1988) 87–91, https://doi.org/10.1088/0143-0815/9/4a/015. Suppl. A.

[94] T. Rahman, M. Hasan, A. Farooq, M.Z. Uddin, Extraction of cardiac and respiration signals in electrical impedance tomography based on independent component analysis, J. Electr. Bioimpedance 4 (1) (2013) 38–44, https://doi.org/10.5617/jeb.553.

[95] B. Lobo, C. Hermosa, A. Abella, F. Gordo, Electrical impedance tomography, Ann. Transl. Med. 6 (2018) 26, https://doi.org/10.21037/atm.2017.12.06.

[96] B.H. Brown, Electrical impedance tomography (EIT): a review, J. Med. Eng. Technol. 27 (2003) 97–108, https://doi.org/10.1080/0309190021000059687.

[97] J.L. Davidson, R.A. Little, P. Wright, J. Naish, R. Kikinis, G.J.M. Parker, H. McCann, Fusion of images obtained from EIT and MRI, Electron. Lett. 48 (2012) 617–618, https://doi.org/10.1049/el.2012.0327.

[98] C. Jácome, J. Ravn, E. Holsbo, J.C. Aviles-Solis, H. Melbye, L.A. Bongo, Convolutional neural network for breathing phase detection in lung sounds, Sensors 19 (8) (2019) 1798, https://doi.org/10.3390/s19081798.

[99] C.D. Smallwood, B.K. Walsh, Noninvasive monitoring of oxygen and ventilation, Respir. Care 62 (2017) 751–764, https://doi.org/10.4187/respcare.05243.

Chapter 7

Respiratory data management

Vassilis Kilintzis and Nikolaos Beredimas
Laboratory of Computing, Medical Informatics and Biomedical Imaging Technologies, Medical School, Aristotle University of Thessaloniki, Thessaloniki, Greece

Introduction

Electronic management of health data seems a trivial task until undesired situations manifest its major importance. Now that telemonitoring and auto-mated analysis begins to play a significant role in delivery of health-care services the need for secure and robust services for data management is even more important. When we refer to data management of respiratory data, we include all the operations and provisions that ensure secure and unam-biguous persistent storage and exchange of all the information that pertains to the respiratory function of an individual. The software and hardware that undertake these operations are a respiratory data management system. The managed information includes, apart from demographics, medical history and the objective measurements that are acquired from medical devices, as well as subjective information identified or assumed by the health-care professionals along with probabilistic output of automated analysis systems. The major obstacles, while managing respiratory related information, arise from their diversity, their volume, and continuously expanding range of new concepts derived from modern analytics.

In this chapter, we will summarize the possible pitfalls of respiratory data management system development and review how the current state-of-the-art data management software tries to tackle the identified problems.

What can go wrong?

When failure of the data management framework is materialized, the lack of end service may not be the worst impact. More important than temporal lack of service, is the impact from reports that are compiled of wrong information from clinical, monitored, or analysis generated data, and clinical decision support systems (DSSs) that generate wrong or not in-time outcome, leading thus to a system that may harm the health of the monitored individuals and cannot be trusted by the end users.

Wearable Sensing and Intelligent Data Analysis for Respiratory Management
https://doi.org/10.1016/B978-0-12-823447-1.00009-9
213

While trying to achieve an intended result, unsuccessful attempts are commonly divided into two broad categories: "execution failures" and "planning failures." Even though correspondence in development of data management systems is not straightforward, still the failures during the operation of a system that relies on a medical data management subsystem (e.g., an Hospital Information System, an electronic medical record application, a clinical DSS, or the software accompanying an imaging technology) can be discriminated in problems arising from human errors (either directly or via the use of graphical user interfaces) and in problems originating from the software/hardware itself. Human errors that are not system-dependent (e.g., user fatigue, bad training, etc.) and hardware-based problems are not of interest in this text. The remaining category of failures of the software to achieve its intended role as a data management system can be the result of either any or a combination of: (a) ambiguity of concepts, (b) loss of integrity, (c) unstable or poor performance, (d) inefficiency of interfaces, (e) inadequate data security, and (f) low scalability, and maintainability.

Ambiguity of concepts

We refer to ambiguity of concepts when the concepts that correspond to the information that is managed are not defined unequivocally or when the unequivocal definition is not shared among the interested actors, i.e., users and developers. For example, the definition of the concept regarding the respiratory rate signal simply as "RR" leads to a common ambiguation with the RR interval,[1] resulting in storing/presenting or providing to analysis erroneous data. Ambiguity of concepts is not only related to the concept meaning but also to the supported data types, units of measurement, or restrictions on the actual managed values. For example, the Fraction of inspired oxygen (FiO_2) is usually defined by physicians as a decimal number <1. If the concept was modeled to accept values in terms of percentages (e.g., 30%) but still was accepting decimal numbers, then a manual entry error due to habit would result into erroneous data in the system. Ambiguity of concepts is the result of inefficient data modeling procedure enhanced by bad communication among the system developers. Both conditions are critical since good communication among data modelers and system developers may hide inefficiency in data modeling, at least in the short term.

Loss of data integrity

Data integrity, although is intuitive to be enforced by a data management service, usually relies also on the constraints applied on the user interface (UI) level.

---

1. RR interval: the time elapsed between two successive R-waves of the QRS signal on the electrocardiogram.

Commonly encountered data integrity issues either concern referential integrity, i.e., missing, or erased data that are referred by another piece of data, or domain integrity, i.e., the existence of data with values outside the expected value-set or data values that are contradicting with previous values. Both referential and domain integrity can be managed by the UI level, but such a solution adds complexity, introduces possible security issues, and reduces maintainability of the whole system. While the most common outcome from violation of referential integrity results in loss of information that can be identified by the system or the user, violation of domain integrity in some cases might be more difficult to identify and contradicting information could lead to serious errors when they are consumed by a DSS or analytics service or the end user. An example of such an unfortunate scenario would be the definition of identified allergies of a person including both category of allergen and specific substance as distinct attributes. For a specific person, substance might be defined by mistake as allergy to "corn" (instead of cortisone) and allergy category (since is an autonomous attribute) defined correctly as "allergy to medication." Subsequently, a Computerized Provider Order Entry (CPOE) system or a DSS that queries the database for allergy information related to medication, assuming it considers both fields and not only the wrong one, will possibly refrain from sending a critical warning. Definition and enforcement of data integrity mechanisms are of high importance in data management. Encountered issues usually arise either from inherent inefficiencies of the data management services, i.e., their inability to define and/or to enforce domain integrity rules,[2] or from the lack of formal and commonly accepted modeling of domain integrity rules.

Unstable or poor performance

Data management performance issues, apart from the obvious impact on user experience due to limited system response, may also impact the operation of the rest of the system performance, such as analysis operations or DSS operations. Slow system response might be acceptable in some cases, such as aggregate reports from a group of patients. Nevertheless, it always deteriorates the user engagement and may lead to unused parts of the system or gaps in registered information, since the time that health-care professionals must spend using a system is limited. Performance must be considered in all data management operations, i.e., storing, updating, deleting, and querying. Specifically for querying, performance of common and uncommon (but still needed and supported) queries must be profiled. Unexpected performance, e.g., a data management system that has chaotic behavior in terms of response time (or no response at all), may result in unexpected timeouts and thus

---

2. To be analyzed later: for example, rules on cardinality of accepted answers.

unforeseen issues in other modules. It must be noted that performance issues are not always based on the database management system per se. There are times that inefficient modeling leads to data fragmentation, hindering the performance of the system.

Inefficiency of interfaces

The UI of a system is the part that the end user (e.g., health-care professional, patient, informal caregiver, etc.) interacts with the system. While the goal is to provide effective operation and control of the system, requiring minimum effort on the user's part, this is not always achieved. Often UI design focusses on providing the intended functionality, neglecting characteristics like efficiency, intuitiveness, active user support against errors, and esthetic design. Efficiency refers to the number of actions and time needed to perform an operation and intuitiveness is the ability of UI to provide the desired operation using controls and procedure that the user would expect. Support for data entry errors and hidden or partially presented information (e.g., a UI that does not highlight active diagnoses over past diagnoses) can be particularly important in health-care-related UIs. The lack of any of those characteristics has a negative impact on user acceptance and may result in missing or erroneous input, unclear output thus triggering undesirable results.

Inadequate data security

With the emergence of the electronic health record (EHR), issues related to leakage of personal information or sensitive health information have become of big importance. Although the security measures on the storage of traditional paper-based records were in most cases questionable, the possibility of large-scale data leakage was minimal. On the other hand, data stored in insecure systems may lead to exposure of sensitive information for many people in a small period and in some cases transparently to the data owners. The importance of this subject was identified in the US Health Insurance Portability and Accountability Act (HIPAA) of 1996 that was mainly aiming on the continuity of health care and the portability of EHR, and it is now reflected in the more generic highly acknowledged General Data Protection Regulation (GDPR) (EU) 2016/679, which was put into effect in EU on May 25, 2018. Health data management systems must consider all provisioned requirements that stem from that document and provide authenticated and authorized user access as well as auditing mechanisms.

Lack of sustainability

Finally, provision on the sustainability of a system is important. A system during deployment might fulfill all user requirements, may be user friendly, may be responsive and robust but, with no provision for implementing changes, may result in either a withdrawn system or a companion system. This situation is common in public hospitals or private practice clinics, where the software for managing health records or even an acquisition device is well accepted for years but new technologies or even minor change requests cannot be integrated into it. This often leads to the acquisition of new systems that operate in parallel either because they lack the full functionality of the previous system, or even if they provide the same functionality, they are not equally accepted. Issues that must be confronted by design are the possible addition of new managed concepts without ambiguity, possible addition of new functionalities (e.g., support for new query types), standard based connections to external systems, and ways to overcome vendor lock-in situations.

Respiratory disease management data

The central data management is the heart of an integrated care solution that entails multiple sources of data and information along with multiple access points, a strong temporal aspect, as well as different computational workflows. This implicates cloud data storage, management and security, ability to support storing and retrieving of multiple, long-term streaming sensor data, other biosensor data and user reports along with numerous calculated features, metadata, and provenance information, as well as DSS outputs.

A respiratory data management system must be capable of managing all types of data related to diagnosis and treatment of respiratory conditions. Apart from basic EHR information commonly shared among all health-care practices such as demographics and administrative information, existing allergies, previous and current diagnoses, medical history and family history, medication, diagnostic imaging, etc., modern respiratory disease management requires the acquisition and assessment of data from specialized medical devices. For many years, acquisition of data from devices measuring oximetry, spirometry, electrocardiogram (ECG), and auscultation was visit based and consisted of either single numerical values or distinct small binary files (e.g., the ECG signal of a periodic visit to the private office of a doctor). Nowadays, medical devices used in everyday health-care practice, as well as portable unobtrusive devices, are transforming patient's home to a health-care environment and can measure periodically or continuously not only basic vital signs but also complex signals and multimedia-like ultrasound recording, multi-lead ECG, electrical impedance tomography (EIT), and multichannel auscultation. These devices require efficient and robust management of more complex data than those stored in traditional paper-based health records.

Apart from the data that are acquired by devices or entered to the system because of the health-care professional's observations or assumptions, there is one more category of data that is growing rapidly not only regarding their volume, but also in the variety of concepts that need to be managed.

This category includes the data that are asserted by software after analyzing either the raw data corresponding to a single observation of a patient or data from a huge number of complete patient records. These practically include biosignal analytics results, clinical decision support outcome (i.e., recommendations, alerts, proposed workflows, etc.), results of data mining and what-if analysis, machine learning (ML)/artificial intelligence (AI) training features, algorithm parameters and resulting labels and classification, as well as trace back and information and explanation for all asserted by software information.

Systematized Nomenclature of Medicine Clinical Terms (SNOMED CT) is a systematically organized computer processable collection of medical terms in hierarchical form. SNOMED CT is considered to be the most comprehensive, multilingual clinical health-care terminology in the world [1]. Regarding respiratory data, two main branches of terms are identified in SNOMED CT. The first is the branch of respiratory observables (SCTID:364048003), which includes the related clinical history and examination set of observable entities. This branch comprises 284 observable entities that include standard questionnaire's scores (e.g., the *Asthma Control Questionnaire score*), measurable concepts (e.g., $FEV1^3/FVC^4$ ratio), or examination observables (e.g., diaphragmatic excursion). There are 182 measurable concepts organized as subconcepts of the 43 direct children of *Respiratory measure* (SCTID: 251880004) concept presented in Table 7.1. The second is the branch of respiratory tract evaluation procedures (SCTID:386043001), which includes 319 terms mainly about auscultation and imaging of chest and respiratory tract. The 53 direct children of *Respiratory tract evaluation* concept are presented in Table 7.2.

The presented list of concepts, along with the rest of relevant SNOMED CT concepts, must be considered when selecting supported concepts for a respiratory data management system. Additionally, as stated above, concepts outside of SNOMED CT that may correspond to new observation procedures, such as chest EIT or new features identified by image processing or ML/AI approaches, may need to be included, depending on the context.

Modeling respiratory data

A data model is the definition of how the data corresponding to the relevant entities of a domain should be organized and how those entities, and thus the corresponding data, relate to each other. The development of the data model, although it is maybe the most common task while developing a health information system, is a procedure that requires several critical decisions to be made, since it is one of the cornerstones of the final system.

---

3. FEV: Forced Expiratory Volume.
4. FVC: Forced Vital Capacity.

TABLE 7.1 Direct children of respiratory measure (SCTID:251880004).

SNOMED CT ID	Concept
699092002	Spirometric lung age
442465007	Respired oxygen tension
440031006	Adequacy of ventilation and alveolar gas exchange
440030007	Adequacy of cardiac output and alveolar gas exchange
417214003	Number of breaths
404996007	Airway patency status
404988002	Respiratory gas exchange status
395111008	Substance level in breath
364065007	Distribution of ventilation:perfusion
313221000	Respiratory percentage
251955000	Alveolar—capillary membrane conductance
251954001	Pulmonary capillary blood volume
251953007	Alveolar volume
251952002	Transfer coefficient (respiratory measure)
251949005	Transfer factor (respiratory measure)
251942001	Respiratory ratio
251908005	Respiratory volume
251907000	Respiratory pressure
251902006	Nitrogen washout measure
251901004	Maximum voluntary ventilation
251899008	Expired gas concentration
251891006	Respiratory shunt
251890007	Alveolar-arterial oxygen tension difference
251883002	Thoracic compliance
251882007	Airway conductance
250823005	Total dynamic compliance
250818005	Ventilation cycle time
165033004	Respiratory flow rate
89919001	Transmural pressure

Continued

TABLE 7.1 Direct children of respiratory measure (SCTID:251880004).—cont'd

SNOMED CT ID	Concept
89624001	Partial expiratory flow-static recoil curve
86377001	Ventilation–perfusion ratio
86290005	Respiratory rate
79063001	Gas flow rate (v)
75098008	Flow history
72947005	Maximum expiratory flow-static recoil curve
70778006	Gas transport time peak following bronchodilator
67461000	Alveolar pressure
61017008	Lung relaxation pressure
43224003	Pleural pressure
35706003	Isovolume pressure–flow curve
30367009	Lung clearance index
6304003	Airway resistance
5222008	Pressure difference between alveoli and mouth

TABLE 7.2 Direct children of respiratory tract evaluation (SCTID:386043001).

SNOMED CT ID	Concept
16321411000119100	MRI of nasopharynx without contrast
720389008	Expiratory CT
719897001	Fluoroscopy-guided insertion of bioabsorbable stent into airway
709641006	Ultrasonography of nasal sinus
709580008	MRI of nasopharynx with contrast
449264008	Auscultation of lower respiratory tract
449195005	Endoscopic ultrasonography and biopsy of bronchus
445212008	Nasal potential difference test

Continued

TABLE 7.2 Direct children of respiratory tract evaluation (SCTID:386043001).—cont'd

SNOMED CT ID	Concept
441677009	Imaging of lung
440369006	Evaluation of oral stage deglutition and pharyngeal stage deglutition
439939004	Endoscopic ultrasonography of bronchus
438623001	Bronchoscopic procedure using ultrasound guidance
438592006	Percutaneous embolization of nasal blood vessel using fluoroscopic guidance with contrast
426755009	Diagnostic fiberoptic endoscopic examination of lower respiratory tract and lavage of lesion of lower respiratory tract
426196002	Diagnostic fiberoptic endoscopic examination of lower respiratory tract and brush cytology of lesion of lower respiratory tract
418876000	Fluoroscopic bronchography
359552007	Exploration of ethmoid sinus
303668008	CT of intrathoracic respiratory structures
287306002	Exploration of bronchus
271312008	Soft tissue X-ray of lung/bronchus
268436004	Contrast radiography larynx/trachea
265039004	Diagnostic endoscopic examination of trachea using rigid bronchoscope
265037002	Diagnostic fiberoptic endoscopic examination of pharynx and larynx
252573007	Taub test
252572002	Laryngeal stroboscopy
241612008	MRI of larynx
241610000	MRI of nasopharynx
241609005	MRI of paranasal sinuses
232615005	Exploration of tracheostomy
232521000	Diagnostic antroscopy via middle meatus
232520004	Diagnostic antroscopy via inferior meatus

Continued

TABLE 7.2 Direct children of respiratory tract evaluation
(SCTID:386043001).—cont'd

SNOMED CT ID	Concept
173147000	Diagnostic fiberoptic endoscopic examination of trachea and biopsy of lesion of trachea
173122002	Diagnostic endoscopic examination of bronchus and biopsy of lesion of lung using rigid bronchoscope
173121009	Rigid endoscopic examination and biopsy below trachea using rigid bronchoscope
169033008	Tomography—larynx/trachea
168812006	Selective bronchography
164772003	Examination of larynx
164771005	Examination of pharynx
112790001	Nasal sinus endoscopy
87829000	Incision and exploration of trachea
87338009	Tracheography
85895003	Contrast bronchogram
84492002	Radiography of nasal sinuses
67244001	Incision and exploration of lung
58011003	Diagnostic radiography of larynx
46857000	Determination of resistance to airflow by plethysmographic method
41785006	Diagnostic radiography of pharynx
27032005	Nose and throat examination
18044005	Endoscopy of trachea
16608009	Endoscopy of lung
16020002	Laryngeal function studies
10847001	Bronchoscopy
2811005	Cineradiography of pharynx

The most used types of data models are the geographic data model, mainly used in geographic information systems (GISs), and the entity–relationship model (ERM), which is the common practice in software engineering when the storage of the data resides in Relational Database Management System (RDBMS). Another type of a conceptual data model is the semantic data model. In a semantic data model, the model describes the meaning of its instances and allows the interpretation of the meaning directly from the instances without the need to know the meta model. This enhancement that a semantic model offers to the domain model definition, along with the advanced expressivity provided by the use of semantic tools, makes ontologies a modern way of describing a domain data model. Semantic interoperability allows computers to share, understand, interpret, and use data without ambiguity [1]. Modern big data problems require more and more unattended analysis procedures to be performed. To enable such functionalities in the future, a system must provide semantic interoperability of data through linked data principles. Those principles dictate the use of Uniform Resource Identifiers (URIs) to uniquely identify resources and the definition of exchanged data in a way that they include not only actual values but also the relationships among them.

Ontological semantic representation of domains is very common in health-related domains. There are several attempts to use ontologies as data model to support EHR data [2], personalized care of chronic diseases [3], and DSSs [4−6]. Health ontologies and terminologies are widely used in health-care information systems. One of the most easy-to-use and freely accessible repositories of health-related ontologies is Bioportal [7] of the National Center for Biomedical Ontology. It incorporates search and representation mechanisms for several health ontologies and terminologies such as SNOMED Clinical Terms, International Classification of Diseases (ICD), Logical Observation Identifier Names and Codes (LOINC), and World Health Organization's Anatomical Therapeutic Chemical (ATC) classification system, among others.

The Web Ontology Language (OWL) is a family of knowledge representation languages for authoring ontologies. Ontologies are a formal way to describe taxonomies and classification networks, essentially defining the structure of knowledge for various domains [8]. OWL Description Logic (OWL DL) is a sublanguage of OWL and it was defined to support those users who want the maximum expressiveness[5] without losing computational completeness (all entailments are guaranteed to be computed) and decidability (all computations will finish in finite time) of reasoning systems.

Health Level 7 (HL7) Fast Healthcare Interoperability Resources (FHIR) is a new HL7 standard for exchanging EHRs. It builds upon previous HL7 data format standards, but also leverages more modern technologic concepts and approaches, aiming to be more developer friendly. Examples of these modern approaches include a RESTful API built upon standard HTTP operations, and a choice between JSON and XML for data representation. HL7 FHIR models

---

5. OWL DL includes all OWL constructs with restrictions such as type separation (a class cannot also be an individual or property, a property cannot also be an individual or class).

the whole health data management domain via a distinct set of concepts (i.e., FHIR Resources) and their properties. Although the original goal was to tackle interoperability and data exchange issues, as successor of previous HL7 standards, FHIR encompasses the collective knowledge of the HL7 Community in identifying the semantics of the health domain resulting into a model that must be considered both for large-scale and for small-scale health domain modeling attempts, either by adopting the standard as-is, by expanding/specializing, or by pruning not applicable properties.

A semantic data model for respiratory data management

Overview

The benefits of integrated medical data management in a standardized and semantically enhanced manner are highlighted in various papers [9—11]. Even so, such an approach is not followed by modern systems since data management per se is convincingly addressed by current state-of-the-art standards and frameworks. The review by Ref. [12] on using Ontologies and Semantic Integration Methodologies to Support Integrated Chronic Disease Management in Primary and Ambulatory Care reveals that there is big potential but still little evidence in this area. Similarly, in Ref. [13] the use of semantics for data quality in integrated chronic disease management is drawn promising but still very immature.

An example of a semantic model for the data management of patients with COPD and comorbidities has been implemented by Ref. [14] as an OWL DL ontology. The implemented ontology had a dual purpose: the first was to define unequivocally the entities in the domain of integrated care of patients with COPD and comorbidities along with their semantics; and the second was to act as the thesaurus for the scaffolding data management engine's interfaces that were auto-generated and used for storing/exchanging of the actual data. In this scope, the entities were defined as OWL classes, while the actual recorded, reported, extracted, and computed data were stored as ontology instances. To enhance semantic interoperability and knowledge sharing, the defined entities were linked with Bioportal-based classes. The connection was based on Bioportal's PURLs (Persistent Uniform Resource Locators) to provide a semantic link to existing acknowledged medical terminologies, such as SNOMED CT or LOINC. Finally, the ontology classes were grouped based on their conceptual relation inside the domain and based on the common shared properties expressed via common ancestor FHIR resources—based classes. For example, HeartRate and BreathingRate were grouped under VitalSigns which itself was a descendant of the FHIR:Observation class that defines the corresponding shared properties in accordance with FHIR. Apart from the defined classes also class-specific restrictions were set regarding their accepted data type, cardinality, and where applicable, corresponding fixed value-set.

Semantic data model implementation steps

The implemented ontology comprised semantic layers, defined as hierarchically connected ontologies, where each additional layer enhances the model. The ontologies were developed in the following order. First, an ontology of "FHIR primitive and complex data types," then the *"FHIR resources and properties"* ontology defining the required HL7 FHIR resources and their properties in OWL, and on the top layer, the "Domain specific data entities" ontology defining the required entities (along with their domain/deployment-specific restrictions) to model the domain of integrated care services for COPD multimorbid patients. The instances of the OWL classes defined in this ontology correspond to the actual data managed by the framework.

Apart from the definition of FHIR semantics and domain-specific entities, additional optional definitions might be required. Using the defined data model to store actual data as instances, additional semantics are needed to address referential integrity constraints. In Ref. [14], such an ontology is proposed, which defines properties to be assigned on specific classes of "Domain specific data entities ontology" so that can be used by the storage engine to enforce referential integrity. For better maintenance and reusability of the model, these properties are defined in a separate ontology: the "ServerSpecificProperties".

Representing primitive data types

Specific design decisions in the definition of the ontologies are interesting. The "FHIR primitive and complex data types" ontology is very important since it defines the method for storing all the data as instances of the ontology classes. Specifically, in FHIR, primitives are defined using XML Schema and building on top of its primitives. New data types can be constructed from XML Schema primitives using restriction (reducing the set of permitted values), list (allowing a sequence of values), and union (allowing a choice of values from several types). Most of the FHIR primitives can be mapped exactly to XSD primitives. For example, fhir:instant maps directly to xsd:dateTime and fhir:string maps to xsd:string. A number of them, however, map to more complex expressions of XSD primitives, like fhir:dateTime which maps to the union of xsd:gYear, xsd:gYearMonth, xsd:date, and xsd:dateTime.

Perhaps, the most basic way to represent FHIR primitive values in RDF is to use RDF Literals [15], where the 'datatype IRI' element will correspond to the appropriate XSD data type without defining FHIR specific data types. An RDF Literal using this approach would have a form like

- "primitive_value_literal"^^xsd:xxx

where xxx is any of the XSD data types.

This approach can be extended further, by subclassing rdfs:Datatype, thereby defining new 'datatype IRIs' that correspond exactly to FHIR primitives.

The second approach would yield literals in the form of

- "primitive_value_literal"^^fhir:yyy

where yyy can be any of the FHIR primitive data types.

Both approaches produce "well-formed" RDF syntax. It is not, however, trivial to check the value legality of RDF Literals with a non-XSD data type IRI. It is important to note that even the use of the well-defined XSD data types is not required, but only recommended by the relevant RDF standard [15].

So, while at first glance, defining new data type IRIs for the FHIR primitives is an intuitive approach, the reality of the matter is that it introduces additional burden to the implementers. While restricting the value space of an xsd data type is trivial, the definition of unions of xsd data types is not straightforward without applying library-specific interfaces, to validate legal values. Value legality is not the only concern when using custom data type IRIs. Developers need not only to validate legal values for primitives, but also operate on these values. How does one compare and sort a list of dates (fhir:date) when some items on the list include only the year while others the full date?

The third approach, which was selected by Ref. [16], is to define OWL classes that correspond to the FHIR data types. An already defined property (rdf:value) exists with the sole intent of representing structured values in RDF. The definition of these classes would include additional semantic information (e.g., in the form of OWL Restrictions) that a reasoner could exploit to validate the legality of assigned values.

After examining the various, different options, we chose to represent the various FHIR data types as distinct OWL classes. To represent any primitive value, one needs only to instantiate the relevant class, and assign it a value using the *rdf:value* property.

The definition starts with the base class *fhir:PrimitiveType*. An appropriate *owl:Restriction* is inserted in the definition (Fig. 7.1), to constraint the use of the *rdf:value* property to a maximum cardinality of 1 per instance.

Subsequently, each FHIR primitive data type is defined as a subclass of *fhir:PrimitiveType*. Additional property restrictions are defined per class, to constraint the Literal values on the *rdf:value* property, to the relevant XSD data types. Regardless of whether an FHIR data type maps directly to one or more (Fig. 7.2) XSD data types, this enables a consistent restriction pattern to be applied. Using this pattern, one can easily extend the ontology to define new primitive types. In addition, the core use of XSD-only data types makes it implementer-friendly since most RDF libraries can validate the legality of XSD values.

```
:PrimitiveType   a   owl:Class ;
    rdfs:label        "Primitive Types"^^xsd:string ;
    rdfs:subClassOf :Element ;
    rdfs:subClassOf [ a owl:Restriction ;
                      owl:maxCardinality 1"^^xsd:nonNegativeInteger ;
                      owl:onProperty     rdf:value
                    ] .
```

FIGURE 7.1   The fhir:PrimitiveType class definition.

```
:dateTime    a        owl:Class ;
    rdfs:label        "A date, date-time or partial date (e.g. just year
                      or year + month) as used in human communication. If
                      hours and minutes are specified, a time zone SHALL be
                      populated. Seconds may be provided but may also be
                      ignored. Dates SHALL be valid dates."^^xsd:string ;
    rdfs:subClassOf :PrimitiveType ;
    rdfs:subClassOf [ a owl:Restriction ;
                      owl:allValuesFrom [ a  owl:Class ;
                      owl:unionOf ( xsd:dateTime xsd:date
                      xsd:gYearMonth xsd:gYear ) ] ;
                      owl:onProperty rdf:value
                    ] .
```

FIGURE 7.2   The definition of fhir:dateTime.

Representing HL7 FHIR complex data types

The same approach, of different and distinct OWL classes, was followed at the definition of FHIR Complex Types. Each Complex Type is represented as an OWL class with appropriate cardinality restrictions in place, as per the FHIR specification. A strict approach was selected when defining the various RDF properties that assign child elements to each complex type. In reference to the "*value*" example used previously, distinct properties were defined, depending on the context of the "value" i.e., *fhir:Quantity.value* (Fig. 7.3), *fhir:ContactPoint.value*, etc. This strict approach is representative of the FHIR specification followed by *rdfs:domain* and *rdfs:range* axioms, defined for each property, to better represent the FHIR definitions.

```
     :Quantity.value   a      owl:ObjectProperty ;
            rdfs:comment   "Comments: The implicit precision in the value
                           should always be honored. Monetary values have
                           their own rules for
                           handling precision"^^xsd:string , "Requirements:
                           Precision is handled implicitly in almost all
                           cases of measurement."^^xsd:string ;
            rdfs:domain    :Quantity ;
            rdfs:label     "The value of the measured amount. The value
                           includes an implicit precision in the
                           presentation of the value."^^xsd:string ;
            rdfs:range     :decimal .
```

FIGURE 7.3   Quantity.value property definition.

Representing HL7 FHIR resources

The "The FHIR resources and properties" ontology was the representation of
FHIR resources, their properties, and the restrictions on properties in OWL.
FHIRResources:Resource was defined as an *owl:class* and acted as a parent
class for all the rest definitions. To ensure the scalability and extensibility of
the ontology a strict methodology was followed for representing all FHIR
resources and their properties. Properties were defined as *Resourcename.-
propertyname* (e.g., Device.type) and were assigned to the specific resource.
For example, the Device.type property was specifically assigned (via
rdfs:domain) only to the resource (owl:class) Device and it is not shared with
other resources. The property range (i.e., allowed value-set) was defined using
rdfs:Range in the assigned property. To define cardinality restrictions (i.e., how
many values the property may have for a specific instance) an *owl:restriction*
was defined in the corresponding FHIR resource class.

Adding the semantics for data integrity

Having the data model as an ontology, additional semantics are needed to
address and enable enforcement of referential integrity constraints. This is
required as ontologies traditionally follow the "Open World Assumption"
(OWA) approach [17]. In OWA the absence of a given fact is interpreted as
unknown information and cannot imply whether that possible assertion is true
or false. In contrast, in the Closed World Assumption (CWA) approach
anything not known to be true is presumed to be false. In the context of the
specific approach, FHIR resources are represented as RDF documents and

references between them are expressed using URIs. By nature, except from validating that it is well formed, a URI is just a string. There is no guarantee that it is dereferenceable, that it identifies any valid resource, or about the type of resource it identifies. As a simple example of the above, for a 'subject' URI inside an Observation resource, using just the OWA, it would be impossible to distinguish between a valid 'subject' URI (e.g., an existing patient) and an invalid one (e.g., a nonexisting resource URI). To achieve this distinction, a property:hasForeignKey was defined. The system can examine any incoming Observation resource and validate that the 'subject' property value (defined as a foreign key in the Observation definition; see Table 7.3) contains the URI of an existing resource. It is only in the case of referential integrity (foreign keys, unique keys) where the OWA approach of RDF/OWL fails to provide any meaningful expression mechanism.

Having the semantic data model as an ontology, a method for expressing and validating constraints on the values of the actual data is required. As already mentioned, for the definition of type conformity, rdfs:Range can adequately describe the constraint, but for specific restrictions, in distinct classes defined in the domain, such as "Every instance of Observation*BodyTemperature* in our domain must have a value less than 44°C," there is no straightforward way to define and validate restrictions. Of the available solutions for RDF validation, SPIN (SPARQL Inferencing Notation) framework [18] provides a basis for constraint checking. SPIN is available as a W3C Member Submission, and a Java-based open-source implementation already exists for it. It is also one of the main technologies that the RDF Data Shapes Working Group is basing its design on (alongside Stardog ICV, Shape Expressions, and Resource Shapes). SPIN constraints are serialized in RDF, therefore easily distributed inside an ontology, in contrast to other approaches such as SWRL [14]. Finally, since it is based on the formulation of SPARQL queries, it is relatively easy for third parties to transform SPIN constraints to other technologic frameworks.

TABLE 7.3 Code excerpt from the Observation class definition showing the use of the:hasForeignKey property.

```
FHIRResources:Observation

rdf:type owl:Class ;

[...]

ServerSpecificProperties:hasForeignKey
(<FHIRResources:Observation.subject>) ;

[...]
```

```
spin:constraint  ⬚ ▽
✱    ASK WHERE {                                              ▽
        ?this :Quantity.code ?code .
        FILTER NOT EXISTS {
          ?this :Quantity.system ?system .
        }.
     }
```

FIGURE 7.4   spin:constraint for the validation of the FHIR logical constraint on Quantity.code and Quantity.system.

Using SPIN, the example data model ontology is populated with constraints that can be evaluated to validate common restriction patterns such as cardinality constraints. In addition, since it is based on SPARQL, it is easy to validate and enforce practically any complex constraint, such as the logical constraints imposed by FHIR on the complex data types (Fig. 7.4) or value level constraints, for example, constraint on the property *Observation.value* (the property about the actual measurement value) for instances of *Body-Temperature* class. This constraint is enforced by the SPIN rule shown in Fig. 7.5. In a functional system, these types of constraints can be evaluated every time a specific resource is created or updated ensuring data integrity and logical coherence. Kilintzis et al. [14] have successfully used SPIN to constraint check against OWL Restrictions and emulate common restriction patterns found in relational databases, such as foreign keys and unique values. In addition, one can use SPIN rules to auto-generate property values and provide dynamic property values that change as the stored data changes, ensuring valid, good quality data.

The downside to selecting SPIN to perform constraint checking is that this operation cannot be natively conducted on the storage engine, but must be located one layer above it, between the storage engine and the Communication API. Although this is a drawback, it is a standard approach, even in relational databases, to check input data for validity and quality before delegating them to the storage engine.

Transition to SHACL

In July 2017, the W3C has ratified the Shapes Constraint Language (SHACL) as an official W3C Recommendation. SHACL was strongly influenced by

```
# Temperature out of bounds(too high)
ASK WHERE {
?this (FHIRResources:Observation.value/FHIRct:Quantity.value)/rdf:value
?value .
FILTER (?value > 44) .
    }
```

FIGURE 7.5   SPIN constraint for high BodyTemperature.

```
FHIRResources:QuantityShape
rdf:type shacl:NodeShape ;
shacl:sparql [
     rdf:type shacl:SPARQLAskValidator ;
     shacl:select """SELECT $this
WHERE {
    ?this FHIRResources:Quantity.code ?code .
    FILTER NOT EXISTS {
        ?this FHIRct:Quantity.system ?system .
    } .
}""" ;
    ] ;
    shacl:targetClass FHIRct:Quantity ;
    .
```

FIGURE 7.6   shacl:NodeShape for the validation of the FHIR logical constraint on Quantity.code and Quantity.system.

SPIN and can be regarded as its legitimate successor [19]. SHACL is a data modeling language that has been developed by a W3C Working Group. As more people work with data coming from a variety of sources, especially for data integration projects, SHACL gives them a way to describe the "shapes" of the data that they are working with so that applications can take better advantage of that data [20]. In Fig. 7.6, the spin:constraint for the validation of the FHIR logical constraint on Quantity.code and Quantity.system is represented as an SHACL NodeShape.

Every SPIN feature has a direct equivalent in SHACL. While SHACL improves over the features explored by SPIN and although SPIN will continue to get support from TopQuadrant, SHACL must be considered for future implementations since it is the current official recommendation.

Persistent storage and data integrity

The current common practice in eHealth systems is to use RDBMS to store medical data since, for many years, this type of systems was used and proved to be robust and efficient. Triplestores, although still not widely used, present significant advantages in terms of modeling complex, semantically enriched information and can also be easily integrated to web service architectures, since the communication is carried over HTTP. On the other hand, exchanging information as triples (data entities composed of subject-predicate-object definitions) adds overhead compared to the exchange of database records for use in an RDBMS. Therefore, the capability to handle large amount of data and system performance are critical issues that must be addressed by a triplestore in order to be selected as a viable alternative to RDBMS.

Relational databases are by far the most widely used solution to save data in enterprise systems; however, they also have several disadvantages. It is not easy to maintain an evolving data model in a relational database. In addition, they suffer from what has been referred to as the "impedance mismatch" problem, that is, the difference between the relational data model that organizes data into tables and rows, and the in-memory data structures an application developer uses when handling those data [21]. Furthermore, relational databases have no "out-of-the-box" support for semantic web technologies, which means that we would have to implement semantic web technologies on a layer out of storage. The ability to directly integrate semantic web technologies to the data model would be a valuable asset to the developing of the DSS module. Finally, it should be noted that, although SQL, the main query language used in relational databases, is standardized, a variety of SQL dialects have evolved over the years for the various implementations that exist.

NoSQL is a term encompassing various families of database technologies that differ from traditional relational databases in the mechanism used to store and retrieve data. These families include key-value stores, key-document stores, column-family stores, and graph stores [22]. NoSQL databases tend to be open-source projects, cluster-friendly, and schema-less.

Most NoSQL solutions lack model definition and integrity checking mechanisms. Furthermore, they implement custom, nonstandardized ways to access data leading to vendor lock-in.

There exists, however, the specific case of RDF Triple Stores. Those are graph databases that store triples of information, a triple being a data entity consisting of subject-predicate-object. Triples can be imported and exported using queries, RDF, and other formats.

Triplestores offer a variety of advantages that are of particular interest. There is a standardized, vendor independent way of accessing data, SPARQL for querying, and RDF serializations for exchanging data. Triplestores can be queried over HTTP, making it easy to fit them inside a service-oriented architecture. They allow easier maintenance of an evolving data model, while, at the same time, natively embedding semantic interoperability. Being, however, a relatively modern technology, RDF Triple Stores are not widely used. In the past, triplestores have been characterized by poor read/write performance [23]. Triplestores show their competitive edge when used to store/retrieve RDF/XML graphs. However, most benchmarks available focus on the performance of triplestores when loading in memory serialized RDF from static files. Although this provides an insight to the engineering skill of the developers of each system, it cannot be used to extrapolate performance in other scenarios. Kilintzis et al. [24] have tested the performance of a basic RESTful web service that stores and retrieves numeric and text data about a patient exchanged in RDF/XML format using three different setups, PHP/MySQL, JENA API/Fuseki triplestore, and JENA API/Virtuoso triplestore, concluding that the current generation of triplestores appears to be ready, at least performance-wise, for developing production-grade systems.

Regardless of their origin, scope, or size, all the data that a storage engine must handle can be classified into the following two broad categories:

- Structured Information, and
- BLOBs (Binary Large Objects, mainly biosignals) along with their metadata that belong to the first category.

The issue of storing BLOBs is a trivial one, technology-wise. A variety of solutions exist, ranging from the storage of flat files, accessed through a basic custom-built web service, to more sophisticated approaches of using an existing, commercial, file storage cloud solution such as Amazon S3, Microsoft Azure BLOB Storage, or Google Cloud Storage.

Achieving compliance to regulations

Although electronic management of health data was one of the first uses of computers dating back to early 1960s, regulatory laws regarding the electronically stored data became active relatively late. For several years, electronically stored health data were regulated by the same principles as orally or written health data.

The two most acknowledged regulations for electronically stored/transmitted health data are the US HIPPA and the EU GDPR.

The HIPAA of 1996 is a federal law that required the creation of national standards to protect sensitive patient health information from being disclosed without the patient's consent or knowledge. While the HIPAA Privacy Rule safeguards protected health information (PHI), the Security Rule protects a subset of information covered by the Privacy Rule. This subset is all individually identifiable health information a covered entity creates, receives, maintains, or transmits in electronic form. This information is called "electronic protected health information" (e-PHI).

To comply with the HIPAA Security Rule, all covered entities must do the following:

- Ensure the confidentiality, integrity, and availability of all e-PHI
- Detect and safeguard against anticipated threats to the security of the information
- Protect against anticipated impermissible uses or disclosures
- Certify compliance by their workforce

GDPR or Regulation (EU) 2016/679 on the protection of natural persons regarding the processing of personal data and on the free movement of such data is an essential step to strengthen individuals' fundamental rights in the digital age and facilitate business by clarifying rules for companies and public bodies in the digital single market.[6]

It must be noted that, unlike HIPAA, GDPR is not exclusively limited to health data. Regarding health data, which are considered sensitive data, particularly strict rules apply. The main data protection issues that arise from the regulation are:

- Data quality − not to process more personal data than necessary.
- Right of information − to inform the citizens about their rights and the purpose of the use of their health-related data.
- Right of access − to provide access on the medical files upon request.
- Retention period − health data must not be kept for longer than necessary.
- Data security − data should be processed only by health professionals or administrative staff reminded of their confidentiality obligations. Furthermore, where necessary, specific security measures on access control and management of all the information processed in the context of health data must be developed.

As it is evident, apart from the ethical and legal aspects, these regulations/laws also describe or imply requirements for health data management systems and must be considered while designing and developing health data management systems.

Conclusions

As mentioned in 2008 by Böcking and Trojanus in *Encyclopedia of Public Health* [25], the efforts at national and international level are concentrating to harmonize standards for data exchange, protect database access and patient identification, define the user roles for health data management and the EHR, including plans for long-term preservation, and evaluate conformity to existing norms and standards on an international level. Almost 15 years after that remark, the problem is yet to be tackled convincingly.

In this chapter we have presented the possible pitfalls of respiratory data management; we have identified the relevant aspects and presented examples of managing respiratory data, enabling their dual use as clinical and research data via their semantic enhancement. This approach can adhere to standards, and generate datasets that can be leveraged by AI, providing also the flexibility needed for adding future concepts.

Abbreviations

AI Artificial Intelligence
API Application Programming Interface

---

6. https://ec.europa.eu/info/law/law-topic/data-protection/data-protection-eu_en.

ATC Anatomical Therapeutic Chemical
BLOB Binary Large Object
COPD Chronic obstructive pulmonary disease
CPOE Computerized Provider Order Entry
CWA Closed World Assumption
DSS Decision Support System
e-PHI electronic Protected Health Information
ECG Electrocardiogram
EHR Electronic Health Record
EIT Electrical Impedance Tomography
ERM Entity—Relationship Model
FEV Forced Expiratory Volume
FHIR Fast Healthcare Interoperability Resources
FiO$_2$ Fraction of inspired oxygen
FVC Forced Vital Capacity
GDPR General Data Protection Regulation
GISs Geographic Information Systems
HIPAA Health Insurance Portability and Accountability Act
HL7 Health Level 7
HTTP Hypertext Transfer Protocol
ICD International Classification of Diseases
IRI Internationalized Resource Identifier
JSON JavaScript Object Notation
LOINC Logical Observation Identifier Names and Codes
NoSQL Non-SQL or Not only SQL
OWA Open World Assumption
OWL Web Ontology Language
OWL DL OWL Description Logic
RDBMS Relational Database Management System
RDF Resource Description Framework
REST REpresentational State Transfer
SHACL Shapes Constraint Language
SNOMED CT Systematized Nomenclature of Medicine Clinical Terms
SPARQL RDF Query Language
SPIN SPARQL Inferencing Notation
SQL Structured Query Language
SWRL Semantic Web Rule Language
URI Uniform Resource Identifier
XML eXtensible Markup Language
XSD XML Schema Definition

References

[1] T. Benson, G. Grieve, Principles of Health Interoperability, third ed., Springer International Publishing, Cham, 2016.
[2] C. Martínez-Costa, S. Schulz, Ontology content patterns as bridge for the semantic representation of clinical information, Appl. Clin. Inform. 5 (3) (2014) 660—669.
[3] T. Mallaug, K. Bratbergsengen, Long-term temporal data representation of personal health data, in: Advances in Databases and Information Systems: 9th East European Conference, ADBIS 2005, Tallinn, Estonia, September 12—15, 2005. Proceedings, Springer, Berlin, Heidelberg, 2005, pp. 379—391.

[4] N. Lasierra, A. Alesanco, S. Guillén, J. García, A three stage ontology-driven solution to provide personalized care to chronic patients at home, J. Biomed. Inform. 46 (3) (2013) 516–529.

[5] D. Riaño, et al., An ontology-based personalization of health-care knowledge to support clinical decisions for chronically ill patients, J. Biomed. Inform. 45 (3) (June 2012) 429–446.

[6] S.R. Abidi, S. Hussain, M. Shepherd, Ontology-based modeling of clinical practice guidelines: a clinical decision support system for breast cancer follow-up interventions at primary care settings, Stud. Health Technol. Inform. 129 (Pt 2) (2007) 845–849.

[7] P.L. Whetzel, et al., BioPortal: enhanced functionality via new Web services from the National Center for Biomedical Ontology to access and use ontologies in software applications, Nucleic Acids Res. 39 (Suppl. 2) (July 2011) W541–W545.

[8] H. Knublauch, D. Oberler, P. Tetlow, E. Wallace, A Semantic Web Primer for Object-Oriented Software Developers, W3C Editor's Draft, 2006 [Online]. Available: https://www.w3.org/2001/sw/BestPractices/SE/ODSD/.

[9] C.W. Kelman, A.J. Bass, C.D.J. Holman, Research use of linked health data - a best practice protocol, Aust. N. Z. J. Public Health 26 (3) (2002) 251–255.

[10] M. Aranguren, J. Fernandez-Breis, M. Dumontier, Special issue on linked data for health care and the life sciences, Semant. Web 5 (2) (2014) 99–100.

[11] B. Tilahun, T. Kauppinen, Potential of linked open data in health information representation on the semantic web, in: SWAT4LS, 2012, pp. 3–6.

[12] H. Liyanage, et al., The evidence-base for using ontologies and semantic integration methodologies to support integrated chronic disease management in primary and ambulatory care: realist review. Contribution of the IMIA primary health care informatics WG, Yearb. Med. Inform. 8 (1) (2013) 147–154.

[13] S.T. Liaw, et al., Towards an ontology for data quality in integrated chronic disease management: a realist review of the literature, Int. J. Med. Inf. 82 (1) (2013) 10–24.

[14] V. Kilintzis, I. Chouvarda, N. Beredimas, P. Natsiavas, N. Maglaveras, Supporting integrated care with a flexible data management framework built upon linked data, HL7 FHIR and ontologies, J. Biomed. Inform. 94 (June 2019).

[15] RDF 1.1 Concepts and Abstract Syntax, n.d. [Online]. Available: https://www.w3.org/TR/rdf11-concepts/. (Accessed 19 September 2021).

[16] N. Beredimas, V. Kilintzis, I. Chouvarda, N. Maglaveras, A reusable ontology for primitive and complex HL7 FHIR data types, in: Proceedings of the Annual International Conference of the IEEE Engineering in Medicine and Biology Society, vol. 2015, EMBS, November 2015, pp. 2547–2550.

[17] R. Reiter, On closed world data bases, in: Readings in Artificial Intelligence, Elsevier, 1981, pp. 119–140.

[18] W3C, SPIN - Overview and Motivation, 2011 [Online]. Available: https://www.w3.org/Submission/spin-overview/.

[19] From SPIN to SHACL, n.d. [Online]. Available: https://spinrdf.org/spin-shacl.html. (Accessed 26 June 2021).

[20] SHACL Tutorial: Getting Started — TopQuadrant, Inc, n.d. [Online]. Available: https://www.topquadrant.com/technology/shacl/tutorial/. (Accessed 26 June 2021).

[21] P.J. Sadalage, M. Fowler, NoSQL Distilled: A Brief Guide to the Emerging World of Polyglot Persistence, Pearson Education, 2013.

[22] P. Bednar, M. Sarnovsky, V. Demko, RDF vs. NoSQL databases for the semantic web applications, in: Sami 2014 - IEEE 12th Int. Symp. Appl. Mach. Intell. Informatics, Proc, 2014, pp. 361–364.

[23] D.J. Abadi, P.A. Boncz, S. Harizopoulos, Column-oriented database systems, Proc. VLDB Endow. 2 (2) (August 2009) 1664–1665.

[24] V. Kilintzis, N. Beredimas, I. Chouvarda, Evaluation of the performance of open-source RDBMS and triplestores for storing medical data over a web service, in: 2014 36th Annual International Conference of the IEEE Engineering in Medicine and Biology Society, EMBC 2014, 2014, pp. 4499–4502.

[25] W. Kirch, Encyclopedia of Public Health: Volume 1: A-H, vol. 2, Springer Science & Business Media, 2008, pp. I–Z.

Part IV

Current challenges in respiratory management systems

Chapter 8

The edge-cloud continuum in wearable sensing for respiratory analysis

Anaxagoras Fotopoulos[1], Pantelis Z. Lappas[1], Alexis Melitsiotis[2]

[1]EXUS AI Labs, Athens, Attika, Greece; [2]EXODUS SA, Athens, Attika, Greece

Introduction

The evolution of health care over the latest years has been significant regarding the use of new technologies such as artificial intelligence (AI), big data, miniaturization of sensors, and the high computational power offered by cloud technologies. A certain number of publications have shown that the vision for a predictive, preventative, personalized, and participatory (known as P4) health care is becoming rather a norm than a vision. The transformation of health care in the digital era involves: (a) the use of both medical and nonmedical devices (consumer wearables), (b) the -omics and the whole genome analysis of human DNA to discover mutations and genetic predisposition, (c) the big data analysis to unravel linkages in data unknown at first sight to human eye, (d) the data mining in (open) scientific knowledge, (e) the association with other type of data of importance, such as the weather or the social media in line with Geospatial Information Systems (GISs) information, (f) lab-on-a-chip technologies for rapid test analysis, (g) personal and family health history and daily habits information (more especially on exercise and nutrition) and others. IBM predicts that during a person's lifetime more than 1100 TB of data are generated, with 60% of those considered to be exogenous (behavioral, socioeconomic, environmental, weather, habits, family history, education, satisfaction, etc.), 30% considered as genomics related, and 10%, clinical information [1].

This chapter aims at presenting a holistic overview of the edge-cloud continuum concept in the wearable sensing. We carefully report recent works in the health-care domain to highlight the benefits of wearable sensing and the P4 health-care revolution and explain in detail the basic terms of edge and cloud computing to help readers understand the recent trends in the area of

Wearable Sensing and Intelligent Data Analysis for Respiratory Management
https://doi.org/10.1016/B978-0-12-823447-1.00002-6

241

AI. Furthermore, special emphasis is placed on respiratory systems. In this direction, several works from the literature have been carefully selected to present the contribution of present AI technological components and tools in the respiratory analysis.

The remainder of this chapter is organized as follows. The P4 healthcare revolution and its main principles are presented in The P4 health-care revolution section. Section Edge-cloud continuum describes the concepts of edge and cloud computing, whereas recent artificial intelligence trends for the edge-cloud continuum are presented in Recent artificial intelligence trends for the edge-cloud continuum section. Section AI-based solutions for respiratory analysis provides readers with recent AI-based solutions for respiratory analysis, while conclusions are given in Conclusions section.

The P4 health-care revolution

Health care is becoming more *participatory* in the sense that people research their conditions online to better understand them, and analyze through (nonmedical usually) wearables key metrics like heart rate level, stress level, SpO2 levels, weight, breathing disturbances during sleep, and other. There is growing tendency from people not only to understand their health status but also to improve it via exercise and healthy nutrition. Apart from "measuring" features, thousands of mobile health-care applications provide a series of workout programmes and healthy food ideas. This target-group of enthusiasts that are willing to transform their lives with a series of corrective actions and increased knowledge can provide high volumes of crowdsourced data. Such data can be used for the development and further optimization of algorithms that will/can act as the basis for the breakthrough improvement of current (nonmedical) wearables.

The human organism is a complex system, and the individual characteristics differ from person to person. As the whole genome sequencing is getting more affordable, it becomes the enabling factor for precision, preventative, and personalized medicine, through the genetic analysis of each person's health and diseases stratification [2]. Health care is becoming more *personalized*, individualizing treatment, based on the analyzed physiological states of each person by optimizing treatment approaches. The cancer research is the sector that has already moved from theory to practice, where optimal chemotherapy regimen is selected based on individual gene expression profile to significantly improve immune response [3]. Systems biology plays a vital role in disease mechanism elucidation, in disease risk estimation, and setting the right treatment strategy. Disease pathways are analyzed in bibliography; however, we are not in a mature level considering the low number of existing pathways related to the number of physiological functionalities, diseases, and the recent advances from noncoding DNA. Regarding the latter, a series of publications have revealed that the information may be hidden to noncoding DNA that represents nearly the 98% of our DNA. The scientific community is excited calling it an unexplored goldmine that can be used for diagnosis and prognosis and serve as potential therapeutic targets [4]. In another aspect, personalized information can be fused with other heterogeneous information from open databases for unraveling and minimizing the risks of side effects for specific treatments. In the future, genome information will be part of the patient's "conventional" Electronic Health Record, while creating omics profiles will become a routine. At first, a series of challenges will need to be addressed, with cost being the most prominent. The development, testing, and approval of personalized drugs could potentially lead to higher costs than drugs targeting at the wider population. Cloud technologies, harnessing the needed of computational power combined with AI, shall speed up research and eventually reduce overall costs.

Although AI has been widely used in health-care research, there are quite few AI applications that are approved from FDA (U.S. Food and Drug Administration) or other similar authorities. The potential of AI in diagnosis, management, and controlled treatment has been published in more than 12,500 articles up to 2020. Health care is becoming more *predictive* by utilizing AI technologies to forecast future events in near real time to improve the health of individuals. AI in health care is based on prior knowledge and historical data and is used in disease diagnosis and prognosis, while it could predict and control biological system activities [4]. However, there are still challenges into applying AI-based algorithms and operating medical devices in daily clinical practice. A recent publication [5] analyzed the existing 64 AI-based medical devices and algorithms that are FDA approved. In the article, it was summarized that the officially approved AI applications have benefited specific medical specialties such as radiology, oncology, ophthalmology, and medical decision-making. AI enabled the time reduction, improved medication adherence, optimized insulin dosages, and supported the analysis of magnetic resonance images [5]. It should be noted that the fact that AI algorithms have managed to outperform radiologists in spotting malignant tumors [6] is a glimpse from what we expect in the future from AI applications in health care.

There is a long way before AI tools tend towards replacing physicians in clinical diagnosis for a series of reasons such as ethics, accuracy, and complexity. Most importantly, people's lives cannot be dependent from AI, since potential wrong decisions may lead to unexpected losses. At present, AI in diagnostics (mainly) is used rather as a tool in the hands of physicians to improve decision-making. In addition, AI can play a drastic role in the hands of health-care policymakers, for predicting the risk of certain population groups in diseases [7] and hospital readmission [8], reducing overall health-care expenditure at macroscopic level. Currently "machines" and physicians are at a stage that struggle to utilize the entirety of data — that continue to grow exponentially in size and different channels (omic-based, nonmedical wearables, medical devices, EHRs, weather, genome, etc.) — for improved clinical decisions. It is evident that the future of *predictive* health care involves more holistic approaches from AI system-of-systems that will aggregate diverse information and predictions from other AI systems. However, the health-care "AI of AI-systems" needs thorough validation, regulation, approval, and continued research. Physicians need thorough training to be able to "digest" information from AI systems that fuse information from heterogenous sources to be able to adopt AI in clinical workflows. Only a limited number of EHR products offer AI functions in their features [9]. For widespread adoption to take place, AI systems must be integrated within EHR systems, in a standardized and regulated manner, taught adequately to clinicians [6]. A regulatory shortcoming is the fact that FDA approved AI solutions with "locked" algorithms that provide same result for each time the same input is entered [5]. In contrary, this prevents the adoption of AI adaptive algorithms such as Reinforcement Learning algorithms that change predictions and adapt over time based on new input. This regulatory gap shall be addressed toward more inclusive strategies for AI.

In predictive health care the major question that AI algorithms try to answer is on whether a catastrophic phenomenon (from clinical aspect) can be predicted prior to its occurrence. Considering an accurate and successful prediction, corrective actions from physicians can save the life of a (potential) patient. Similarly, if through personalized health care a genetic predisposition is unraveled, a set of prophylactic actions can be taken, ranging from improved daily habits, drug prescription to proactive surgeries (e.g., for the case of breast cancer). Furthermore, participatory health care can alter the well-being of people fitting their daily habits, exercise, and nutrition to minimize factors that could potentially affect their health care. The 3Ps (predictive, participatory, and personalized) feed the fourth pillar of *preventative* health care in the broader sense of proactive intervention. In preventative health care, the primary focus shall be on the causes rather than the symptoms of disease, which will enable the intervention to occur earlier in the disease propagation, preventing possibly the disease existence in first place [74].

There are two sectors that contribute indirectly to preventative health care: social networking and gamification. Social networking starts to introduce a new "type" of health care: the *peer-to-peer health care*. According to a survey, 23% of social network users have followed their friend's personal health suggestions. It is yet hard to measure and understand on whether this has overall benefit or harm to the health of those that followed the suggestions, since health suggestions should be made only from physicians to avoid potential danger. Dyson et al. conducted a review on self-harm acts from the social media use [10] indicating the penetration of new technologies in medicine. On the contrary, Giustini et al. [11] published a systematic review on the effective use of social media in public health and medicine. Some cases have showcased statistically important positive effects in adopting healthy behavior through social networks [12,13]. In addition, social media are also used for educating general public and vulnerable populations in risk factors and avoiding infectious agents [11]. The so-called serious games for health is another sector on the rise involved in transforming daily habits but also in some cases used in rehabilitation. Serious games have been used mainly for physical activity and nutrition motivation in achieving individual targets, stroke rehabilitation [14], behavioral change, and disease education [15]. However, game development demands large amounts of time and cost to achieve a professional polish similar to large software house entertainment games. Gamification in the broader sense involves the embedding of selected motivational features to an existing application. Such features could be, for example, point scores, badges, levels, comparison via leaderboards, competitions, challenges, or continuous progress goals [16]. Such narratives provide positive emotional changes in individuals through repetitive small "wins." These small wins may create addiction (depending on how the game is designed) that can lead toward the achievement of a bigger goal. It is worth mentioning that the overall health of a person is affected from various factors in the daily habits in homes and workplaces. There is a need for maintaining or even improving the health and the wellness of individuals far away from hospitals in homes and workplaces. Vast amount of information needs to be visualized and analyzed in (near) real time, indicating, in an intelligent manner, actions that may prevent the hospitalization. With the rise of computational power in smartphones, tablets, and wearables the analysis of information is transitioning from hospital to home.

Edge-cloud continuum

The industrial Internet of Things (IIoT) is associated with the extension and use of the Internet of Things (IoT) in industrial sectors and applications (e.g., smart manufacturing, smart cities, smart health care, etc.), based on three-tier architecture: edge tier, fusion tier, and cloud tier [17]. The edge tier focuses on gathering information via variety of data sources. The fusion tier provides connectivity between "things" and the cloud part of the IoT solution, whereas it enables data preprocessing and filtering before moving them to the cloud and transmits control commands going from cloud to "things". Furthermore, the cloud tier performs large-scale data computation to produce insights that generate value. In this direction, big data technologies and AI have received a great deal of attention from academics, consultants, and practitioners yielding numbers of revolutionary applications and services that are primarily driven by high-performance computation and storage facilities in the cloud (e.g., data lake containing data in their natural format, big data warehouse consisting of cleaned, structured, and matched data, machine learning (ML), and data analytics). One of the major challenges in many IoT systems is the high-energy consumption due to the transmission of data to the cloud [18]. On one hand, cloud computing is centrally deployed on a global scale and has become an essential, necessary, and crucial part of processing IoT data. However, it faces several difficulties such as transmission latency, bandwidth constraints, and high-energy consumption [19]. On the other hand, edge computing is seen as a "decentralized cloud" that drives the computing power near or at the source of data to improve energy efficiency and security, as well as decrease latency. The following sections analyze the concepts of edge and cloud computing, as well as the main principles of a potential ML-based fusion engine.

Cloud computing

Cloud computing is an innovative technology based on the principle of virtualization, where third parties (i.e., cloud service providers) rent out their computing resources (i.e., network devices, storage devices, software applications, platforms, or services) via the network [20]. In particular, cloud computing allows users to use clouds (i.e., servers that are accessed over the Internet, as well as the software applications and databases that run on those servers) on demand, regardless of location and time. The cloud may be (a) public when the computing resources are accessible via Internet and made available by a cloud service provider, who manages his infrastructure, (b) private when the cloud is reserved for the exclusive use of a single organization, or (c) hybrid that is a mix of public and private clouds. Furthermore, a cloud can be characterized as an internal or an external when it is a cloud internal to the consumer organization or not, respectively. With respect to a pricing policy, a cloud can be also characterized as a commercial or a noncommercial cloud. Several commercial and noncommercial cloud services are reported in Table 8.1. It is considered that, nowadays, the most utilized service in the cloud is the cloud storage. As a result, too many cloud storage providers exist (Google Drive, MediaFire, Sync.com, OneDrive, pCloud, Dropbox, MEGA, iCloud, Amazon Drive, etc.).

Although cloud computing seems to be a new technological trend in nowadays' information age, the concept behind it is very old. Cloud computing is directly related with Complexity Theory. As Melanie Mitchell

TABLE 8.1 Commercial and noncommercial clouds.

Commercial clouds	Noncommercial clouds
- **AWS**: The cloud computing solution provided by Amazon. - **GCP**: The cloud platform provided by Google. - **Azure**: The cloud computing service provided by Microsoft.	- **OpenNebula**: The cloud management platform to support the growing needs of developers and DevOps admins across the cloud. - **OpenStack**: A set of software components that provide common services for cloud infrastructure.

mentioned in her book entitled *Complexity: A Guided Tour*, complex systems have been studied by humanity for thousands of years [21]. She traced an interesting journey back to Aristotle in Ancient Greece and his thoughts of how thinking and scientific discovery are influenced by the dynamical systems theory until the 16th century and Galileo's studies on motion. More recently, in the 20th century, the need for high-performance computing has been strongly reported by scientists in the fields of Applied Mathematics, Physics, and Computer Science. In 1955, John McCarthy dreamed up the concept of computing time-sharing (i.e., the beginning of the cloud concept), whereas J.C.R. Licklider developed ARPANET (Advanced Research Projects Agency Network) in 1969 (i.e., the basis of the Internet). Moreover, in 1972, IBM developed its mainframe "Virtual Machine/370" system, while 3 years later (in 1975), John Holland and his collaborators at the University of Michigan published their studies in Genetic Algorithms [22]. In the 1970s, soft computing started to be especially mentioned by scientists who tried to solve complex problems. Soft computing is in contrast to hard computing which refers to exact and numerically rigorous calculations. It should be mentioned that soft computing is associated with heuristic algorithms which aim at solving nondeterministic polynomial (NP)-hard problems (i.e., problems from which a solution cannot be obtained in a polynomial time) [23].

In addition to this, through the 1990s (1990–98), the Internet grew exponentially, and in 1999 the cloud computing concept started to be promoted via the launch of Salesforce.com. Our century, the 21st century, is related mainly with the following milestones: the launch of (a) Amazon Web Services in 2002, (b) Amazon Elastic Compute Cloud and Google Apps cloud-based services in 2006, (c) Apple iCloud storage solution in 2011, as well as (d) Google Drive in 2012. It should be noticed that the use of more powerful mobile devices as well as better networks and faster access to the Internet via VHDSL (Very High Bitrate Digital Subscriber Line) technology will result in a considerable cloud boom in upcoming years.

Cloud computing (or serverless computing) can be distinguished into two classes: (a) computing models and (b) cloud services (Fig. 8.1) [24,25].

Computing models

Computing models can be assumed as the predecessors of cloud computing and can be categorized into three levels. The first categorization is based on computing models for software architecture design, the second categorization is associated with the concept of high-performance and soft computing, and the third categorization is related to utility computing.

Several software architecture designs have been introduced in the literature, from the single-tier architecture and peer-to-peer (P2P) models to

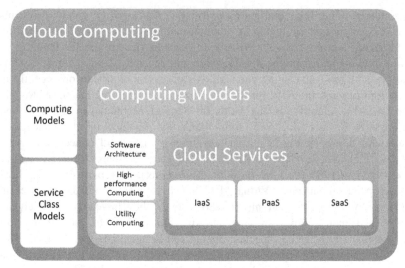

FIGURE 8.1 Cloud computing.

nowadays' client–server models (Fig. 8.2). In single-tier architecture the business logic and database access code are combined into a single program and deployed in a single physical machine, while a P2P model is consisting of two or more physical machines that are connected and share resources without using a separate server. Contrary to P2P models, the client–server models follow a master–slave configuration, in which a separate dedicated server serves as a master to which slaves (i.e., clients) are connected.

Depending on the requirements of scalability, elasticity, availability, and performance, several client–server models can be introduced by modifying the available number of tiers (e.g., two-tier, three-tier, N-tier client–server computing). Considering the potential complexity of business logic in enterprise applications (e.g., asynchronous messaging services, e-mail services, etc.) different layers can also be assumed constructing thus N-layer, N-tier client–server models.

As far as the high-performance computing concept is concerned, the notions of clusters and grids have been introduced in the literature. Particularly, terms such as "cluster computing" (Fig. 8.3) and "grid computing" (Fig. 8.4) are often used to facilitate the process of solving complex problems in terms of time complexity (i.e., how long a computation takes to execute) and space complexity (i.e., how much storage is required for a computation). A typical cluster consists of several computers (called nodes). One of the nodes is called head or master node that is responsible to access the cluster and distribute tasks in other nodes (called compute nodes). Depending on the need of clusters several types of clusters exist such as load-balancing clusters, high-availability clusters, fault-tolerant clusters, and high-performance clusters.

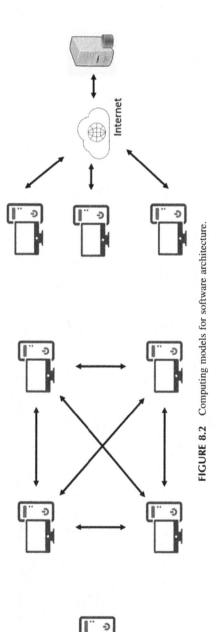

FIGURE 8.2 Computing models for software architecture.

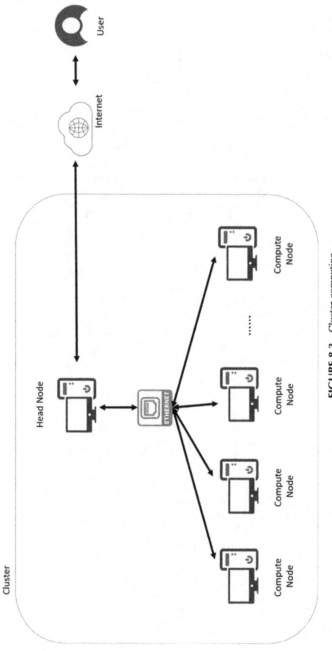

FIGURE 8.3 Cluster computing.

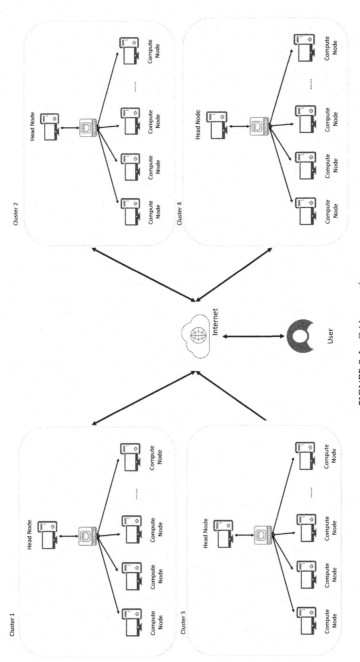

FIGURE 8.4 Grid computing.

When a single cluster is not enough to satisfy specific computational requirements, grid computing takes place. Grid computing assumes that different autonomous, heterogeneous, and geographically dispersed computational resources are combined, via the network, into a single virtual entity so as to solve complex problems (i.e., function as a supercomputer). It is of interest to note that high-performance computing is highly connected with several forms of parallel computing (i.e., bit-level, instruction-level, data, and task parallelism) where many computational executions are carried out simultaneously in compute nodes of one or more clusters (sharing thus memory and storage to achieve a specific computational goal).

Soft computing and heuristics are also popular in high-performance computing domain (Fig. 8.5). "Heuristic" comes from the Greek word "ευρίσκω" which means find or discover and refers to a family of approximation algorithms. Approximation algorithms are optimization algorithms that intend to find near-optimal solutions for NP-hard problems [26,27]. They can be distinguished into (a) classic heuristics, (b) meta-heuristics, (c) math-heuristics, (d) sim-heuristics, and (e) learn-heuristics. Examples of classic heuristics are the construction heuristic algorithms (i.e., myopic algorithms) and the local improvement heuristics. The constructive heuristics are suitable for generating valid initial solutions for constrained optimization problems. Actually, these algorithms start with an empty solution and repeatedly extend the current solution until a complete solution is obtained. In contrast, the local improvement heuristics start with a complete solution and then try to improve the current solution further via local moves. However, these algorithms are potential to stuck in a local optimum (minimum or maximum). This disadvantage is addressed by meta-heuristic algorithms which can (i) escape from local optimum by applying clever strategies and (ii) tackle large-size problem instances by delivering satisfactory solutions in a reasonable time.

Generally, meta-heuristics can be classified into two classes: (1) single-point search meta-heuristics and (2) population-based search meta-heuristics. Single-point search meta-heuristics (e.g., Tabu Search, Multi-start Local Search Method, Greedy Randomized Adaptive Search Procedure, Iterated Local Search, Guided Local Search, Variable Neighborhood Search, Adaptive Method, Smoothing Method, Simulated Annealing, etc.) manipulate and transform a single solution during the search, whereas population-based optimization algorithms consist of a population of candidate solutions to some problem, and via iterations, the population evolves to a better solution to the problem. The population-based search meta-heuristics are often reported as evolutionary optimization algorithms since they are biologically and nature-inspired optimization algorithms. Examples of evolutionary algorithms are the following: Genetic Algorithms, Evolution Strategies, Evolutionary Programming, Genetic Programming, Differential Evolution, Ant Colony

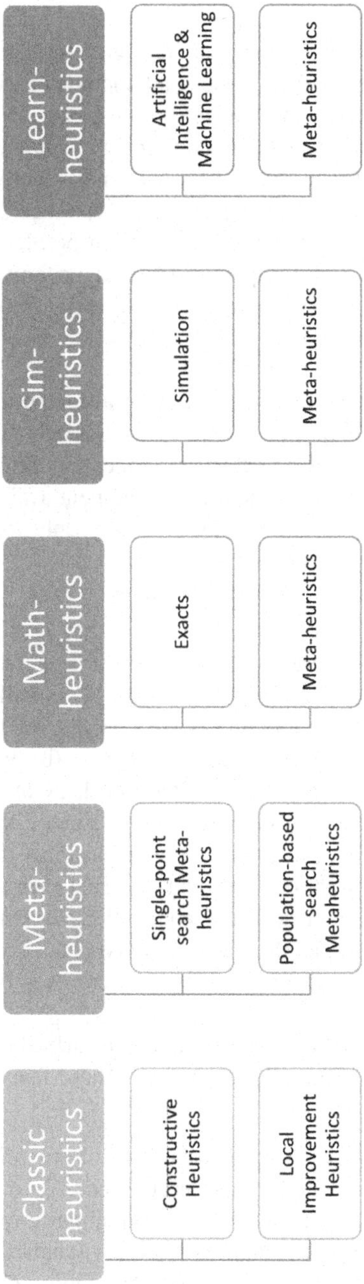

FIGURE 8.5 Heuristics.

Optimization, Particle Swarm Optimization, Artificial Immune Systems, Artificial Bee Colony Algorithm, Bacterial Foraging Optimization, Shuffled Frog Leaping Algorithm, etc.

The recent literature has shown an increased interest in so-called math-heuristic methods that combine exact and meta-heuristic approaches (e.g., decomposition approaches, column generation-based approaches, etc.) [28]. Additionally, sim-heuristics aim at combining Monte Carlo simulation or Markov Chain Monte Carlo Simulation with meta-heuristics so as to tackle stochastic variants of optimization problems [29]. Regarding learn-heuristics, the main goal is to solve large-scale data-driven problems by combining AI approaches, supervised and unsupervised ML algorithms with meta-heuristic optimization algorithms (e.g., Swarm Intelligence, Evolutionary Learning, Evolutionary Fuzzy, Evolutionary Neural Networks, Evolutionary Clustering, Evolutionary Decision Trees, etc.) [30,31].

Finally, utility computing can be seen as a computing model that helps providers to sell their computing resources. Particularly, utility computing is a service provisioning model in which a cloud service provider makes his computing resources and infrastructure management available to the customer as needed. As a result, customers pay providers only for the computing resources and computing power they use.

Cloud services

It should be noted that several types of cloud services exist based on what an individual or a company demands (e.g., storage, e-mail access, music and video streaming, gaming, remote access, etc.). In general, there are three major categories: (a) Infrastructure-as-a-Service (IaaS), (b) Platform-as-a-Service (PaaS), and (c) Software-as-a-Service (SaaS). IaaS allows distributed virtualized computing resources such as servers, storage devices, network devices, and virtual machines to be shared, whereas PaaS is the combination of both hardware and software development, offering thus to developers a platform to code and test their software applications on various hardware and operating systems. As far as the SaaS is concerned, cloud service providers offer completely developed applications to users over the Internet via subscription policies.

The aforementioned cloud services are characterized by some features that indicate the benefits of using cloud services. These features are summarized below:

- Fast elasticity and scalability: the capability of providing scalable services rapidly (elasticity) and hence meet users' needs by reducing or extending the supply of resources, as well as handling growing loads without compromising the existing quality of service standards (scalability).
- Self-service on-demand provisioning: the ability of a user to request through an online interface of his/her own any available computing

resource from the catalogue of the cloud service provider, without his/her intervention.

- Resource pooling: the ability of cloud service provider to handle the needs of multiple tenants at the same time, and their variances in demands.
- Pay-as-Per-Use pricing: the ability of a user to pay only what he/she really consumes (utility computing).

While cloud computing offers multiple benefits to users (e.g., flexibility, automatic updates, reduced costs, access from anywhere, competitiveness, document control, etc.), security risks due to a potential unavailability of infrastructure or privacy attacks (hacking) that may result in theft or loss of sensitive data still remain [32]. Therefore, trustworthy cloud computing has received an increased attention from researchers who are interested in investigating how the goals of trustworthy computing (i.e., security, privacy, reliability, and integrity) are represented in computer science research [33].

Data and information fusion

How to capture reliable, valuable, and accurate information in massive data is one of the most significant research topics nowadays [34]. Big data are accompanied by difficulties and challenges in a data-driven service provision due to its "5Vs" (i.e., volume, velocity, variety, veracity, and value) [35]. Traditional data fusion techniques include probabilistic fusion (e.g., Kalman Filter and its extensions, Sequential Monte Carlo, Markov Chain Monte Carlo, etc.), evidential belief reasoning fusion (e.g., Dempster–Shafer theory and complexity reduction approaches), fuzzy set theory–based fusion, and rough set theory–based fusion [36]. Data fusion challenges such as data imperfection, data inconsistency, data confliction, data alignment/correlation, data type heterogeneity, fusion location, and dynamic fusion indicate that raw data cannot be extracted as information with high value by once. Several data fusion architectures have been proposed in the literature to deal with data fusion challenges like (a) the Joint Directors of Laboratories (JDL) architecture, (b) the Luo and Kay's architecture, and (c) the Dasarathy's architecture [37].

Furthermore, it should be mentioned that information fusion is becoming a major need in data mining and knowledge discovery in databases [38]. Hence, identifying the most significant features that mostly contributed to decision-making through decision support systems is fundamental to tackle management problems in a feature selection context. Feature selection can be seen as a dimensionality reduction technique to mitigate the noise induced by the large number of parameters of interest [39,40]. Several linear and nonlinear dimensionality reduction approaches have been proposed in the literature depending on whether variable transformations take place or not.

On one hand, the focus is placed on (a) exploring high-dimensional data with the goal of discovering structure or patterns that lend to the formation of statistical hypotheses, (b) visualizing the data using scatterplots when dimensionality is reduced to 2D or 3D, and (c) analyzing the data using statistical methods, such as clustering, smoothing, probability density estimation, or classification. Actually, the Principal Component Analysis, the Singular Value Decomposition, the Nonnegative Matrix Factorization, the Fisher Discriminate Analysis, the Factor Analysis, and Random Projections are used to transform the data to a new set of variables that are linear combination of the original variables (linear models). Additionally, various models for a nonlinear mapping between the high-dimensional space and the low-dimensional space such as the Isometric Feature Mapping and Auto-encoders (deep learning approach) can be considered (nonlinear models). On the other hand, filter-based, wrapper-based, and hybrid feature selection techniques are available to discover knowledge from raw data by selecting the most appropriate subset of features that maximizes a chosen performance measure. Classic filter—based approaches such as the Pearson Correlation (linear correlations) or the Spearman Correlation (nonlinear correlations), ranking filter-based approaches like the Mutual Information or Chi-square statistic, as well as wrapper-based approaches such as the Forward Feature Selection, the Backward Feature Elimination, or Meta-heuristic Algorithms are some examples of feature selection approaches [41,42].

An exploratory data analysis approach is very often used to reveal patterns and features that help the data analyst to understand, analyze, and model the data [43]. Graphical presentation formats and computational intelligence techniques may be combined to convey meaningful insights into the problem. Two types of data tours are supported in the exploratory data analysis context. The first data tour assumes that data should first be explored without assumptions about probabilistic models, error distributions, and number of groups or relationships between the features for the purpose of discovering what they can tell a data analyst about the problem he/she is investigating. Missing values or outliers can be also identified. The second data tour is based on dynamic graphical methods that allow a data analyst to interact with the display to uncover structure. Data transformation techniques (e.g., normalization, standardization, binarization) are available to produce effective visualization or more informative analysis. In addition to this, data partitioning (training, validation, and test sets) is recommended to evaluate a chosen performance measure, decide a reasonable model for the data, find a smoothing parameter in density estimation, estimate the bias and error in parameter estimations, and so on. The distribution of the original dataset indicates whether the dataset is imbalanced or not. In case of imbalanced dataset, undersample or oversample techniques might be used with respect to the problem domain. Furthermore, feature selection techniques can be examined in a computational intelligence framework where advanced ML techniques and evolutionary optimization algorithms such as the Genetic Algorithm are combined to select the most appropriate set of features. This approach may effectively influence the ML life cycle by suggesting optimal ML strategies in a model selection context. It should be mentioned that before applying a feature selection technique, automated feature engineering may take place for building and deploying accurate ML models by handling necessary but tedious tasks so data scientists can focus more on other important steps of the ML workflow. Depending on whether unstructured or structured transactional and relational datasets occur, deep learning approach [44] or deep feature synthesis approach [45] can be applied, respectively, for automating feature engineering/extraction tasks.

Edge computing

According to Cao et al. [46], edge computing reflects enabling technologies that allow computing at the edge of the network. As a result, edge computing can be seen as a new computing paradigm for performing calculations closer to the user and closer to the source of data [47]. It should be noticed that edge computing is not going to replace cloud computing [48]. It is required to share the pressure of the cloud by storing and processing data on edge devices, reducing thus the load of network bandwidth and the energy consumption of these edge devices [49].

Edge computing architecture can be seen as a federated network structure that extends cloud services to the edge. Some current paradigms within the domain of edge computing are the following: (a) fog computing, (b) cloudlet, (c) ad hoc clouds, and (d) mobile edge computing [50,51]. Fog computing extends cloud services closer to IoT devices by acting as an intermediate layer between the traditional cloud and edge devices. Moreover, cloudlet is a decentralized architecture to augment the computation capabilities of mobile devices by reducing communication delays, saving energy, and minimizing the application response time. As far as the ad hoc clouds are concerned, they aim at delivering of computing services over the network with the benefits of improved manageability in terms of resource management, routing, security, and privacy. Mobile edge computing aims at solving the delay-sensitive and context-aware applications for the proximity of mobile devices subject to constraints related to characteristic of wireless networks.

Several tools and platforms are available for (1) sharing data at the edge through virtualization mechanisms for computation (e.g., Hyper-V, Xen, Fusion, VirtualBox, KVM, etc.) and networking (e.g., OpenSwitch, Open-Flow, etc.), (2) simplifying resource management on heterogeneous and resource-constrained edge devices (e.g., Kubernetes, Docker, etc.), (3) executing ML and computer vision models (e.g., TensorFlow, OpenCV,

Torch, OpenNN, Caffe, Theano, Deeplearning4j, etc.), as well as (4) simplifying devices programming (e.g., AWS Greengrass, Apache Edgent, Amazon IoT, Microsoft Azure IoT, IBM Watson IoT, Google Cloud IoT, etc.)

Edge computing is an emerging technology which has not been extensively researched in the literature [52]. As a result, the area could be further investigated in order to (i) design and provide suitable mechanisms to guarantee the integrity, usability, interoperability, adaptability, and availability of selected services, (ii) proceed to the fusion of edge and cloud computing by learning from the previous experiences in the fusion of big data gathered due to the edge-cloud collaboration, and (iii) define security strategy guidelines in the edge environment.

Recent artificial intelligence trends for the edge-cloud continuum

It is widely recognized that AI is expected to play a crucial role in the success of edge-cloud continuum. The objective of this section is to present the most promising areas of AI that can further improve cloud computing, edge computing, and multisensor fusion.

Trustworthy and explainable artificial intelligence

In recent years, AI systems based on ML excel in many fields. AI has become a key enabling technology for the sciences and industry. This is due to the success of AI technologies and automatic decision-making that allow the development of increasingly robust and autonomous AI applications. AI systems become particularly essential for their users, for the people who are affected by AI decisions, and for the researchers and developers who create the AI solutions. However, many of these AI systems are not able to explain their autonomous decisions and actions to human users [53]. Most of the AI applications are based on the analysis of historical data which encodes historical biases, whereas ML models are developed based on the experience recorded in these datasets to make decisions. The role of explanations in data quality in the context of data-driven ML models is very crucial especially when it is needed to address the issues of transparency as linked to AI from sociolegal and computer scientific perspectives [54,55].

Recently, some ideas and insights have been provided by the research community as far as the explainable AI is concerned, focusing not only on the AI models, but also on the data, that support the model development [56]. The main focus is placed on the definition, identification, and explanation of errors in data and the appropriate repair actions. Moreover, a set of processes have been introduced to mitigate and manage general classes of bias in AI algorithms [57]. These biases are generally divided into those associated with the mapping of the business intent into the AI algorithm, those that arise due to the distribution of training samples, and those related to individual input samples. Therefore, understanding the reason why a decision has been made by a machine is crucial to grant trust to a human decision-maker [58]. The goal is to explain the algorithmic decisions of AI solutions with nontechnical terms in order to make these decisions trusted and easily understandable by humans. Especially the latter is very important in the context of legal decision support systems which should guaranty the causality and explainability of the AI since humans and the law often desire or demand answers to the questions "Why?" and "How do you know?" [59]. Explainable AI should also deal with the question of "How to evaluate the quality of explanations?" given by an explainable AI system. The System Causability Scale, as well as the Interpretable Confidence Measures, have been introduced in the literature to measure the quality of explanations [60,61].

A recurrent concern about AI that is based on ML algorithms is that they operate as "black boxes" without explaining their computational results in a language close to a human expert [62]. Especially, in the context of legal decision support systems, this difficulty to identify how and why the AI algorithms reach particular decisions, recommendations, or predictions reflects the nowadays judicial demand for explainable AI and rational judicial reasoning. Considering this issue, causal inference and fuzzy logic–based systems are strongly recommended in the literature since they can hold great promise for enabling current AI approaches to become trusted and explainable [63,64]. Moreover, neuro-fuzzy and evolutionary fuzzy systems can be used as computational methods that are able to combine computational intelligence and human-understandable linguistic inference model in practical applications. Thanks to this hybridization (neural networks and fuzzy logic; evolutionary optimization and fuzzy logic), superb abilities can be provided to fuzzy modeling in many different scenarios combining expert knowledge and data science practices under seven aspects: (a) the ability to perceive rich and complex information, (b) the ability to discover causal information from observational data, (c) the ability to learn in a particular context, (d) the ability to abstract, (e) the ability to create new meanings/concepts, (f) the ability to reason for decision-making, and (g) the ability to explain the prediction/ decision outcome.

Artificial intelligence in adversarial environments

AI and especially ML technology have become a mainstream in a large amount of business domains. While these ML models are being increasingly trusted for decision-making purposes, the safety of systems using such models has become an increasing concern. Specifically, in the context of IIoT in which a heterogeneous network is assumed (i.e., the large number of connected devices and the heterogeneity of the exchanged data), monitoring this kind of networks is quite challenging and hard, whereas potential attacks against them (e.g., denial of service attack, false data injection attack, time delay attack, replay attack, spoofing attack, 0-day attack, etc.) are increasing [65]. On one hand, AI technologies such as Deep Belief Networks, Recurrent Neural Networks, Convolutional Neural Networks, and Generative Adversarial Networks (GANs) have been introduced into cyber security to construct smart models for implementing malware classification and intrusion detection approaches [66,67]. On the other hand, AI models face various cyber threats, which will disturb their sample, learning, and performance [68]. Adversarial machine learning (AML) sits at the intersection of artificial intelligence and cyber security, and it is typically outlined as the study of effective intelligent techniques against associate adversarial opponent. Although AML has been extensively studied in the area of image recognition [69,70], in other domains such as intrusion and malware detection, the exploration of such methods is still growing [71]. Adversarial attack strategies such as Box Constrained Limited-memory Broyden—Fletcher—Goldfarb—Shannon, Fast Gradient Sign Method, Jacobian-based Saliency, Deepfool, Carlini, and GANs have been applied to intrusion and malware attack scenarios. In addition to this, many adversarial defenses have been introduced to counteract many of the attack strategies like adversarial training, gradient masking, defensive distillation, feature squeezing, universal perturbation defense method, MagNet, ML classification over encrypted data, and safe distributed ML systems.

Nowadays, AI and ML models, as well as the information systems that use them, are not implemented in a way to ensure safe operation when deployed. To maximize the security level of AI and ML-oriented systems, it is urgent to design and develop novel and special AI adversarial algorithms and models, respectively. Hence, a novel cyber security protection that can cope with diversified and sustainable security threats toward AI and ML should be investigated. In this direction, a potential adversarial model can be designed considering the following four dimensions: (a) the adversarial goal using both the expected impacts and the attack specificity of security threats (i.e., targeted attack or indiscriminate attack), (b) the adversarial knowledge examining whether or not an attacker knows training data, features, ML algorithms and their parameters, objective functions, and feedback information (i.e., black-box, gray-box, or white-box attacks), (c) the adversarial capability referring to the attacker's capability of controlling training and test datasets (i.e., causative or exploratory attacks), as well as (d) the attacking strategy depicting specific behaviors of manipulating training and test datasets to achieve an attacker his/her goals [72]. In the context of the CIA triad of confidentiality (i.e., the protection of personal information), integrity (i.e., the process of maintaining and assuring the accuracy and completeness of data over their entire life cycle), and availability (i.e., the information must be available when it is needed), three major types of AI and ML attacks are generally recognized: privacy attacks (confidentiality based), poisoning attacks (integrity and availability based), and evasion attacks (integrity based). Adversarial models based on computational intelligence approaches used in classification, clustering, or hybrid fashion under intrusion and malware scenarios can be examined to model and evaluate security threats against the training phase (i.e., backdoor poisoning attacks) and the inferring phase (i.e., evasion attacks) of the ML life cycle. In addition to this, cryptographic technology based on differential privacy or homomorphic encryption can be examined to protect data privacy (i.e., privacy attacks). Finally, game-theory approach in which the AML model can be seen as a noncooperative game (defender: ML algorithm) using machine and evolutionary learning can be examined to evaluate the ML threats.

Artificial intelligence for blockchain

The concept of decentralized artificial intelligence (D-AI) that is based on the combination of AI and blockchain has been recently emerging [73]. D-AI aims at facilitating data processing, analytics and decision-making on trusted, digitally signed, and secure shared information that is stored, retrieved and transfered in the context of a distributed and decentralized blockchain architecture [74,75]. Several blockchain applications have already been implemented in several fields such as financial, data management, business and industry, privacy and security, education, governance, integrity verification, and health [76]. Furthermore, examples of decentralized applications are related to autonomic computing, planning, optimization, knowledge discovery, knowledge management, learning, search, reasoning, and perception. In this direction, several technologies based on D-AI have been presented such as snipAIR, Chain intel, MedRec, KEEL, and Nebula genomics.

Although various applications and technologies have been provided, the concept of D-AI is a complex and challenging task. Several AI approaches can be proposed for optimizing the already powerful technology of blockchain. AI can be involved in improving blockchain system's resilience by creating robust ciphers and improving the system's attack—defense process (e.g., increase the efficiency of intrusion detection system by applying swarm intelligence). In addition to this, AI can be used to figure out the scalability from decentralized data sources and provide optimal answers for the blockchain system (e.g., increase the efficiency of blockchain system that is limited by scalability issues such as response time, initiation time, cost brought forth per confirmed transactions, etc.). AI can also enhance the efficiency by predicting the utility of blockchain node for fulfilling mining (e.g., execute fast and active learning to speed up resource calculation and enhance overall system performance). Furthermore, AI methodologies can be applied to enhance mass scale system such as power system planning and operation (e.g., provide optimized energy consumption from the large-scale complex system perception). Finally, with the increase in the amount of private data in blockchain system, the data encryption becomes an issue for assuring user's intimacy. Different AI approaches can be used for finding the secret keys for tackling this problem.

ML-based fusion engine

Combining fusion concepts with ML aims at addressing data fusion challenges effectively with respect to criteria such as efficiency, quality, stability, robustness, and extensibility. A potential ML-based fusion engine can be composed of three fusion modules and adversarial learning (AL) models (Fig. 8.6).

In particular, there are three fusion-based modules which provide a toolbox of AI algorithms for exploring data in depth (see Explorable Data Analysis Fusion, EDAF), extracting features and discovering knowledge (see Explainable Information Fusion, EIF), as well as supporting decision-making processes (see Explainable Decision Fusion, EDF). The aforementioned fusions, namely the data fusion, the information fusion, and the decision fusion, are data-driven AI algorithms based on ML practices which are evaluated in adversarial environments to secure the ML-based fusion engine from potential ML attacks (i.e., backdoor poisoning attacks, evasion attacks, and privacy attacks). The latter depicts the AL module which is responsible for the definition and the application of appropriate ML risk strategies. Analytically, the AL module will support all the aforementioned modules by managing potential diversified and sustainable security threats toward AI and ML tasks.

Explorable and explainable AI for data quality and causal reasoning

Inadequate data acquisition processes can lead to incomplete, inconsistent, biased, or low predictive value data. To avoid these errors, mechanisms can be established to guarantee the quality of the data, which is an essential component to reduce the uncertainty of the data-based modeling. Hence, a set of metrics to quantify the quality of the data can be explored. Advanced data tours and data processing techniques can be designed for exploring data in depth, in terms of quality, and discovering hidden patterns. These functionalities are part of the EDAF module (see Fig. 8.6).

To begin with, SQL (e.g., MySQL) and NoSQL (e.g., MongoDB) databases can be available for storing and managing structured transactional and relational datasets, as well as unstructured data (e.g., documents, images, video, etc.), respectively. The option of a distributed file system like the Apache Hadoop File System can be examined for holding very large amount of data and providing easier access. The latter can be achieved by (a) storing files across multiple machines in redundant fashion so as to reduce the system from possible data losses in case of failure and (b) making applications available to

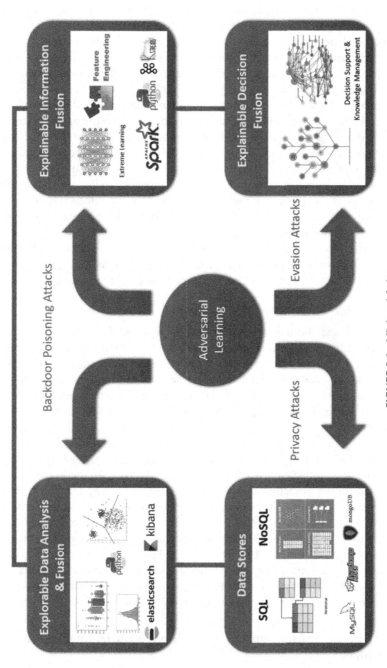

FIGURE 8.6 ML–based fusion.

parallel processing. As far as the EDAF module is concerned, cutting-edge technologies such as Elasticsearch can be used for managing raw data flows from a variety of sources, including logs, system metrics, and web applications. In addition to this, a data visualization and management tool for Elasticsearch such as Kibana can be used for providing real-time histograms, line graphs, pie charts, and maps. The EDAF supports an exploratory data analysis process based on ML algorithms for classification, clustering, regression, and anomaly detection. The identification of missing values and outliers as well as the exploration of high-dimensional data for discovering structure or patterns that lead to the formation of statistical hypotheses are assumed to be the two basic features of the EDAF for exploring potential sources of bias, calculating quality metrics, and providing explanations in data quality. Advanced statistical and ML approaches such as ranking approaches and/or wrapper-based feature selection techniques can be used to reduce the dimensionality of a potential problem, find the most significant variables of the dataset, identify correlations, and in general explore data in depth. In addition to this, ML approaches such as clustering techniques, neural networks, or anomaly detection algorithms can be applied to investigate similarities, predict risk levels, or design an intelligent and accurate fault detection process.

Explainable information and decision fusions

As it has already been mentioned, the EIF module (see Fig. 8.6) is associated with the information fusion which is responsible for transforming data into information. Actually, this module is very important since cleaned, unbiased, and balanced datasets for the data-driven decision-making process should be provided. In the context of the EIF module, cutting-edge technologies such as the Apache Spark and the Apache Kafka can be examined for (a) designing a fast feature engineering and feature selection process based on parallel computing and (b) establishing a steam processing approach for storing, reading, and analyzing streaming data, respectively. AI methods based on evolutionary optimization, supervised and unsupervised learning algorithms may support the extraction of new features from the existing ones, as well as the feature importance evaluation, finding thus the most significant decision variables for prediction or causal analysis purposes. Furthermore, web services can be used for exploiting the EIF's outputs for decision support. As a result, the EDF module (see Fig. 8.4) depicts an explainable human in the loop AI-based decision support process allowing experts and decision-makers to perceive rich and complex information, learn in a particular context, reason for decision-making, and easily explain the decision outcome.

Tiny machine learning and energy-efficient artificial intelligence algorithms

Although the computation capability of edge devices has increased extremely during the past decade, it is still challenging to perform sophisticated AI algorithms in these resource-constrained edge devices for energy-efficient processing at the edge, as well as optimal distribution of resources and tasks along the edge-cloud continuum [77,78]. This open problem reflects a recent field of study on ML and embedded systems known as tiny machine learning (TinyML) [79]. As a result of increased computational load and energy consumption, novel approaches based on cutting-edge computational intelligence methods can be proposed to simultaneously achieve smartness, security, and energy efficiency. In particular, ensemble tree-based models such as Extreme Gradient Boosting, Random Forests, or Adaptive Boosting can be used to investigate architectural characteristics of embedded systems for filtering high-volume sensor data before further processing. In addition to this, ML inference approach based on evolutionary learning and deep neural networks on the IoT edge node can be used to reduce the workloads and energy consumption of resource-constrained edges by exploiting the trade-off between the data accuracy requirement of applications, sensing data prediction quality, and the energy efficiency of edges.

AI-based solutions for respiratory analysis

The main function of a respiratory system is to supply oxygen to the blood and expel waste gases (of which carbon dioxide is the main constituent) from the body through breathing [80]. Breathing monitoring and pattern recognition with wearable sensors [81] are very crucial for detecting of sleep apnea and identifying abnormal behaviors of the respiratory system [82]. In this direction, P4 (predictive, preventive, personalized, and participatory) health care [83] based on Bayesian inference and time series ML approaches seems to be very promising by exploring personalized health data (gathered from wearable sensors), analyzing them as time series, and constructing accurate and high predictive models.

To begin with, digital signal processing techniques have been widely used for analyzing breath and lung sounds which are assumed as important indicators of respiratory health and disease [84]. Artificial neural networks and support vector machines are often used to detect and classify respiratory patterns into normal, obstructive, restrictive, and mixed patterns using spirometry data [85]. Early and correct diagnosis of respiratory system abnormalities is vital to patients with chronic heart failure who often develop breathing patterns [86]. Hence, identifying periodic breathing patterns and nonperiodic breathing patterns is a crucial step for discriminating patients.

Supervised ML algorithms such as the K-nearest neighbor, decision trees, random forests, and support vector machines were also used as decision-aid tools to categorize airway obstruction severity in chronic obstructive pulmonary disease (i.e., a respiratory disease characterized by a chronic airflow limitation), enhancing thus medical services for patients [87,88]. It is worth mentioning that the need for optimization in terms of time and cost is often highlighted in the literature when computer-aided systems are used for disease classification purposes. Recently, respiratory function tests such as spirometry, impulse oscillometry, and body plethysmography were proposed to be combined with AI algorithms like artificial neural networks and fuzzy logic to optimize time resources and reduce the costs of medical device usage needed for testing of a patient and costs of medical professional attending the measurement [89].

In addition to this, it should not pass unnoticed that several limitations exist when data from wearable sensors are used for respiratory analysis purposes. The process of extracting valuable information from wearable sensors is still a challenging task. Many research areas have focused on ML-based data fusion techniques for better understanding of data and facilitating experts to evaluate the state of human actions and diagnose the illnesses. Swarm Intelligence approaches have been introduced in the literature to address challenges such as the runtime complexity, analyze the variations in the periodic signals, and reach an acceptable accuracy level [90].

Although it is widely recognized that AI has many potential applications in respiratory medicine, particularly with assisting diagnosis from examination, bedside tests, and imaging [91], it remains to be validated in clinical settings, exploring simultaneously the ethical implications of their application in primary care practice [92].

Conclusions

This chapter explored the concept of the edge-cloud continuum in the wearable sensing for respiratory analysis. Through a systematic review, the main principles of the P4 health care in the big data era have been analyzed, whereas AI technological components and ML tools have been described in the context of the cloud and edge computing, respectively. Additionally, recent trends in the fields of AI and respiratory medicine have been highlighted to indicate the potential of AI in respiratory analysis. As it can be observed, AI has many potential applications in respiratory medicine, particularly with assisting diagnosis from examination, bedside tests, and imaging. Big data challenges, information fusion from wearable sensors, and high predictive modeling are still open issues in the literature for classifying patients with respect to their breathing patterns.

References

[1]   L. Latts, The Age of Big Data and the Power of Watson, IBM Watson Health, 2016.
[2]   L. Hood, M. Flores, A Personal View on Systems Medicine and the Emergence of Proactive P4 Medicine: Predictive, Preventive, Personalized and Participatory, New biotechnology, 2012.

[3] K. Yu, Q.A. Sang, P.Y. Lung, W. Tan, T. Lively, C. Sheffield, M.J. Bou-Dargham, J.S. Liu, J. Zhang, Personalized chemotherapy selection for breast cancer using gene expression profiles, Sci. Rep. (2017), https://doi.org/10.1038/srep43294. PMID: 28256629; PMCID: PMC5335706.

[4] P. Qi, X. Du, The long non-coding RNAs, a new cancer diagnostic and therapeutic gold mine, Mod. Pathol. 26 (2013) 155—165, https://doi.org/10.1038/modpathol.2012.160.

[5] S. Benjamens, P. Dhunnoo, B. Meskó, The state of artificial intelligence-based FDA-approved medical devices and algorithms: an online database, npj Digit. Med. 3 (2020) 118, https://doi.org/10.1038/s41746-020-00324-0.

[6] T. Davenport, R. Kalakota, The potential for artificial intelligence in healthcare, Fut. Healthcare J. 6 (2) (2019) 94—98, https://doi.org/10.7861/futurehosp.6-2-94.

[7] A. Rajkomar, E. Oren, K. Chen, et al., Scalable and accurate deep learning with electronic health records, Dig. Med. 1 (2018) 18. www.nature.com/articles/s41746-018-0029-1.

[8] A. Nait Aicha, G. Englebienne, K.S. van Schooten, M. Pijnappels, B. Kröse, Deep learning to predict falls in older adults based on daily-life trunk accelerometry, Sensors 18 (5) (2018).

[9] L.L. Low, K.H. Lee, M.E. Hock Ong, S. Wang, S.Y. Tan, J. Thumboo, N. Liu, Predicting 30-day readmissions: performance of the LACE index compared with a regression model among general medicine patients in Singapore, BioMed Res. Int. (2015) 169870, https://doi.org/10.1155/2015/169870.

[10] M.P. Dyson, L. Hartling, J. Shulhan, A. Chisholm, A. Milne, et al., A systematic review of social media use to discuss and view deliberate self-harm acts, PLoS One 11 (5) (2016) e0155813, https://doi.org/10.1371/journal.pone.0155813.

[11] D. Giustini, S.M. Ali, M. Fraser, M.N. Kamel Boulos, Effective uses of social media in public health and medicine: a systematic review of systematic reviews, Online J. Public Health Inform. 10 (2) (2018) e215, https://doi.org/10.5210/ojphi.v10i2.8270.

[12] L. Laranjo, A. Arguel, A.L. Neves, A.M. Gallagher, R. Kaplan, et al., The influence of social networking sites on health behavior change: a systematic review and meta-analysis, J. Am. Med. Inf. Assoc. 22 (1) (2015) 243—256, https://doi.org/10.1136/amiajnl-2014-002841.

[13] C.A. Maher, L.K. Lewis, K. Ferrar, S. Marshall, I. De Bourdeaudhuij, C. Vandelanotte, Are health behavior change interventions that use online social networks effective? A systematic review, J. Med. Internet Res. 16 (2) (2014) e40.

[14] A. Chytas, D. Fotopoulos, V. Kilintzis, T. Loizidis, I. Chouvarda, Upper limp movement analysis of patients with neuromuscular disorders using data from a novel rehabilitation gaming platform, IFMBE Proc. 76 (2020) 661—668.

[15] D. Johnson, S. Deterding, K.A. Kuhn, A. Staneva, S. Stoyanov, L. Hides, Gamification for health and wellbeing: a systematic review of the literature, Inter. Interven. 6 (2016) 89—106, https://doi.org/10.1016/j.invent.2016.10.002.

[16] K. Seaborn, D.I. Fels, Gamification in theory and action: a survey, Int. J. Hum. Comput. Stud. 74 (2015) 14—31.

[17] S. Munirathinam, Industry 4.0: industrial internet of things (IIOT), Adv. Comput. 117 (1) (2020) 129—164.

[18] J. Azar, A. Makhoul, M. Barhamgi, R. Couturier, An energy efficient IoT data compression approach for edge machine learning, Future Generat. Comput. Syst. 96 (2019) 168—175.

[19] M. Shafique, T. Theoharides, C.-S. Bouganis, M. Abdullah Hanif, F. Khalid, R. Hafiz, S. Rehman, An overview of next-generation architectures for machine learning: roadmap, opportunities and challenges in the IoT era, in: Paper Presented at the Proceedings of the 2018 Design, Automation & Test in Europe Conference & Exhibition (DATE), Dresden, 2018, pp. 827–832.

[20] M. Zbakh, M. Essaaidi, P. Manneback, C. Rong, Cloud Computing and Big Data: Technologies, Applications and Securities, Springer, Cham, Switzerland, 2019.

[21] M. Mitchell, Complexity: A Guided Tour, Oxford University Press, Oxford, United Kingdom, 2009.

[22] J. Holland, Adaptation in Natural and Artificial Systems, United States of America: University of Michigan Press, Ann Arbor, Michigan, 1975.

[23] C. Papadimitriou, K. Steiglitz, Combinatorial Optimization: Algorithms and Complexity, Dover Publications, Inc, Mineola, New York, United States of America, 1982.

[24] C. Surianarayanan, P.-R. Chelliah, Essentials of Cloud Computing: A Holistic Perspective, Springer, Cham, Switzerland, 2019.

[25] J. Anel, D. Montes, J. Iglesias, Cloud and Serverless Computing for Scientists: A Primer, Springer, Cham, Switzerland, 2020.

[26] E.-G. Talbi, Metaheuristics: From Design to Implementation, John Wiley & Sons, Inc, New Jersey, United States of America, 2009.

[27] X.-S. Yang, Nature-inspired Optimization Algorithms, Elsevier, London, United Kingdom, 2014.

[28] V. Maniezzo, T. Stutzle, S. Voβ, Matheuristics: Hybridizing Metaheuristics and Mathematical Programming, Springer, New York, United States of America, 2009.

[29] A. Juan, J. Faulin, S. Grasman, M. Rabe, G. Figueira, A review of simheuristics: extending metaheuristics to deal with stochastic combinatorial optimization problems, Oper. Res. Perspect. 2 (2015) 62–72.

[30] L. Calvet, J. de Armas, D. Masip, A. Juan, Learnheuristics: hybridizing metaheuristics with machine learning for optimization with dynamic inputs, Open Math. 15 (1) (2017) 261–280.

[31] Z.-H. Zhou, Y. Yu, C. Qian, Evolutionary Learning: Advances in Theories and Algorithms, Springer, Singapore, 2019.

[32] F. Al-Turjman, Trends in Cloud-Based IoT, Springer, Cham, Switzerland, 2020.

[33] T. Lynn, J. Mooney, L. van der Werff, G. Fox, Data Privacy and Trust in Cloud Computing: Building Trust in the Cloud through Assurance and Accountability, Palgrave Macmillan, Cham, Switzerland, 2021.

[34] P. Rashinkar, V.S. Krushnasamy, An overview of data fusion techniques, in: Conference Paper Presented in Proceedings of the 2017 International Conference on Innovative Mechanisms for Industry Applications (ICIMIA), Bangalore, 2017, pp. 694–697.

[35] T. Meng, X. Jing, Z. Yan, W. Pedrycz, A survey on machine learning for data fusion, Inf. Fusion 57 (2020) 115–129.

[36] B. Khaleghi, A. Khamis, F. Karray, S. Razavi, Multisensor data fusion: a review of the state-of-the-art, Inf. Fusion 14 (2013) 28–44.

[37] H.B. Mitchell, Data Fusion: Concepts and Ideas, Springer, New York, United States of America, 2012.

[38] V. Torra, Information Fusion in Data Mining, Springer, New York, United States of America, 2003.

[39] T. Hastie, R. Tibshirani, J. Friedman, The Elements of Statistical Learning: Data Mining, Inference, and Prediction, Springer, New York, United States of America, 2008.

[40] H. Liu, H. Motoda, Feature Selection for Knowledge Discovery and Data Mining, Springer Science and Business Media, New York, United States of America, 2012.

[41] C. Dhaenens, L. Jourdan, Metaheuristics for Big Data, ISTE Ltd and John Willey & Sons, Inc, London, United Kingdom, 2016.

[42] S. Wang, J. Tang, H. Liu, Feature selection, in: C. Sammut, G. Webb (Eds.), Encyclopedia of Machine Learning and Data Mining, Springer, New York, United States of America, 2015, pp. 503–511.

[43] W. Martinez, A. Martinez, J. Solka, Exploratory Data Analysis with MATLAB®, CRC Press, New York, United States of America, 2011.

[44] I. Goodfellow, Y. Bengio, A. Courville, Deep Learning, MIT Press, Massachusetts, United States of America, 2016.

[45] J.-M. Kanter, K. Veeramachaneni, Deep feature synthesis: towards automating data science endeavors, in: 2015 IEEE International Conference on Data Science and Advanced Analytics (DSAA), Paris, France, 2015, pp. 1–10.

[46] J. Cao, Q. Zhang, W. Shi, Edge Computing: A Primer, Springer, Cham, Switzerland, 2018.

[47] K. Cao, Y. Liu, G. Meng, Q. Sun, An overview on edge computing research, IEEE Access 8 (2020) 85714–85728.

[48] D. Sabella, A. Reznik, R. Frazao, Multi-access Edge Computing in Action, CRC Press, New York, United States of America, 2020.

[49] W. Yu, F. Liang, X. He, W.-G. Hatcher, C. Lu, J. Lin, X. Yang, A survey on the edge computing for the internet of things, IEEE Access 6 (2018) 6900–6919.

[50] P. Habibi, M. Farhoudi, S. Kazemian, S. Khorsandi, A. Leon-Garcia, Fog computing: a comprehensive architectural survey, IEEE Access 8 (2020) 69105–69133.

[51] T. Wang, Y. Lu, Z. Cao, L. Shu, X. Zheng, A. Liu, M. Xie, When sensor-cloud meets mobile edge computing, Sensors 19 (2019) 1–17.

[52] H. Bangui, S. Rakrak, S. Raghay, B. Buhnova, Moving to the edge-cloud-of-things: recent advances and future research directions, Electronics 7 (2018) 1–31.

[53] A.-B. Arrieta, N. Diaz-Rodriguez, J. Del Ser, A. Bennetot, S. Tabik, A. Barbado, S. Garcia, S. Gil-Lopez, D. Molina, R. Benjamins, R. Chatila, F. Herrera, Explainable artificial intelligence (XAI): concepts, taxonomies, opportunities and challenges toward responsible AI, Inf. Fusion 58 (2020) 82–115.

[54] T. Miller, Explanation in artificial intelligence: insights from the social sciences, Artif. Intell. 267 (2019) 1–38.

[55] B. Verheij, Artificial intelligence as law, Artif. Intell. Law 28 (2020) 181–206.

[56] L. Bertossi, F. Geerts, Data quality and explainable AI, ACM J. Data Manag. Inform. Qual. 12 (2) (2020) 1–9.

[57] D. Roselli, J. Matthews, N. Talagala, Managing bias in AI, in: L. Liu, R. White (Eds.), WWW '19: Companion Proceedings of the 2019 World Wide Web Conference, 2019, pp. 539–544 (San Francisco, USA).

[58] D. Gunning, M. Stefik, J. Choi, T. Miller, S. Stumpf, G.-Z. Yang, XAI – explainable artificial intelligence, Sci. Robot. 4 (37) (2019) 1–2.

[59] L. Branting, C. Pfeifer, B. Brown, L. Ferro, J. Aberdeen, B. Weiss, M. Pfaff, B. Liao, Scalable and Explainable Legal Prediction, Artificial Intelligence and Law, 2020, https://doi.org/10.1007/s10506-020-09273-1.

[60] A. Holzinger, A. Carrington, H. Muller, Measuring the quality of explanations: the system causability scale (SCS), KI 34 (2020) 193–198.

[61] J. Van der Waa, T. Schoonderwoerd, J. van Diggelen, M. Neerincx, Interpretable confidence measures for decision support systems, Int. J. Hum. Comput. Stud. (2020), https://doi.org/10.1016/j.ijhcs.2020.102493.

[62] O. Loyola-Conzalez, Black-Box vs. White-Box: understanding their advantages and weaknesses from a practical point of view, IEEE Access 7 (2019) 154096−154113.

[63] K. Kuang, L. Lian, Z. Geng, L. Xu, K. Zhang, B. Liao, H. Huang, P. Ding, W. Miao, Z. Jiang, Causal inference, Engineering 6 (2020) 253−263.

[64] R. Chimatapu, H. Hagras, A. Starkey, G. Owusu, Explainable AI and fuzzy logic systems, in: D. Fegan, C. Martin-Vide, M. O'Neill, M. Vega-Rodriguez (Eds.), Theory and Practice of Natural Computing, Springer, Cham, Switzerland, 2018.

[65] N. Tuptuk, S. Hailes, Security of smart manufacturing systems, J. Manuf. Syst. 47 (2018) 93−106.

[66] D. Berman, A. Buczak, J. Chavis, C. Corbett, A survey of deep learning methods for cyber security, Information 10 (4) (2019) 1−35.

[67] Y. Hong, U. Hwang, Y. Yoo, S. Yoon, How generative adversarial networks and their variants work: an overview, ACM Comput. Surv. 52 (1) (2017) 1−43.

[68] A. Handa, A. Sharma, S. Shukla, Machine Learning in Cyber Security: A Review. WIREs Data Management and Knowledge Discovery, 2018, https://doi.org/10.1002/widm.1306.

[69] J. Agnese, J. Herrera, H. Tao, X. Zhu, A survey and taxonomy of adversarial neural networks for text-to-image synthesis, WIREs Data Manag. Knowl. Discov. (2019), https://doi.org/10.1002/widm.1345.

[70] H. Xu, Y. Ma, H.-C. Liu, D. Deb, Adversarial attacks and defenses in images, graphs and text: a review, Int. J. Autom. Comput. 17 (2) (2020) 151−178.

[71] N. Martins, J. Cruz, T. Cruz, P. Abreu, Adversarial machine learning applied to intrusion and malware scenarios: a systematic review, IEEE Access 8 (2020) 35403−35419.

[72] B. Biggio, G. Fumera, F. Roli, Security evaluation of pattern classifiers under attack, IEEE Trans. Knowl. Data Eng. 26 (4) (2014) 984−996.

[73] D. Dillenberger, P. Novotny, Q. Zhang, P. Jayachandran, H. Gupta, S. Hans, D. Verma, S. Chakraborty, J. Thoma, M. Walli, R. Vaculin, Sarpatwar, Blockchain analytics and artificial intelligence, IBM J. Res. Dev. 63 (2/3) (2019) 5−14.

[74] T. Dinh, M. Thai, AI and blockchain: a disruptive integration, Computer 51 (9) (2018) 48−53.

[75] S. Kuang Lo, Y. Liu, S. Yen Chia, X. Xu, Q. Lu, L. Zhu, H. Ning, Analysis of blockchain solutions for IoT: a systematic literature review, IEEE Access 7 (2019) 58822−58835.

[76] F. Casino, T. Dasaklis, C. Patsakis, A systematic literature review of blockchain-based applications: current status, classification and open issues, Telematics Inf. 36 (2019) 55−81.

[77] K. Tamilselvan, P. Thangaraj, Pods − a novel intelligent energy efficient and dynamic frequency scalings for multi-core embedded architectures in an IoT environment, Microprocess. Microsyst. 72 (2020) 102907.

[78] L. Valerio, M. Conti, A. Passarella, Energy efficient distributed analytics at the edge of the network for IoT environments, Pervasive Mob. Comput. 51 (2018) 27−42.

[79] P. Warden, D. Situnayake, TinyML: Machine Learning with TensorFlow Lite on Arduino and Ultra-low-power Microcontrollers, O'Reilly Media, Inc, Boston, United States of America, 2020.

[80] J. Batzel, F. Kappel, D. Schnediz, H. Tran, Cardiovascular and Respiratory Systems: Modeling, Analysis and Control, SIAM, Philadelphia, Unites States of America, 2007.

[81] N. Dey, A. Ashour, S.-J. Fong, C. Bhatt, Wearable and Implantable Medical Devices: Applications and Challenges, Elsevier, Inc, London, United Kingdom, 2020.

[82] Z. Moussavi, Fundamentals of Respiratory Sounds and Analysis, Morgan & Claypool Publishers, Williston, North Dacota, United States of America, 2006.

[83] M. Flores, G. Glusman, K. Brogaard, N. Price, L. Hood, P4 medicine: how systems medicine will transform the healthcare sector and society, Pers. Med. 10 (6) (2013) 565–576.

[84] R.-M. Lee, N.-C. Tsai, Dynamic model of integrated cardiovascular and respiratory systems, Math. Methods Appl. Sci. 36 (2013) 2224–2236.

[85] S. Jafari, H. Arabalideik, K. Agin, Classification of normal and abnormal respiration patterns using flow volume curve and neural network, in: Paper in the Proceedings of 5th International Symposium on Health Informatics and Bioinformatics, 2010, pp. 110–113, https://doi.org/10.1109/HIBIT.2010.5478898.

[86] J.C. Rodriguez, C.J. Arizmendi, C.A. Forero, S.K. Lopez, B.F. Giraldo, Analysis of the respiratory flow signal for the diagnosis of patients with chronic heart failure using artificial intelligence techniques, in: I. Torres, J. Bustamante, D. Sierra (Eds.), VII Latin American Congress on Biomedical Engineering CLAIB 2016, Bucaramanga, Santander, Colombia, 2017, pp. 481–484.

[87] J. Amaral, A. Lopes, A. Faria, P. Melo, Machine learning algorithms and forced oscillation measurements to categorize the airway obstruction severity in chronic obstructive pulmonary disease, Comput. Methods Progr. Biomed. 118 (2015) 186–197.

[88] E. Ladanza, V. Mudura, A decision support system for chronic obstructive pulmonary disease, in: L. Lhotska, L. Sukupova, I. Lackovic, I.G. Ibbott (Eds.), World Congress on Medical Physics and Biomedical Engineering, Springer, 2019, pp. 321–324. IFMBE Proceedings.

[89] A. Badnjevic, L. Gurbeta, M. Cifrek, L. Pecchia, Pre-classification process symptom questionnaire based on fuzzy logic for pulmonary function test cost reduction, in: A. Badnjevic (Ed.), CAMBEBIH 2017, Springer, 2017, pp. 608–616. IFMBE Proceedings.

[90] M. Janidarmian, A.R. Fekr, K. Radecka, Z. Zilic, A novel algorithm to reduce machine learning efforts in real-time sensor data analysis, in: P. Perego, A. Rahmani, N. TaheriNejad (Eds.), Wireless Mobile Communication and Healthcare. MobiHealth2017, Springer, 2018, pp. 83–90. Lecture Notes of the Institute for Computer Sciences, Social Informatics and Telecommunications Engineering.

[91] C. Lovejoy, E. Philips, M. Maruthappu, Application of artificial intelligence in respiratory medicine: has the time arrived? Respirology 24 (2019) 1136–1137.

[92] S. Boers, K. Jongsma, F. Lucivero, J. Aardoom, F. Buchner, M. de Vries, P. Honcoop, E. Houwink, M. Kasteleyn, E. Meijer, H. Pinnock, M. Teichert, P. van der Boog, S. van Luenen, R. van der Kleij, N. Chavannes, SERIES: eHealth in primary care: Part 2: exploring the ethical implications of its application in primary care practice, Eur. J. Gen. Pract. 26 (1) (2020) 26–32.

Chapter 9

Strategies for long-term adherence

Laura Romero Jaque[1] and Vicente Traver Salcedo[2]
[1]Universitat Politecnica de Valencia, Valencia, Spain; [2]ITACA — Universitat Politecnica de
Valencia, Valencia, Spain

Introduction

Adherence is defined by the WHO as "The degree to which a patient's behaviour, about taking medication, following a diet or modifying lifestyle habits, conforms to the recommendations agreed upon with the professional sanitary" This term seeks to frame the patient as a subject of the treatment itself and not as an object, that is, patients must be active partners with health professionals in their care and such good communication between them is an essential requirement for effective clinical practice.

- Adherence can be divided or classified according to the duration of treatment:
  o Adherence to short-term treatment: treatment lasting less than 6 months.
  o Adherence to long-term treatment: treatment lasting more than 6 months, and its objective is to avoid relapses (continuation treatment) and prevent recurrences (maintenance treatment), prevent complications, and reduce morbidity and mortality rates.

On the other hand, lack of adherence can be divided or classified according to different criteria:

- According to the intention:
  o Intentional lack of adherence: characterized by the patient's willingness to show clearly that he/she will not take the medication.
  o Lack of unintentional adherence is the result of involuntary forgetfulness related to the loss of memory or autonomy of the patient, the complexity of the treatment, the lack of creation of daily routines and habits, etc.
- Depending on the point at which the treatment was discontinued [1,2]:
  o Primary adherence: occurs when after the prescription of the treatment, the patient does not even start it by carrying out a direct desertion.

Wearable Sensing and Intelligent Data Analysis for Respiratory Management
https://doi.org/10.1016/B978-0-12-823447-1.00011-7
273

o Secondary adherence: occurs when the prescribed treatment is not properly followed; common actions in this scenario include dosing with amounts and frequencies that are not correct, forgetting the dose, in general, accompanied a posteriori by extra doses to compensate for the loss, increased dose frequency, or premature discontinuation of treatment.

To quantify the lack of adherence, observable values are used that seek to establish the proportion of the dose taken concerning the theoretical prescribed one, obtaining the following types according to such relationship:

o Partial noncompliance: The patient only follows the treatment at specific times.
o Sporadic noncompliance: Occasionally the patient does not perform the action requested by the treatment, generally due to forgetting a dose.
o Sequential noncompliance: The patient abandons the treatment when observing a slight improvement but readheres when the symptoms reappear.
o Compliance with a white coat: The patient only adheres when consultation or review is going to happen soon in the health center. This behavior is observed especially related to chronic diseases.
o Complete noncompliance: Indefinite discontinuation of treatment.

Adherence is a complex and multifactorial process whose result depends on the interrelation between the different agents that include patients, health professionals, the health system, and the direct environment of the patient. Among all these relationships, according to [2], the most significant is the doctor—patient relationship, since even the best-conceived prescription can be useless, if clear and fluid communication is not established with the patient through regular information flows for setting expectations or goals in treatment. It is also important to distinguish between short- and long-term adherence as it is important to provoke adherence in the patient but even most important, to guarantee a long-term adherence that will impact in patient health outcomes. Table 9.1 summarizes all the other factors.

General practice reviews estimate that about 50% of chronic disease medications are not taken as prescribed. Adherence rates tend to be even lower for patients with asthma, COPD, and AR with estimates ranging widely between 22% and 78% [3].

Failure to adhere to regular asthma, COPD, or AR action plan causes poor control of the disease, causing exacerbations and a reduction in patient quality, which translates into more hospitalizations and visits to the emergency room. These events have economic consequences, such as increased hospitalization and visits to the emergency department, which unnecessarily increase the negative consequences of healthcare and health systems.

For example, currently, the total cost of asthma as a single chronic condition comprises 1%—2% of healthcare expenses where precisely persistent

TABLE 9.1 List of factors influencing adherence to treatment.

	Factors
The system and the health-care team	- Lack of knowledge and training of health-care providers in the management of treatment and inadequate understanding of the disease - Lack of training in modifying the behavior of patients who do not adhere
Illness	- Inadequate understanding of the disease
Treatment	- Complex treatment regimens - Prolonged duration of treatment - Frequent doses - Adverse effects of treatment
Patient	- Forgetfulness - Misunderstanding of medication instructions - Poor parents' understanding of medications for childhood asthma - Patients' lack of perception of their own vulnerability to the disease - Lack of information from patients about prescribed daily doses and misconceptions about the disease and treatments - Persistent misunderstandings about side effects - Drug abuse

Source: O. Ibarra Barrueta, R. Morillo Verdugo, Lo que debes saber sobre la adherencia al tratamiento. Badalona: Euromedice Vivactis, 2017. http://www.biblioteca.cij.gob.mx/Archivos/Materiales_de_consulta/Drogas_de_Abuso/Articulos/libro_ADHERENCIA.pdf, Long Term Adherence to Treatment. Word Health Organization, Geneva 2004 https://www.paho.org/hq/dmdocuments/2012/WHO-Adherence-Long-Term-Therapies-Spa-2003.pdf.

clinical conditions cause systematic and disproportionately high hospitalization and urgent care processes, with a ratio of almost 1: 1 between direct and indirect costs. If we compare such data with the costs of outpatient and pharmaceutical services, the impact on health systems is less despite being used with great frequency [2].

The consequences could be summarized in:

o People's quality of life is worsening.
o It makes difficult the control of the disease.
o Greater probability of relapses and exacerbations appears.
o Side effects or poisoning might appear.
o It can drive the chances of death.
o It involves an increase in social and health resources.

Therefore, the major goal of this chapter is to present the key concepts about medium- and long-term adherence for respiratory diseases that can help to get the best health outcomes, following the value-based health-care approach.

This chapter not only presents such key concepts, but also introduces the different dimensions that need to be considered and adapted to each specific case in order to maximize treatment adherence, as personalization is key to achieve it. Different use cases and tools will be presented, looking for the reader's empowerment that can help him/her to prescribe the best adherence strategy to be followed based on age, digital health literacy, culture, context, health status, or other factors.

Materials and methods

This section details the materials and methods that have been necessary for the development of this study. A nonsystematic review has been carried out in different databases of scientific articles (Scopus, Cochrane Library, PubMed, Web of Science, and Google Scholar) based on the search strategy. The PRISMA methodology was selected to perform a search-structured bibliography.

The phases of the search strategy were as follows:

1. **Initial search.** We obtained studies and articles published between January 2017 and November 2020 in four different databases: PubMed, Scopus, Web of Science, and Cochrane Library. The main search terms performed were as follows:

   (respiratory disease OR chronic respiratory disease OR asthma OR CPOD OR rhinitis) AND adherence AND (medication OR drug) AND (enhancing OR improving OR improve) AND (intervention OR strategies OR tools) AND (e-health OR digital technologies OR smart medicine OR TICS) AND (review OR systematic review).

   In addition to this query, searches were also carried out with simple terms or keywords as:

   Adherence, long-term, chronic disease, digital technologies, asthma, wearables, strategies, management, e-health, chronic obstructive pulmonary disease, enhancing interventions, telemedicine, interventional tools, smart medicine, device gadgets, apps, and device, adherence, monitoring, management, medication, patients, inhaler, technologies.

2. **Screening.** The results obtained yielded 71 results, of which, eliminating duplications, 68 articles remained. After a reading of the articles, 25 of them were discarded since the title and the subject matter were outside of our application framework, due to different reasons.

3. **Suitability.** As a result of the screening, 16 articles were obtained on which an exhaustive analysis was carried out, looking for those reviews that provided us with more information and established comparisons and relationships based on the following reasons:

   a. The topic should be focused on adherence strategies for chronic respiratory diseases (CRDs): asthma, COPD, or AR-based on ICTs.

b. The objectives of the work should be to propose improvements or recommendations to the existing ones, establish the limitations and current barriers in the clinical application, identify indicators that guarantee the correct follow-up of the patient, or describe all the tools and strategies based on ICTs concerning their behaviors.

4. **Inclusion.** Finally, the articles involved in the analysis were 16 articles where systematic and nonsystematic reviews were included. Concerning the nonsystematic reviews case, data synthesis was limited to a narrative summary of the included evidence as a pooling of quantitative data was not possible due to the evident heterogeneity among the identified studies.

This heterogeneity is due to the complex and multifactorial nature of the adherence process, which is represented in the different selection criteria for the target population, medical conditions, and adherence measures. This poor methodological quality in the reviews means that specific analysis cannot be carried out on a particular population group to which different intervention strategies may have been applied. The strategies vary so much between them to cope with the different determinants of nonadherence that it is difficult to be able to compare them directly (Fig. 9.1).

Intervention strategies

From these 16 reviewed articles, different intervention strategies were extracted, which we will use as a common analysis toolkit for any possible ICT.

Next, a list of types of intervention strategies is described based on Refs. [1,2] that establishes these standards and is the most widely propagated and used in publications.

- Interventions based on the simplification of treatment.
- Educational interventions: They provide information by all available means so that patients who understand their condition and their treatment are more informed and easier to take their medication well. These questions include answers to: how and when to take the medicine, as well as the storage conditions of these. Education about the disease will focus on the problems caused by noncompliance, as well as the benefits of the following treatment appropriately.
- Therefore, the informed patient can participate in the early recognition of seizures and their management.
- Behavioral interventions. They modify or reinforce behaviors and empower patients to participate in their care. This group includes everything from relatively simple interventions, such as adapting the treatment to the patient's routines or the use of pillboxes, to much more complex interventions, such as motivational interviewing.

  It is essential to provide tips and tricks to help you remember the shot, as well as to adapt the shot to your daily routine to the complexity of the treatment regimen.

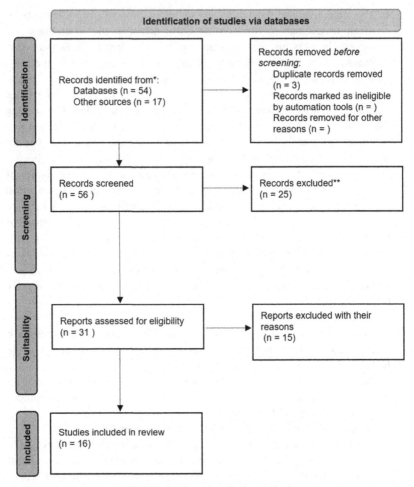

FIGURE 9.1   Flow chart for item selection.

- Interventions based on feedback with the patient on adherence obtained through electronic devices and aspects of their disease.
- Interventions based on medication reminders through technology.
- Reward-based interventions: This could be to recognize the efforts that the patient makes to improve adherence and reduce the number of visits to the doctor or pharmacy if adherence is adequate.

Thus, to develop any type of intervention to improve adherence, a series of basic questions must be considered [2]:

- Support the patient, do not judge or blame him.
- Individualization of interventions.

- Generate bonds of sufficient trust between the patient and the professional to be able to express their doubts, difficulties, and concerns, especially in pediatrics.
- Accept the principles of patient autonomy and shared decision-making.
- Possible repercussions on the patient since the interventions consume resources and could have negative effects (lack of privacy and autonomy of the patient, increase in adverse effects, etc.).
- Multidisciplinary collaboration between health-care professionals and the health-care team.
- Establish reminders to make patients aware of the importance of taking medication.
- Adherence is a dynamic process that will require a variable intervention strategy to improve it.

Results

In this section, ICT-based strategies through the meta-analysis from 16 studies, including literary reviews, systematic reviews, and literary reviews, are analyzed. The methodology followed is the identification of the technological modality used (mHealth, wearable, or electronic monitoring devices), the intervention and control tools used, the effect on adherence, and the health of the patient, as well as the indicators related to the improvement in adherence detected by the authors. Once this identification and description of the reviewed strategies have been carried out, a comparison is made between the different plans described in each review, seeking to find which indicators of all those described in the literature show greater effectiveness and efficiency.

Tools	Description strategy	Effect on adherence	Examples
	mHealth		
Messaging services	Alert reminders by: • record symptoms and medication on an e-health platform • take medication	Effect on erratic nonadherence	SMS
Social media	Educational intervention	No effects at first glance	Twitter
Patient Diaries/Reports	Recording symptoms and medication on an e-health platform	Effect on erratic nonadherence	*My asthma journal*
Informative Videos/Text	Educational intervention	Effect on unconscious nonadherence	AsthmaCare
Medication taking schedule/schedule	Reminders in collaboration with messaging services	Effect on erratic nonadherence	
Applications/ websites with integrated tools	Gamification, gratuities system, use previous tools	Effect on erratic, intelligent, or inconsistent adherence	Scripps Asthma Coach

Applications/websites with integrated tools

The web page strategy generally obtained an increase in adherence by reducing the number of emergency visits and the number of exacerbations along with other parameters in several studies carried out. Incorporating technology to advance asthma controller adherence, William C. Anderson in 2017 [4] carried out a type of web page strategy based on shared use by children and parents, the former as subjects of treatment and the latter as guarantors that the treatment is being carried out correctly. The key factor of these improvements was the involvement level that parents had when using the website and communicating with the doctor, as well as the usability degree of the application.

Other relevant aspects were the study of specialized apps for the adult population such as Scripps Asthma Coach or AsthmaCare in the child population, both included together with disease control tools such as daily reminders for the use of medications and personalized prevention of crisis triggers, new tools to improve the use of the application such as gamification systems and reward systems. The use of these new tools resulted in better adherence to ICS, with more than 58% of patients achieving the inhaled corticosteroids (ICSs) adherence goal of at least 50%, Asthma Control Test (ACT) scores improved with reduced use of short-acting inhaled beta2-agonist (SABA).

Novel methods for device and adherence monitoring in asthma [5], on the other hand, were based on applications for smartphones specifically designed to improve asthma control through alerts and actions initiated by the patient, such as information inquiries and data entry with the ACT as the primary endpoint to assess disease control over the intervention. Its secondary goals included replacing check-ups, advising when to seek medical care, monitoring asthma over time, collecting data to present to health-care professionals, providing educational materials and instructions, and recording side effects.

The reception of this management system was positive for both patients and health-care professionals. The patients showed a significant improvement in ACT scores and forced expiratory volume in the values of the first second (FEV1) during the study period and the results of the questionnaire showed that people with asthma applied more frequently than mobile health systems monitor asthma over time (72%) and collect data to present to your health-care teams (70%).

A study [6] showed the use of applications in combination with integrated sensors, which were based on currently available inhaler-based monitoring devices that use various approaches to measure adherence, including the capture of the time and date of medication use, audio recording during inhaler use, and telemonitoring capability based on remotely captured spirometry measurements.

Messaging services

In general, it was found that the use of SMS messaging used to go in combination with other tools such as electronic medical device (EMD) and web pages [4]. The combination of SMS and EMD reminders using a web page as an intermediary generally presented improvements in adherence when electronic reports were not used, substituting their value for measurements of the physiological variables of the patients when they were being used, showing improvements in asthma control, quality of life, and reduction in exacerbations.

On the other hand, Ref. [7] reported that by receiving SMS alerts inviting the patient to register their symptoms and medication in an e-health platform, results of self-reported adherence were obtained, the rate of clinical care and improvement of symptoms was slightly higher in the groups that used them versus those that did not.

Informative Videos/Text

Producing engaging animation-based videos has become much more cost-effective in the recent years due to the improvement of text-to-speech technology. The results of several studies demonstrate that animation-video-based technology is more engaging and can be used to supplement in-person consultations. While computer-based learning programs that improve chronic disease self-management skills and personalized approaches for asthma management are being implemented, it is critical to determine whether providing health-care knowledge in an interactive manner is an effective adaptive treatment strategy for addressing the needs of transition based.

Medication taking schedule/schedule

The effectiveness of a treatment depends on both the efficacy of a medication and patient adherence to the therapeutic regimen. On general basis, simplify medication taking on patient characteristics at the first level of drug use along with a schedule appropriate follow-up, is a common indirect method to measure adherence.

Social media

Social media can play a role in the surveillance and management of allergic diseases as it represents a useful method to quickly disseminate health-related information [7]. The objective of this study was to explore the potential of Twitter as a source of information on the prevalence of allergic diseases, based on the observation that Rhinitis allergic sufferers can write tweets that include symptoms and names of drugs they take for treatment. The obtained results showed a high correlation both with the tweets that reported RA complaints and with the tweets that mentioned an antihistamine.

Electronic medical devices (EMDs)			
Device type	Description strategy	Effect on adherence	Examples
Electronic inhalers	• Record the date and time of activations • The total number of activations used, and doses omitted • Send data to app	Effect on unconscious nonadherence	Propeller Health, SmartTrack
Sensors	• Record movement, heart rate, and oxygen levels • Inhalation technical analysis for respiratory noises	Effect on unconscious nonadherence	SmartTrack
Wearables	Constant event detection and monitoring	Effect on unconscious nonadherence	Health and Environmental Tracker

EMDs are now considered the gold standard for controlling medication, as they provide objective data and are not biased by patient self-report. Studies in SmartTrack [8] in children with audiovisual reminders or EMDs only showed how the mean percentage of adherence was 84% in the group intervention compared to 30% in the control group. The ACT score improved by more than three points in the intervention group. Despite these improvements, there were no differences between the two groups concerning missed school days, lung function, or emergency department visits. In a real-world study by Merchant et al. patients were randomly assigned to EMDs (Propeller Health) with or without comments on the use of SABA (short-acting bronchodilator). Both studies demonstrated the benefit of using DME to improve adherence and provide feedback to the patient and the physician. The potential limitations of using EMDs are the cost and time of implementation in the clinic and the possible lack of benefit in the technique of drug administration.

Electronic tracking devices that provide patient-directed reminders and personalized feedback messages have shown modest results in improving adherence to asthma medications, particularly devices that provide feedback to health-care providers to encourage discussion. Around these patients, the review 'New Concepts and Technological Resources in Patient Education and Asthma Self-Management' [9] highlights the project based on the Propeller Health Asthma Platform application to reduce the use of SABA, improve asthma control, and facilitate asthma self-control by providing personalized, data-driven feedback. The Propeller Health Platform includes an FDA-approved sensor that measures the use of inhaled drug. The device records the date, time, and the use of inhalers, being further transmitted via Bluetooth to a paired smartphone app. The application records the location of the event and transmits the data to secure remote servers that patients and providers can access. Favorable results in SBA and ACT were obtained at the same time while the daytime and night time symptoms decreased significantly. No significant change in activity was observed. Participants also reported increased awareness and understanding of asthma patterns, level of control, SABA use (time, location, and triggers), and best preventive practices.

Discussion

Recommendations

Following the entire literature analyzed, we can conclude which of all the strategies reviewed are capable of better-guaranteeing adherence. To carry out this conclusion, we will do so following certain parameters of personal elaboration that, considering the complexity and multifactorial nature that surrounds adherence, can allow us to have an overview of all the strategies and allow us to establish comparisons despite having different tools. These parameters would be:

- **Decision aid,** that is, how the information provided by the strategy makes it possible to foresee a behavior that shows signs of low adherence and to act on the patient to resume correct compliance, as well as the ability to prevent exacerbations.
- **Personalization** either through information on physiological parameters or tests written by the patient such as tests or medication diaries, as well as being able to adapt the dynamics of the strategy to the specific conditions of the patient such as their age or knowledge about their disease.
- **Patient participation,** that is, if the role to be fulfilled in the strategy will be passive or active, depending on the degree of involvement of the patient in taking control of their treatment, long-term adherence could be affected positively or negatively.

Plato's myth of the cave is a metaphysical explanation of how one arrives at knowledge, at the truth, and allows us to explain the proposed strategic plan based on the three previous pillars.

1. **Activation:** thus, patients who are chained to unreliable beliefs and sources can only see in the dark the objects that are shown to them (the doctor's indications). To get to this state, it is first necessary to make patients aware of what their disease is, what repercussions it has on their life, and how to improve their quality of life through an activation with educational and reminder strategies (videos/texts/SMS). Then, the patients start the treatment.
2. **Take over:** the ascent from where the patients are chained to the exit means walking with enormous effort uphill. In this phase, patients assume control. Thanks to all the information obtained and learned, patients feel not only capable of understanding their disease, but also of influencing its

development. Therefore, patients take responsibility for their treatment backed by feedback strategies on health status integrated into an interactive app, for example.

3. **Participation:** leaving the cave and facing the sunlight and seeing what exists is getting knowledge, which is putting what has been learned into practice. In this last phase, patients can then carry out actions that require greater involvement, such as web portals, community and electronic newspapers.

The strategies based on mHealth made use of combined strategies of smart devices such as electronic inhalers for the monitoring of physiological variables in combination with the SMS delivery of reminders to take medication, writing electronic diaries about their symptoms, and training on the use of inhalers or nebulizers. These strategies force somehow patients to be more involved in the treatment, as they are suggested to write a brief report on their symptoms whereas they are also allowed to see an evolution of their health status; therefore, based on this self-reported information, patients can make decisions in treatment or in their daily life to change or improve the trends detected in the reports.

Tool	Decision support	Customization	Patient involvement
SMS sending	Acts on intentional nonadherence	Frequency, alerts, individualized notifications	Passive and reactive paper
Patient diaries/Reports	Retrospective analysis	-------	Active role with positive feedback
Medication taking schedule	The event calendar should be combined with other tools	Adaptation to the patient's routine	Passive paper
Informational videos/Text	Awareness acts on intelligent and unconscious adherence	Feedback based on recorded symptoms or searches	Passive paper predisposes to an active state
Wearables/EMD	Real-time monitoring	-------	Passive paper
Apps/Web portals	Interaction environment	Gamification and usability	Active role

SMS sending

- The greatest functionality presented by this strategy is the ability to intervene on unintentional lack of adherence due to forgetfulness about the amount and frequency of taking the medication.
- The frequency can be adapted according to the assigned treatment and the form of the message showing words or motivational phrases that are user-friendly.

- The patient has a passive and reactive role. It frees the patient from the burden of remembering when and how much medication to take. Therefore, the error of taking the treatment not indicated is reduced. The effects are mainly seen in the short term and even though adherence decreases in the long term, it is still higher than when this tool is not available.

Patient diaries/Reports

- The greatest advantage of this tool is in its retrospective analysis, since both the patient and the health-care provider can see a history of the patient's symptoms, which allows the patient to have a greater perspective of his/her condition whereas the doctor can follow his/her evolution.
- Personalization is low, since regardless of the characteristics of the patient, the strategy does not change the way he/she acts.
- This tool requires the patient to play an active role, since, without the patient's involvement in writing his/her symptoms regularly, the diary loses its purpose. Although the writing and subsequent reading of the written entries could generate positive feedback on the patient, it is also probable that the patient will be exhausted due to the continuous effort of entering the entries.

Medication taking schedule

- Little help with the decision, it is about the calendar of events, it does not provide more information.
- Passive participation does not imply constant action by the patient to carry out the strategy.

Informational videos/text

- They allow patients to be trained and informed about the characteristics of their disease, raising awareness of the repercussions of improper treatment, as well as providing knowledge on how to deal with the symptoms and what they mean.
- The basic customization is low, since they are videos accessible to the entire population. However, a feedback system could be used to recommend videos and articles according to the symptoms that patients have had and reported. Furthermore, the values of the electronic monitoring could be applied to filter the type of video that is recommended to them.
- Patient participation in the strategy is passive. However, it predisposes the patient to an active state and therefore better recognizes their health status and feels a greater commitment that translates into good adherence in the future provided it be combined with other strategies.

Wearables/EMD

- These are the main ones that will monitor the physiological variables, so through this registry, adherence and health status can be measured for different diseases [10]. Based on this information, the health-care provider can take decisions and even seek new strategies that can improve the adherence to treatment.

Apps/web portals

- There has been a 515% increase in the number of recorded adherence apps since 2012 [11]. However, there continues to be variability in the quality and functionality of adherence apps, which would decrease the likelihood of consumers finding the best adherence app on their first try. Providers should be familiar with several quality adherence apps to recommend to patients as an additional tool for patients to use in addressing medication nonadherence.

Use cases

The growth of CRDs in recent decades has changed morbidity and mortality trends and has become a problem worldwide, which affect people of all ages. Among the main risk factors identified for CRDs are tobacco smoking, air pollution (indoor and outdoor), allergens, and occupational risks and vulnerability.

Many patients with a chronic respiratory condition will require long-term treatment and health-care support with the main goal of care focusing on reducing symptoms and improving quality of life. The burden of disease and treatment of chronic respiratory conditions disrupts patients' structures of everyday life and imposes limitations on activities. Patients can achieve well-being in these circumstances by accepting their limitations and adjusting to them, replacing former activities with new meaningful activities they can enjoy and by taking advantage of good days and emotionally adapting to bad days. Self-capacity, trustful care with continuous care relationships, and access to medications have also been identified as essential to well-being.

However, the pandemic forced the adoption of digital technology and communications; one of the biggest fields that aided by ICTs during the pandemic is health care. Technology has played a major part in improving standards health organizations around the world, hospitals, testing facilities, and laboratories. Digital health has become an essential part of the medical sector and plays a key role during the current health crisis. COVID-19 has given impetus for countries to accelerate the adoption of digital health. Digital adoption had to be done at an unprecedented speed—telemedicine for the delivery of health care really was scaled up.

TABLE 9.2 Population characteristics for the three different use cases.

Youth
- Between 14 and 26 years
- High autonomy
- Digital natives
- Little patience
- Immature stage
- Sensitive to the environment

Adults
- Between 27 and 59 years
- Autonomy
- Maturity
- Manual and digital dexterity
- Manifesto of negative consequences on health due to lifestyle

Elders
- Between 60 and 80 years
- Cognitive and functional deficiencies
- Comorbidity
- Polypharmacy
- Concern about adverse effects

Three use cases are presented with different features as example to present how the different steps can be established to maximize the long-term adherence. Therefore, after introduction of each use case, a strategy is drafted, presenting the best suitable tools for the different steps: activation, take over, and participation (Table 9.2).

Use case 1: young person

Characteristics:
- Student
- Mild persistent asthma
- Low self-esteem and high stress
- Sedentary
- Daily treatment: Inhaled ICS

Strategy

When compared with medications used for other chronic diseases, most of the medications used for treatment of asthma have very favorable therapeutic ratios. The pharmacological options for long-term treatment of asthma fall into the following three main categories [12]:

- Controller medications: these are used to reduce airway inflammation, control symptoms, and reduce future risks such as exacerbations and decline in lung function. In patients with mild asthma, controller treatment may be delivered through as-needed low-dose ICS-formoterol, taken when symptoms occur and before exercise.
- Reliever (rescue) medications: these are provided to all patients for as-needed relief of breakthrough symptoms, including during worsening asthma or exacerbations. They are also recommended for short-term prevention of exercise-induced bronchoconstriction. Reducing and, ideally, eliminating the need for reliever treatment is both an important goal in asthma management and a measure of the success of asthma treatment.
- Add-on therapies for patients with severe asthma: these may be considered when patients have persistent symptoms and/or exacerbations despite optimized treatment with high-dose controller medications (usually a high dose ICS and an LABA) and treatment of modifiable risk factors.

Once asthma treatment has been commenced, on-going treatment decisions are based on a personalized cycle of assessment, adjustment of treatment, and review of the response. For each patient, in addition to treatment of modifiable risk factors, controller medication can be adjusted up or down in a stepwise approach to achieve good symptom control and minimize future risk of exacerbations, persistent airflow limitation, and medication side effects. Once good asthma control has been maintained for 2−3 months, treatment may be stepped down in order to find the patient's minimum effective treatment.

Step 1: activation

When dealing with a young avatar, the use of social networks such as Instagram, Facebook, Twitter, and WhatsApp will be the first sources of information that the patient uses even before having been diagnosed with the disease. Therefore, through the presence of a feed where publications on asthma prevention and control campaigns are prioritized to search for content and strategies, according to the preferred network used by the patient.

These social networks will work as social reminder and awareness for the patient. However, as an extra input, SMS reminders are also welcome in order to not forget the timing for medication taking.

Step 2: take over

With programs easily downloadable to the cell phone, which try to help both the doctor and the patient through gamification, they seek to help in the task of controlling asthma.

With the development of wireless communication and smart phones, social network is going mobile, which promotes the emergence of mobile social network. Therefore, social factors have to be taken into the consideration in the design of health-care services. For instance, a mobile persuasive social game, named Movipill, is proposed to improve the medication adherence based on the social competition. Movipill encourages medication adherence behavior by giving more points to users that take their medication very close to or at the prescribed time [13].

This will enable a self-management of chronic disease for the patient during their everyday activities, such as status logging for capturing various problems or symptoms met, and social sharing of the recorded information within the patient's community, aiming to facilitate disease management.

Step 3: participation

In general, people are willing to have social interaction among their friends, relatives, or persons with the same interest. The social reminders from friends, relatives, or persons with the same interest are more effective than that from the strangers. Therefore, to leverage the efficacy of social reminders, a dynamic community that contains friends, relatives, and people with the same interest based on property similarity and topology similarity could encourage social interaction among community members of asthma patients and improves the medication adherence, thanks to the contagion effect.

Use case 2: adult

Characteristics:
- Middle-class employee
- Severe allergic rhinitis
- High stressed
- Sedentary
- Daily treatment: Immunotherapy, corticosteroids, antihistamines, and decongestants

Strategy

In patients with perennial symptoms attributable to indoor allergens (e.g., dust mites, furry animals, indoor molds, cockroaches), avoidance of the allergen is a critical first step in treatment. Environmental control programs should always be based on accurate assessments of both sensitization (by skin or in vitro testing) and exposure. These strategies are particularly helpful to patients who are sensitized only to indoor allergens with no evidence of allergy to pollens or outdoor molds [14].

- **Dust mites:** House-dust mites *Dermatophagoides farinae* and *Dermatophagoides pteronyssinus* are ubiquitous throughout the world (except in dry or alpine regions), and approximately 30%−40% of patients with allergic rhinitis are allergic to allergens produced by these mite species. The mite avoidance must be aggressive to be effective and should include encasement of pillows and mattresses in impermeable covers, washing of all bedding in hot (>130°F) water, and elimination of carpeting in favor of tile or hardwood floors.

- **Animals:** Virtually any furry pets may result in allergic sensitization and ultimately symptoms of rhinitis. However, most of the available clinical data regarding the efficacy of animal avoidance measures come from studies of indoor cats and their major allergen.
- **Indoor mold:** Identification of homes with mold growth is often difficult. Indoor mold spores from species of *Aspergillus* and *Penicillium* are most likely to emanate from potentially damp areas such as crawl spaces (because of defective plumbing or poor drainage), attics (because of roof leaks), and under sinks. The presence of a musty smell and the visual presence of mold confirm the problem. Occasionally, however, wall spaces, carpet backing, and other areas with limited access may harbor mold growth, and identification of the mold may be delayed or even missed. Complete correction of all plumbing, drainage, and construction defects must be undertaken to eliminate significant mold problems. In some cases of extensive mold growth, major portions of a house may have to be rebuilt. More limited measures, such as the application of bleach or fungicides, have not been shown to be beneficial.
- **Cockroach:** Although cockroach allergy has been most heavily implicated as a pathogenic factor in children with asthma, it may play a substantial role in perennial allergic rhinitis as well. The best indicator of a significant cockroach infestation is the presence of emanations on the floor. Cockroach exposure is usually not limited to the kitchen or dining room but may affect all living areas because allergen is passively transferred on shoes and clothing. Pesticide application is only temporarily effective, and problems will recur unless food and garbage are appropriately packaged and handled.
- **Outdoor seasonal allergens:** Plant pollens and outdoor molds (e.g., species of *Alternaria* and *Cladosporium*) are responsible for the symptoms of seasonal allergic rhinitis and are generally very difficult to avoid completely. During indoor activities, keeping all windows and doors shut and the use of an air-conditioner eliminate most pollen from the inside of the house. Certain mold spore counts tend to be highest late in the evening or early in the morning, especially in damp climes, and this may be a consideration for patients who are mold allergic. However, altering schedules and activities is undesirable for most patients, and, for this reason, avoidance measures play a limited role in allergic rhinitis caused by outdoor allergens.

Step 1: activation

Patients are unlikely to be adherent to their AR treatment unless they have a clear understanding of their illness and the treatment, including correct expectations about treatment cost, duration, benefit, and potential side effects. Patients who are confused or doubtful about their treatment are likely to stop taking it. In turn, when patients are provided with the knowledge and skill to self-manage their illness, they will be more adherent.

In the treatment of AR, the health-care provider focus has been directed at choosing an effective medication based on the patient's history and symptoms in a process that can be provider-centric while failing to determine what the patient would prefer. Although providers are often directed to consider patient preferences, for example, by asking patients whether their willingness to use a specific AR medication would be impacted by the medication's cost, odor, or taste, such approaches do not fully embrace patient-centeredness. Adherence is most likely to improve when patients are engaged in a process of shared decision-making where the provider attempts to reach concordance with the patient about treatment choices and goals, promoting greater involvement of the patient in deliberations about treatment options.

Step 2: take over

Once a patient leaves the clinic, providers have limited opportunity to influence patient behavior. Still, if effective communication in the office has set the stage for enhancing adherence, subsequent contact can reinforce this motivation. Follow-up communication may occur during return visits to the clinic, telephone calls, and other technology-based interventions such as:

- Online patient diaries/reports
- SMS messaging

Patients are frequently scheduled for a return visit to help evaluate treatment response. These visits also provide an opportunity for follow-up discussion about the patient's perceptions regarding the treatment and to address any new questions or concerns that may interfere with adherence. Patient support through follow-up telephone calls, letters, and a combination of the two can have a positive impact on patient adherence. Communication and health information technology have been recently employed to provide follow-up enhancement of adherence without requiring additional provider time. Such adherence-enhancing approaches include leveraging text messaging and interactive voice recognition technology to activate patients and encourage better disease self-management.

Step 3: participation

- **Apps informing on risk factors for allergic rhinitis:** risk factors for exacerbations of AR and asthma include allergen exposure, climatic factors, and air pollutants. It is therefore of great importance to identify levels of risk factors that can induce symptoms in allergic patients. Among these, pollen exposure is the most important for pollen allergic patients. Therefore, forecasting symptoms of pollen-related AR for the individual patient should improve disease control and plan pharmacological intervention and/ or prevention of exposure.

- **Apps including health data:** many mHealth apps support patients with AR via self-monitoring through an electronic diary (e-Diary), personalized feedback, and/or patient education. They aim to improve patient education and self-management on a daily basis but require an evidence-based evaluation given that the information provided on the app stores is limited, in particular for the apps' validity. This can be done by evaluating the effectiveness of the app with the patients' clinical outcomes.

Use case 3: elderly person

Characteristics:
- Retirement
- Severe COPD
- Comorbidity: Cardiovascular disease (CVD) and depression
- Sedentary and smoker
- Daily treatment: Inhaled Bronchodilators

Strategy

Identification and reduction of exposure to risk factors are important in the prevention and treatment of COPD. All individuals who smoke should be encouraged to quit. The level of FEV1 is an inadequate descriptor of the impact of the disease on patients and, for this reason, individualized assessment of symptoms and future risk of exacerbation should also be incorporated into the management strategy for stable COPD. All patients with COPD with breathlessness when walking at their own pace on level ground benefit from rehabilitation and maintenance of physical activity, improving their exercise tolerance and quality of life and reducing symptoms of dyspnea and fatigue [15].

- Pharmacologic therapy is used to reduce symptoms, reduce frequency and severity of exacerbations, and improve health status and exercise tolerance. Existing medications for COPD have not been conclusively shown to modify the long-term decline in lung function that is the hallmark of this disease.
- Routine follow-up is essential in COPD. The frequency of follow-up visits and type of examinations needs to be individualized. In general, the following aspects need to be considered:
  - Symptoms: at each visit, inquire about changes in symptoms since the last visit, including cough and sputum, breathlessness, fatigue, activity limitation, and sleep disturbances. Questionnaires such as the COPD Assessment Test (CAT) can be performed every 2—3 months. Trends and changes are more valuable than single measurements.

- Smoking status: at each visit, determine current smoking status and smoke exposure. Strongly encourage participation in programs to reduce and eliminate wherever possible exposure to COPD risk factors.
- Lung function: it may worsen over time, even with the best available care. Decline in lung function is best tracked by spirometry performed at least once a year to identify patients whose lung function is declining quickly.
- Pharmacotherapy and other medical treatment: to adjust therapy appropriately as the disease progresses, each follow-up visit should include a discussion of the current therapeutic regimen. Dosages of various medications, adherence to the regimen, inhaler technique, effectiveness of the current regime at controlling symptoms, and side effects of treatment should be monitored. Treatment modifications should be recommended as appropriate with a focus on avoiding unnecessary polypharmacy.
- Exacerbation history: evaluate the frequency, severity, and likely causes of any exacerbations. Specific inquiry into unscheduled visits to providers, telephone calls for assistance, and use of urgent or emergency care facilities is important. Severity of exacerbations can be estimated by the increased need for bronchodilator medication or corticosteroids, by the need for antibiotic treatment, or by documenting hospitalizations.
- Comorbidities: identify and manage them in line with local treatment guidance.

Along with the efforts of health-care providers, elderly persons or their families seek information to meet their needs. They sought the required information from various sources, including "professional and nonprofessional resources" and "the mass media," which includes television, satellite, and the Internet. Illiterate participants used the radio as an information source. Some participants obtained the information from the nurses. However, in many cases, the elders used easily accessible nonprofessionals and peers who, in some cases, can give misleading information to the elderly patient. They also used "their experiences" in "continuing with normal life." In some cases, due to the long history of the disease, the elderly persons had gained valuable experiences in self-management.

Participants started participating in the process of care once they trusted health-care providers. Despite the fact that knowledge about one's condition is necessary, to engage in self-care, in most cases, participants did not have sufficient knowledge about the disease. Moreover, teams of health-care providers believed that, in order to control the disease, knowledge alone is not sufficient and that it is necessary for elderly persons to apply their self-care knowledge.

However, in some cases, the participants did not follow the recommended lifestyle due to some other reasons. An instance of this was the inability by all participants to quit smoking, due to psychological dependence and other deterrent factors. These factors included lack of knowledge, impatience, disability, dementia, drug shortages in the market, and financial problems.

Step 1: activation

In the empowerment process, health-care providers' efforts to involve elderly people in the treatment entailed use of strategies such as "giving information," "striving to reduce dependency," and "participation in disease management." This ensures that health-care providers, especially physicians, employ different ways to inform ill elderly patients about the disease, treatment, diagnosis, physiotherapy and pulmonary rehabilitation, oxygen therapy, and self-care. In some cases, information was provided at the time of diagnosis. Such as:

- Doctor interviews
- Informational videos
- SMS/Calendar's reminders

Step 2: take over

Faced with a complex disease in old age, elderly people were more dependent than at any other time. In addition to helping older family members with the process of striving in the course of life, another important approach was "helping to stabilize whose life" to improve the lives of the older people receiving help. In order to enhance elderly persons' psychosocial capacity while they are striving in the course of life and searching for information, families seek to enhance the elders' psychosocial capacity. In the struggle with COPD, elderly people and their families have to cope with and overcome the disease, due to the disease being the most important concern of the family and affecting the lives of family members. Various attempts are made to ensure the well-being of elderly patients, to provide them with physical, emotional, and financial support, and to ensure that their living conditions are comfortable. Therefore, some tools are used in order to make this happens, such as:

- Web pages
- Treatment simplification
- EMD

They can help in the control and simplification of the process.

Step 3: participation

COPD is a leading cause of hospital readmissions, with reported 30-day readmission rates of 20% [16]. The reasons for readmission are complex, often associated with breathing difficulty, respiratory deterioration, and lack of home care plan. Postdischarge care has therefore been essential to assist patients in addressing health issues. This often involves clinical follow-up and examination, gas exchanges, and care planning in a multidisciplinary team approach.

Therefore, we present as possible solution, the Hospital-at-home (or Home Hospitalization) service that provides active treatment by health-care professionals in the patient's home for a condition that otherwise would require acute hospital inpatient care. Hospital-at-home programs offer an alternative for patients requiring treatment at the hospital or emergency department. It also allows care providers to discharge patients at an early stage, and continue to treat the patients at home. A possible instantiation of a telehealth program to support hospital-at-home for COPD could be provided by electronic diaries and an app.

Conclusions

Long-term adherence is key to guarantee proper health outcomes for the patients' diseases, especially for respiratory ones, whereas the decisions of the patients have a large influence on final outputs, what also affects the sustainability of our health-care system. This chapter has presented an overview of different techniques and tools after a scoping review on the topic, suggesting a framework to deal with this topic and bringing lessons learnt from the existing evidence. Therefore, it's key to be aware that we need a strategy for each case and that has to be personalized based not only on patients' features but also on contextual factors (region, culture, family, etc.). After drafting the strategy, three steps are always needed: activation, take-over participation, and this is something in a permanent and iterative loop that needs to be adjusted as adherence is also related to any of the multiple daily decisions patient (or relatives in some of the cases) is taking so a continuous assessment need to happen. This is not just about using technology, but technological devices and analytic tools should be the key differential element to 1) contextualize the patient environment to define as much as possible a personalized strategy, 2) support the patient and relatives in their decisions through the different steps, and also 3) quantify the long-term adherence to the treatment and measure its impact on patient health outcomes.

Still there is a long way to go, where multidisciplinary teams are addressing this challenge from a multifaceted perspective that needs a holistic approach. Among others, authors suggest three activities that still need investment around this challenge:

- New innovative approaches combining behavior change theories with existing technological devices and tools to personalize this experience for the patient.

- Common data repositories that allow the generation of specific evidence about long-term adherence in a personalized way, allowing the creation of digital twins and going beyond stratification.
- Educational activities addressed to clinicians, patients, and relatives, showing the existing evidence and the need to be aligned among the different stakeholders, always respecting the freedom of choice but at the same time, convincing the patients and relatives about the importance of having a GPS browser like that guides you not on the road but on any decision on your daily life that is affecting your health outcomes.

References

[1]   O. Ibarra Barrueta, R. Morillo Verdugo, Lo que debes saber sobre la adherencia al trata-
       miento. Badalona: Euromedice Vivactis, 2017. http://www.biblioteca.cij.gob.mx/Archivos/
       Materiales_de_consulta/Drogas_de_Abuso/Articulos/libro_ADHERENCIA.pdf.
[2]   Long Term Adherence to Treatment, Word Health Organization, Geneva, 2004. https://
       www.paho.org/hq/dmdocuments/2012/WHO-Adherence-Long-Term-Therapies-Spa-2003.
       pdf.
[3]   New Insights to Improve Treatment Adherence in Asthma and COPD, https://doi.org/10.
       2147/PPA.S209532.
[4]   Incorporating Technology to Advance Asthma Controller Adherence, William C. Anderson,
       2017, https://doi.org/10.1097/ACI.0000000000000343.
[5]   Novel Methods for Device and Adherence Monitoring in Asthma Matteo Bonini and Omar
       S, Usmani, 2017, https://doi.org/10.1097/MCP.0000000000000439.
[6]   Mobile Health and Inhaler-Based Monitoring Devices for Asthma Management, 2019,
       https://doi.org/10.1016/j.jaip.2019.08.034.
[7]   Rhinitis: Adherence to Treatment and New Technologies, 2017, https://doi.org/10.1097/
       ACI.0000000000000331.
[8]   Asthma Management in the Era of Smart-Medicine: Devices, Gadgets, Apps, and Tele-
       medicine, 2018, https://doi.org/10.1007/s12098-018-2611-6.
[9]   New Concepts and Technological Resources in Patient Education and Asthma Self-
       Management, 2020, https://doi.org/10.1007/s12016-020-08782-w.
[10] L. Wedlund, J. Kvedar, Wearables as a tool for measuring therapeutic adherence in behavioral
       health, Npj Digit. Med. 4 (2021) 79, https://doi.org/10.1038/s41746-021-00458-9.
[11] Assessing the Medication Adherence App Marketplace from the Health Professional and
       Consumer Vantage Points, 2017, https://doi.org/10.2196/mhealth.6582.
[12] Bronchial Asthma: Diagnosis and Long-Term Treatment in Adults, 2008, https://doi.org/
       10.3238/arztebl.2008.0385.
[13] R. de Oliveira, M. Cherubini, N. Oliver, MoviPill: improving medication compliance for
       elders using a mobile persuasive social game, in: Proceedings of the 12th ACM Interna-
       tional Conference on Ubiquitous Computing (UbiComp '10), Association for Computing
       Machinery, New York, NY, USA, 2010, pp. 251−260, https://doi.org/10.1145/
       1864349.1864371.

[14] Guidelines for the Diagnosis and Management of Asthma in Children and Adolescents, Clinical Practice Guideline, 2019. https://ginasthma.org/wp-content/uploads/2019/06/GINA-2019-main-report-June-2019-wms.pdf.

[15] Global Initiative for Chronic Obstructive Lung Disease, 2019. https://www.mscbs.gob.es/organizacion/sns/planCalidadSNS/pdf/GOLD_Report_2015_Apr2.pdf.

[16] Digital Health for COPD Care: The Current State of Play, 2019, https://doi.org/10.21037/jtd.2019.10.17.

Chapter 10

Respiratory decision support systems

Ioanna Chouvarda[1], Eleni Perantoni[1], Paschalis Steiropoulos[2]

[1]School of Medicine, Aristotle University of Thessaloniki, Thessaloniki, Greece; [2]Medical School, Democritus University of Thrace, Alexandroupolis, Greece

Introduction

Respiratory diseases are considered, in general, as complex conditions, as their initiation and progression are multifactorial and driven by the interaction between genetic factors, comorbidities, environmental exposures, treatments, etc.

Various respiratory conditions, such as asthma, COPD, cystic fibrosis (CF), sleep apnea, or malignant conditions like lung cancer and respiratory infections (influenza or COVID-19) affect many individuals in all age groups. In addition to acute respiratory care, these conditions pose a challenge for the health system resources, which requires careful and efficient management.

Decision-making, including diagnostic, treatment-related, and optimization of procedures and organization, is the core of clinical routine. The digital transformation in health promises to leverage decision-making.

As mentioned in Sutton et al. [68], "*A computerized clinical decision support system (CDSS) is intended to improve healthcare delivery by enhancing medical decisions with targeted clinical knowledge, patient information, and other health information.*"

Such a system may support clinicians, administrative staff, patients, caregivers, or other health-care providers involved in care, with individualized information specific for each patient and situation that requires a decision, thus helping to make more timely and accurate decisions and to avoid errors or adverse events.

As an example, expert systems (ESs) based on "IF…THEN" rules have been used to implement consistently evidence-based processes, including complex guidelines and protocols. Model-based decision support systems (DSSs) have been used in intensive care unit (ICU) for ventilation support.

Wearable Sensing and Intelligent Data Analysis for Respiratory Management
https://doi.org/10.1016/B978-0-12-823447-1.00008-7
299

Data-driven DSSs have been used when intensive data processing is required, for example, when the decision is heavily based on biosignals and bioimages. Among the numerous examples, many success stories have been identified, as well as risks and pitfalls. Often criticism against a clinical DSS (CDSS) relates to "Alert fatigue" and clinical burnout [31], which happens when the CDSS overwhelms users with redundant or unimportant information, or unacceptable workflow freezes and burdens.

During the last years, the data-driven economy is being incrementally established, pushing for big data principles employment in digital economy applications in health. Several enabling technologies that play a role in this digital transformation can be identified, which play a role in the evolution of respiratory decision support systems (RDSSs).

Electronic medical records with advanced capabilities and computerized clinical workflows are increasingly adopted, thus increasing the volume of digital data to be considered for decision-making and facilitating the incorporation of CDSSs. These factors, together with the responsibility to deliver value-based care and the pressures for optimal resource use in more and more challenging conditions, make the CDSSs essential tools.

In parallel, connected health technologies (CHTs) can facilitate the patient—professional partnership in decision-making and offer opportunities for incorporating DSS in patient self-management [12]. In the previous years, CHT research was mainly focused on monitoring and measurement information, while the provision of feedback recommendation and prescriptive support in the context of CHT seem to be research priorities in the coming years [45].

The increased availability of health data enables the integrated data analysis (spanning from DNA to protein and metabolism, to physiology and external factors), which can address and lead to a better understanding of a range of phenomena, rather than a narrow symptom-based decision. This approach links to the systems medicine (SM) paradigm that considers the respiratory system as a complex system, and focuses on the interactions between the different components within it, functioning at different levels, and the interaction between organs or systems, e.g., between the cardiovascular and respiratory system in specific conditions [53]. SM and in silico modeling are fundamental pillars of information and knowledge for an RDSS.

Finally, AI's role in a DSS in respiratory care cannot be overlooked. AI in medicine, both machine learning (ML) and deep learning, shows potential toward improving diagnosis, and support treatment decisions, provided that adequate quantities of data are available. In emergency situations, like the COVID-19 pandemic, AI-based prediction models and RDSS constitute promising solutions toward alleviating the asphyxiating pressure on health systems, provided that their reliability and trustworthiness can be ensured beyond preliminary results.

This chapter aims to provide an overview of the CDSS approaches and enabling technologies in respiratory care. This integrated view of the aspects that need to be considered when designing for RDSS includes (a) the continuum of decision between acute and chronic care, (b) the wealth and added value of biomedical data to be employed, and (c) the complementarity of methods to be pipelined or aggregated.

A variety of methodological aspects are discussed, along with RDSS cases in chronic and acute care. Challenges in the maturation of RDSS are identified along with future research directions of interest.

Overview of the RDSS domain

A wide range of diseases and conditions implicate the respiratory system and require decision support, spanning in the whole range between acute and chronic care or even transition care.

Emergency and intensive care

The emergency COVID-19 pandemic has put pressure on health systems and led to the fast implementation of prediction models, for diagnosing COVID-19 in patients with suspected infection, for predicting hospital admission, for prognosis of diagnosed patients, and for detecting people at risk of becoming infected. As discussed in a recent review [77], preliminary prediction models may prove unreliable, and future high-quality model development and validation are needed, taking into account quality of reporting, risk of bias, and extended performance evaluation.

Currently, the vast majority of widely applied decision support approaches in ICU are limited to basic alarms, which are both unreliable, considering the false alarms, and insufficient. A characteristic effort to improve arrhythmia alerts in ICU was the CINC challenge 2015 [15].

The gap in the efficiency of critical care DSS is highlighted by the alarm fatigue problem [76], raised by a high number of clinical alarms to which the clinical staff is continuously exposed, and thus desensitized. Although this phenomenon is widely observed, it is not accurately measured with direct metrics. In order to reach a higher accuracy and meet the emerging needs, research is turning to more sophisticated methods, including (a) integration of data, (b) complicated calculations, and (c) continuous analysis of waveforms, rather than sparse data.

Some major issues in a respiratory ICU are described below to better sketch the RDSS needs and challenges.

ARDS diagnosis

Acute respiratory distress syndrome (ARDS) is a critical condition, with substantial mortality. It is a characteristic case of in-hospital patient deterioration. However, its routine diagnosis remains a challenge, due to the variety of risk factors and etiologies, including the lack of diagnosis support tools and the gaps in CT image radiomics. It is considered an underdiagnosed condition in ICU [21], which has negative consequences in treatment decisions. Recently, a number of ARDS diagnostic models were proposed, using, for example, routinely collected clinical variables and information from radiology reports [40] for early prediction, or "sniffer" systems that automatically analyze electronic medical record data recognize ARDS in clinical practice [75].

Mechanical ventilation optimization

Conflicting clinical goals and competing risks that change dynamically for each ICU patient under mechanical ventilation make its management a complicated task. Overall, the aim is to offer adequate ventilation without creating damage. DSSs may help clinicians find the correct balance when complex decisions need to be made.

Important insights on mechanical ventilation CDSS have been previously discussed [34], using a particular example (Real-time Effort Driven ventilator management [REDvent]) to illustrate the concepts. Overall, the main points highlighted are the need to provide simple and explicit instructions, a transparent knowledge base, to understand the decline of recommendations, and provide a reevaluation and adjustment mechanism.

Pediatric ventilation is a particular case. In an older work [69], the CDSS tools for mechanical ventilation in children are distinguished into two categories. The first category refers to commercial CDSS built into specific ventilators, as proprietary fixed rule systems, lacking mechanisms to modify or adapt the rules or report reasons for disagreement with recommendations. Such systems are likely to gradually show decreased acceptability and efficiency. The second one is about ventilator agnostic tools, focusing on a particular ventilation phase (e.g., initiation or weaning) or ventilation mode (e.g., pressure support). The latter can be closed-loop systems. The effectiveness of pediatric RDSS has been variable with no apparent benefits, and, as also suggested in Hartman's DSS evaluation of DSS for weaning mechanical ventilation in children, acceptability is low, and recommendations are not often followed [24]. These facts suggest the need for more maturation steps for improvement and broader adoption.

ARDS treatment

DSS has been considered among the strategies to improve a protective ARDS treatment, i.e., low tidal volume (LTV) mechanical ventilation [66]. The DSS mainly focuses on customizing electronic health records to provide LTV ventilation reminders or alerts at convenient times (i.e., when the respiratory data page is viewed on patients receiving mechanical ventilation, at the time of initiation or change of a ventilator order) or alerts if ARDS patients received injurious tidal volumes for more than 1 hour. Such approaches tend to be useful, although still with variable results and low acceptability. Supporting better ARDS recognition can be combined with the support of more advanced ARDS therapies.

ICU-telemedicine

In recent years, telemedicine ICU programs have been deployed [11], mainly linking a highly expert (potentially academic) ICU to smaller nonacademic ICUs, for patient monitoring, recognition of complex conditions (e.g., ARDS), and staff education purposes. Such systems were considered promising, but many open issues were pinpointed regarding their cost-effectiveness and the broad deployment applicability.

Such issues can potentially be alleviated based on data analysis and decision support tools. Based on data collection from such networks [18], including a significant number of ICU stays, research is conducted to develop and test new predictive and prescriptive analytical solutions and DSSs, that will help to better coordinate care for critically ill patients, and optimize resources and processes of the critical care telemedicine network. Leveraging telemedicine intensive care units (tele-ICUs) may prove particularly useful in the future, in extreme health conditions, like the COVID-19 pandemic.

Chronic care

In chronic care, future digital health systems aim to support predictive, preventive, personalized, and participatory medicine ("P4-medicine"). Decision-making challenges pertain to clinicians, nurses, patients, and caregivers, especially in integrated care settings, including telehealth solutions.

Lung cancer diagnosis and management

Lung cancer is among the most common cancers among men and women in the world, as well as a major cause of death. Early-stage detection can improve survival probability, but it is heavily dependent on quantitative assessment of lung nodules by radiologists, which is time-consuming if manually performed. Therefore, trustful automated detection of lung lesions is a need, and a number of studies have focused on solving this problem, mainly in CT imaging modality [47]. The combination of multiple modalities, including ET/CT and pathology images, has recently gained attention [70] for accurate diagnosis and treatment support.

Considering treatment support, different RDSSs have been proposed, specific to cancer types. The review by Ref. [61] studies RDSS for treatment of metastatic non–small cell lung cancer (NSCLC), which predicts overall survival and/or progression-free survival, for the whole group or subgroups. Gaps are identified regarding performance, coverage of the entire treatment spectrum, incorporation of genetic and biological markers, and evidence on toxicity and cost-effectiveness.

COPD and asthma diagnosis and management

Chronic obstructive pulmonary disease (COPD) is a widespread respiratory disease in the aged population, causing chronic airflow limitations, often interlinked with other chronic conditions. Personalized medicine in COPD has been focused in dealing with heterogeneity of disease trajectories, clinical presentation, and response to existing therapies [19]. However, predicting and/ or modifying the course of the disease (including exacerbations), and predicting response to specific interventions, are gaining attention [30] but are not yet resolved issues. These unmet needs may be answered via a SM approach, integrating data and phenomena at multiple scales in computational models. Such approaches can set the basis of new advanced RDSS for COPD.

Nurses often run patient management in COPD telemedicine programs, and an RDSS has a role in supporting clinical reasoning and avoiding nursing staff from being overwhelmed by the continuous patient monitoring data flow. In Ref. [2] the DSS was based on a combination of symptoms and measurements, and used a rule-based system to classify a patient as "stable patient" indicating change that need follow-up, and "unstable patient" indicating a severe change or a critical condition. This approach was evaluated in practice as useful for identifying health problems and prioritization. However, deeper clinical information regarding each patient's health status is needed to fully promote the understanding of health changes and provide more certainty for a decision on actions.

In the case of asthma self-management, the myAirCoach system [35] has been presented. This is a promising system that monitors several physiological, behavioral, and environmental factors, which are further processed and aggregated to provide short-term prediction of asthma control level for daily and real-time personalized patient guidance, and long-term prediction of exacerbation risks, to support clinical decision-making and alert medical personnel [38]. While this monitoring and decision support approach was found to improve asthma control and quality of life, extended validation and evidence are missing.

Obstructive sleep apnea diagnosis and management

Obstructive sleep apnea (OSA) is a highly prevalent but underdiagnosed disease, with a complex pathophysiology. In order to provide a solution for underdiagnosis, a rule-based DSS based on straightforward questions for pediatric OSA detection in primary care is discussed in Ref. [29] as a feasible approach for automating OSA screening and detection.

In recent years, CHTs, including portable and wearable devices, from polygraphy to smartwatches, have been proposed for OSA diagnosis in the home environment, with various quality and accuracy, suggesting the need for an evaluation framework [54]. OSA continues to be investigated, regarding the various physiological traits, clinical presentations, and biomarkers, and research efforts bring more attention to the need for support of tailored and effective treatments [6].

Methods for respiratory DSS

Biomedical data for respiratory DSS

A variety of signal and sensor-based approaches have been developed and deployed in clinical environments and home monitoring schemes toward accessing information of value to an RDSS. These fall under different categories: estimation of breathing rate and content, lung function and structure, biological and clinical data, and intervention/treatment-related data, as well as data related to systems interaction.

Respiratory rhythm and content

Spirometry is a typical test that can assess lung function by measuring the volume of inhaled and exhaled air and exhalation speed, and its values are of diagnostic value according to medical guidelines (e.g., for asthma or COPD). It can be used in a clinical or ambulatory environment, as is nowadays feasible at homecare settings, although the unsupervised use may pose quality issues [14]. Pulmonary function is measured by spirometry via a challenging maximal breathing maneuver, which is not accurate or suitable for long-term unsupervised monitoring.

The respiratory signal can be used for the diagnosis of clinical state or assessment of treatment and recovery, both at clinical/intensive care and home setting. Continuous sensing by wearables strain sensors has been proposed. PPG-derived respiratory frequency is easily measured, but its accuracy depends on the measurement site and breathing pattern [25]. It is best measured at the forehead and finger, respectively, for normal and deep breathing patterns.

SpO_2 signal can provide detailed information about blood oxygen level [71]. Hypoxia can be related to chronic lung conditions, including COPD and sleep apnea. A simple SpO_2 drop has recently been proposed [58] to early detect COVID-19-related hypoxemia, before the manifestation of more severe symptoms such as shortness of breath. Statistical indices such as the oxygen desaturation index (ODI, number of desaturations per hour) are often used for apnea/hypopnea characterization. Other more sophisticated biosignal analysis methods have been proposed, based on detailed characteristics of desaturation events. Such analysis is vital for the accurate automated classification of OSA, the characterization of events severity, and relation with their clinical status, e.g., prediabetic insulin [55].

As for capnography, this semiperiodic waveform fluctuates between inspiration and expiration and measures how much CO2 a person is exhaling and is typically monitored in intensive care or homecare ventilation. In nonintubated patients, it can be used to assess the pulmonary vessels' ventilation and perfusion, e.g., for asthma classification with ML and capnograph waveform features [65].

New sensors have been proposed as point-of-care diagnostics. In the review of Ref. [26], approaches are discussed to detect volatile organic compounds (VOCs) as biomarkers for respiratory diseases. Such approaches can identify the VOCs distinctive of several respiratory diseases, including COPD, asthma, lung cancer, pulmonary arterial hypertension (PAH), obstructive sleep apnea syndrome (OSAS), tuberculosis (TB), CF, and pneumoconiosis. The underlying sensing mechanisms involve several innovative techniques, including spectroscopy techniques, nanomaterials, chemiresistors, acoustic sensors, colorimetric sensors. On the other hand, electronic noses (e-noses) can distinguish VOCs based on pattern recognition or specific olfactory fingerprints, and human exhaled breath profiling by e-nose can support screening/diagnosis of respiratory diseases [17].

Lung function and structure

Lung sounds (auscultation) and cough have been considered as traditional means for diagnosis [62], as discussed in Chapter 5. Enabled by CHTs [12], the availability of digital lung sounds in broader monitoring conditions has enriched the knowledge and leveraged the use of digital auscultation. Automated lung sound detection is crucial for such analysis [57]. Adventitious sounds have been systematically studied during the last years concerning understanding of exacerbation and postexacerbation recovery, and monitoring of chronic conditions such as COPD across clinical and nonclinical settings. For example, in Ref. [52] it was found that inspiratory crackles seem to persist until 15 days postexacerbation.

In respiratory medicine, imaging is considered as the diagnostic cornerstone (see Chapter 6). A broad spectrum of imaging techniques is available. Chest imaging studies include X-rays, computed tomography (CT), magnetic resonance imaging (MRI), nuclear scanning, ultrasonography, and positron emission tomography (PET). The main modality is still CT, which allows fast and high-resolution assessment of the lung and surrounding structures, e.g., detection of lung nodules. Additionally, MRI and PET are gaining attention with regard to direct functional information, as elaborated in Ref. [41].

As discussed in Chapter 6, electrical impedance tomography (EIT) is a promising, noninvasive, and radiation-free technique with rich information content, applicable in a broad spectrum [20], in intensive and chronic care. EIT has been mostly studied in critical care, as a bedside functional imaging modality for continuous monitoring of lung ventilation and perfusion, mechanical ventilation, and detection of lung ventilation problems [72]. For example, EIT may improve (a) monitoring of lung function during ARDS, (b) assessment of patients' responses to changes in ventilator settings and mode, and optimize mechanical ventilation settings, (c) detecting complications such as derecruitment and pneumothorax, and (d) providing estimates of perfusion distribution. While such studies have taken place, more extended clinical validation studies are expected to explore the technology's full potential and introduce EIT in practice. EIT shows potential in pulmonary function testing in patients with COPD, asthma, and CF in chronic care. Wearable solutions aim to bring chest EIT to home monitoring [13].

Biological and clinical data

The EHR is a useful tool to enable the rapid deployment of standardized processes. Beyond usual care, it has also proven an essential tool in extreme cases, as in supporting the clinical needs of a health-care system managing the COVID-19 pandemic.

In Ref. [60] the design and implementation of EHR-based rapid screening processes, laboratory testing, clinical decision support, reporting tools, and patient-facing technology related to COVID-19 outbreak management are discussed. Multiple COVID-19-specific tools were proposed to support outbreak management, including scripted triaging, electronic check-in, standard ordering and documentation, secure messaging, real-time data analytics, and telemedicine capabilities.

Ref. [75] examines six automated ARDS "sniffer" systems and tools that can automatically analyze electronic medical record data to detect ARDS and discusses their role in improving recognition of ARDS in clinical practice. The reported sensitivity for ARDS detection spans in a wide range (43%−98%), and so does the positive predictive value (26%−90%), while a potentially high risk of bias was estimated. The need for robust evaluation of ARDS sniffer systems and their impact on clinical practice remains ongoing.

Another work [48] proposes an integrated point-of-care COVID-19 Severity Score. Using clinical data, biomarker measurements of C-reactive protein (CRP), N-terminus pro B type natriuretic peptide (NT-proBNP), myoglobin (MYO), D-dimer, procalcitonin (PCT), creatine kinase-myocardial band (CK-MB), and cardiac troponin I (cTnI) combined in a statistical learning algorithm to predict mortality. Based on this, clinical decision support tools for COVID-19 can prioritize critical care in patients at high risk for adverse outcomes.

Intervention and treatment related data and interaction with other organs/signals

In critical care, ventilator data can be of use for smart decisions. In Ref. [39], the continuous ventilation data are used through using ML for the generation of smart alarms, that predict in the short term (next 5 min) the presence of high/low driving pressure of mechanically ventilated patients, therefore suggesting the need for increased/decreased attention and adaptation of ventilation parameters.

Ventilation data together with physiological respiratory parameters can be utilized for tools that support optimizing ventilation, including weaning decisions [53], or minimization of asynchronies [67].

In Ref. [16], rapid learning (Rle) concepts were discussed toward improving treatment decisions. Rle involves reusing clinical routine data to develop models that can predict treatment outcomes, and then clinically applying and evaluating these models via DSSs. The Rle approach in focus deploys a previously developed DSS in a typical clinic for NSCLC patients, and it uses routine care data to validate the system. The prognostic groups are identified based on patient and tumor features, and for each group therapy can be individualized based on the model predictions.

RDSS enabling technologies

Databases and knowledge bases

A number of respiratory sounds databases have been generated and made available for the training of digital lung auscultation diagnostic support [62]. Lung medical imaging collections can be found at the Cancer Imaging Archive,[1] including data, mainly CT, at different time-points the care pathway, diagnosis, pre/during/posttreatment.

In Ref. [56] a freely available ICU database is described. Although not specific to respiratory care, it can help in new knowledge discovery on the dynamics and system interplay. However, it is not easy to find multiparametric data, temporal data, especially data recorded in home environments. An overview of relevant datasets is available in Table 10.1.

When it comes to knowledge bases, several works are dedicated to generating disease ontologies, mainly of diagnostic value. Ryerson et al. [63] proposes a standardized ontological framework for fibrotic interstitial lung disease, aiming to homogenize the diagnostic classification of patients.

This is also important in domains of emerging knowledge, like COVID-19 [27]. The Coronavirus Infectious Disease Ontology (CIDO) is a community-based ontology that supports coronavirus disease knowledge and data standardization, integration, sharing, and analysis.

The COPD Ontology is a biomedical ontology used to model concepts associated with COPD in routine clinical databases.[2]

The approach followed by Ref. [36] provides a representation of semantically enriched EHR data for COPD and comorbidities, based on an OWL ontology built upon HL7 FHIR resources, that can also support SPIN rules and constraints.

---

1. https://www.cancerimagingarchive.net/.
2. https://bioportal.bioontology.org/ontologies/COPDO/?p=summary.

TABLE 10.1 Example datasets for AI model training for RDSS.

Data	Description	References
Respiratory sounds	COPD, pneumonia, Bronchiectasis, Bronchiolitis, Upper/Lower respiratory Tract infection, healthy	https://bhichallenge.med.auth.gr/
Chest imaging	Cancer Pneumonia COVID-19	https://luna16.grand-challenge.org/ https://www.cancerimagingarchive.net/ https://www.kaggle.com/paultimothymooney/chest-xray-pneumonia https://bimcv.cipf.es/bimcv-projects/bimcv-covid19/ https://github.com/ieee8023/covid-chestxray-dataset
ICU/ ventilation	ICU patients including vital signs, ventilation data, and other treatments and events	https://eicu-crd.mit.edu/ https://github.com/AmsterdamUMC/AmsterdamUMCdb
Biosignals	OSA	https://physionet.org/content/ucddb/ https://physionet.org/content/apnea-ecg/ https://physionet.org/content/mimic3wdb-matched/1.0/
Biological (-omics)	Cancer Chronic	https://www.cbioportal.org/ http://pulmondb.liigh.unam.mx/

Systems medicine and computational models in RDSS

A deeper understanding of the respiratory phenomena can be of use in a more personalized RDSS approach. Systems biology or SM is an approach complementary to the classic reductionist approach followed in medicine that focuses on the interactions between the different components within one organizational level (genome, transcriptome, proteome) and among levels. SM [9] contributes to the interpretation and understanding of the pathogenesis and pathophysiology, biomarker discovery, and design of innovative therapeutic targets. This is very relevant in the case of respiratory diseases, as they are generally related to the interaction of multiple factors at different levels. Such opportunities for the adoption of SM approaches in COPD management, focusing on proteomics and metabolomics, are proposed in Ref. [44]. As suggested, SM approaches can be incorporated in an RDSS, and support the identification of disease subclusters, as well as the selection of effective therapies.

In a recent work [74], a comprehensive COVID-19 network (CovMul-Net19) is proposed. It contains all available known interactions involving SARS-CoV-2 proteins, the related diseases and symptoms, and compounds that can potentially target them. While not a DSS, it can be considered a knowledge tool that can support educated treatment decisions, especially in view of a precision medicine approach.

A precision medicine approach requires systems-level understanding, integrating phenomena at multiple scales, ensuring diagnosis and treatment support in an individualized manner. In silico modeling enables the generation of mechanistic hypotheses using patient-specific computational models that incorporate a unique patient's profile (omics, physiological, and anatomical), and can study treatment strategies. In Ref. [33], following this paradigm, an agent-based, asthmatic virtual patient is described that predicted the impact of multiple drug pharmacodynamics at the patient level, paving the way for personalized RDSS.

Connected health technologies in RDSS

The evolution of CHT, including sensors, mobile systems, and cloud computing, has allowed the extension of services and decision-making in the whole continuum between daily life at home, clinical setting, and acute care.

In Ref. [37], the application of CDSSs in a tele-ICU is considered. The use of multiple data streams and cloud computing would be basic components of a tele-ICU, toward supporting the analysis, management, and decision support of multiple remote units. ML algorithms are expected to have an important role as an integral part of tele-ICU CDSS. Such systems are expected to capitalize and grow upon the big volumes of data generated and made available by tele-ICU systems. When policies for sharing data are in place, the higher data availability can be exploited for better learning in ML/AI-based solutions.

With respect to telehealth and self-management of chronic respiratory conditions, RDSS can offer automated treatment advice to the chronic patient without a health-care professional's interference. This approach is presented in Ref. [5], concerning COPD self-management of symptom worsening and exacerbation, based on 12 symptom-related yes/no questions and the measurement of SpO_2, forced expiratory volume in one second (FEV_1), and body temperature. The automated treatment advice is based on expert knowledge and Bayesian network modeling. The system was validated and showed high sensitivity and negative predictive value, and in many cases provided patients with useful advice for day-to-day symptom management.

DSS technology

DSSs can follow a rule-based approach, a model-based approach to optimize a specific treatment, or a data-driven approach.

Model-based DSS

Mechanistic models, including compartmental models, can simulate at physiological organ level, the lung function, the respiratory function, or some aspect of it. Model parameters can be fine-tuned to the individual patient, by fitting the models to real measurements and thus estimating optimal model parameters. The 'what-if' scenarios can be explored via simulation, to support decisions. Both open-loop and closed-loop approaches can be supported.

In Ref. [32], a characteristic computerized model-based DSS is presented and evaluated. It offers advice on inspired oxygen fraction, tidal volume, and respiratory frequency, with personalized parameters. It can automatically provide advice about the ventilation strategy with minimal risks. It is based on physiological models simulating the effect of ventilation strategies and penalty scores to increasing the risk for various adverse effects, e.g., hypoxemia, or acidosis/alkalosis. While such approaches can be valuable, providing accurate and robust advice, it is always important to recognize the model assumptions and limitations, for example, patient parameters in transitional status.

Computerized guidelines and rule-based/expert systems as RDSS

The main advantage of this category of approaches is the incorporation of existing knowledge, interpretability of decisions, and potential applicability in low computational resource settings, for example, in primary care or mobile health settings.

Several technical approaches are applicable. Drools[3] is a Business Rules Management System (BRMS) solution. The semantic knowledge bases can make use of Semantic Web Rule Language (SWRL), or Owl reasoners Jena[4], Hermit,[5] RDfox.[6] Several open-source python-based projects can also be of interest, as lightweight options: like experta[7], or durable_rules[8].

ESs have been considered for chronic diseases screening. In Ref. [7] the different pulmonary diseases that include persistent obstruction of lower airways are considered, under the umbrella term chronic obstructive lung disease (COLD), like chronic bronchitis and emphysema, and specific asthma patterns, all of them often underdiagnosed, especially in primary care. An ES was proposed and validated, aiming at COLD diagnosis support, based on symptoms and standard measurements, and was found as a safe and robust supporting tool for COLD diagnosis in primary care settings.

---

3. https://www.drools.org/.
4. https://jena.apache.org/.
5. http://www.hermit-reasoner.com/.
6. https://www.oxfordsemantic.tech/.
7. https://github.com/nilp0inter/experta.
8. https://github.com/jruizgit/rules.

In Ref. [1], asthma and COPD are considered as major chronic diseases, excessively underdiagnosed. An expert diagnostic system was proposed that can differentiate among patients with asthma, COPD, or a normal lung function based on measurements of lung function and information regarding patient's symptoms. Data from 3657 patients were used to build the system and then independently verified using data from 1650 patients. The system shows a high accuracy for all three classes, which contributed to a 49.23% decrease in demand for conducting additional tests, therefore decreasing financial cost.

Concerning computerized clinical guidelines, a characteristic example is the hybrid system Lung Cancer Assistant that includes guideline rule–based recommendations (implemented with the LUCADA lung cancer ontology) and a probabilistic DSS based on a Bayesian network trained on the English Lung Cancer Audit Database, to aid clinicians achieve more informed treatment selection decisions [64].

As already stated by Luger and Stubblefield [46], typical ESs present some core "deficiencies" They do not include necessarily in-depth knowledge of the domain, in the sense that they cannot explain the underlying mechanisms. They cannot learn from experience to continue evolving. They are rigid, in the sense that they function only in problems contained in their knowledge bases.

ML/AI and data-driven RDSS

AI, and ML as part of it, follows the data-driven research paradigm that capitalizes on the availability of large nonhumanly processable datasets, to generate models of diagnostic or predictive value, to support health-care professionals in making clinical decisions [49]. In an RDSS, these ideas may apply, for example, to chronic conditions with regard to diagnosis, staging, exacerbations, and survival. Such models are particularly useful in cases of gray zones, or cases where current knowledge and evidence do not support thoroughly decision-making, allowing for ML approaches to improve clinical decisions and even minimize patient risk. A data-driven RDSS can involve one or more of the following technologies:

A. Classic approaches, which involve: (i) analysis of respiratory data [50], e.g., wavelet analysis for respiratory sound analysis, (ii) selection of the most informative features, (iii) employment of features in ML-based models (e.g., random forest or support vector machine classifiers), and potentially rule-based systems that employ as high-level concepts the diagnostic outcomes of ML models.

B. Deep learning, with a variety of neural network architectures (with CNNs and LSTMs among the most typical ones) and applications in images, biosignals, text recognition, generation of synthetic data [22,23].

C. Transfer learning, which literally means that experience gained from one domain can be transferred to other domains, and fine-tune part of the model for the specific problem. This is typically used for the detection of structures in imaging, e.g., lymph nodes [80] and for the classification of "objects," and is usually based on previously known and successful architectures (e.g., VGG16) for which pretrained models are available for problem-specific fine-tuning.

D. Reinforcement learning (RL) is a goal-oriented learning method to solve complicated control problems. It uses a state at each time, an action to change the state, a transition probability, and a reward function per state-action. Based on that, it develops a policy, a set of rules for taking actions [43].

Some characteristic cases are discussed below.

New approaches improving scoring and classification

In Ref. [51], an ML method using a shallow neural network is proposed to accurately estimate apnea–hypopnea index (AHI) and ODI using only the continuous measurement of blood oxygen saturation signal (SpO$_2$). This improved estimation with affordable home-measured means can open the way for a more affordable screening of OSA and can help address underdiagnosis.

As clinical information systems may include errors, incorrect labels, missing values, ML systems evolve to cope with these issues. For example, in Ref. [59], an algorithm to detect ARDS is presented that accounts for the uncertainty of training labels existing in real life records.

AI in pandemics and emergency

AI methods have been proposed during the COVID-19 pandemic for identification of positive subjects, treatment, prognosis, and monitoring.

In Ref. [8], different CNN architectures of different depth are compared, as regards five chest related pathologies (the CheXpert dataset) and different types of pneumonia (toward distinguishing the COVID-19-related pneumonia). Interestingly, both shallow networks with a low need for resources and deeper and more complex ones can achieve excellent classification performances.

In Ref. [10] review, it is found that AI was applied to COVID-19 in four areas: diagnosis, public health, clinical decision-making, and therapeutics. However, several methodological and evaluation issues were identified in the proposed methods, including insufficient data for model creation and internal/external validation, as well as ethical, trustful, and efficient use.

Treatment optimization and reinforcement learning

In Ref. [43] survey on RL for clinical decision support in critical care, the RL-based decision support was applied to optimize the choice and dose of medications, the timing of interventions (e.g., ventilation), and for personalization of laboratory values. Several challenges were identified, regarding RL system design (e.g., actions), realistic evaluation metrics (e.g., alternatives to mortality), model choice, and extent of realistic validation.

A characteristic deep reinforcement learning (DRL) pipeline for treatment-related RDSS is described in Ref. [73]. It aims for automated dose adaptation, and specifically automated radiation adaptation protocols for NSCLC patients, to maximize local tumor control at reduced rates of radiation pneumonitis. It includes three components: (a) a generative adversarial network (GAN) to create synthetic data for training from a relatively limited sample size, (b) a radiotherapy model RAE based on a deep neural network (DNN) to enable simulation of transition probabilities between its states when making decisions for adaptation of personalized radiotherapy treatment courses, and (c) a deep Q-network (DQN) for choosing the optimal dose. Careful validation of such systems is critical to their success and acceptance.

A detailed RDSS example

This section presents a generic RDSS framework applicable in Chronic Care that includes a galaxy of tools spanning between primary care, secondary care, and continuous monitoring. Table 10.2 summarizes this framework.

TABLE 10.2 An RDSS framework for chronic care.

Step	Need	RDSS opportunities
Primary care screening	Early diagnosis, to address underdiagnosed cases, manageable and affordable resources	AI/ML model based on daily life data (signals and symptoms) and mobile technology—results to be shared with primary care doctor for shared decisions. Tools for the primary care doctor to easily screen patients
Expert diagnosis and treatment	Precision diagnosis and personalized treatment, not contradicting but enhancing existing guidelines	ML to combine biological and physiological data, recognize disease subtype, and suggest tailored treatment DSS for respiratory and comorbidity treatment
Continuous monitoring	Support patient in lifestyle changes and adherence and self-care support Manage transitions from/to secondary care Manage comorbidity	DSS using multiple sensors for daily lifestyle planning and adaptation DSS for symptom management—including guidelines Detect deterioration and guide patient—combine AI and knowledge systems AI/DSS for guiding patient in the transition from/to hospitalization

An example is further elaborated with regard to OSA, inspired by the work of [6]. Primary care screening for OSA is essential, as it is a widely prevalent and highly underdiagnosed condition. Its early diagnosis can help in better management and will positively impact interrelated conditions, including obesity, cardiometabolic syndromes, as well as other respiratory conditions. The primary care RDSS can have two legs for diagnosis:

(a) Use electronic symptom-based screening tools, customized per age/gender, in the general population, administered via mobile health (mhealth) apps, and
(b) When opted-in, combine with affordable home-based sensors (e.g., wearable oximeter, activity, sound, heart rate) and ML to detect possible apnea.

Accurate diagnosis in secondary care can benefit from an RDSS that follows a personalized approach to OSA. This must consider the sparsity of sleep labs, and the difficulty of sleep lab exam. Thus, OSA diagnosis will need to employ, when acceptable, home-based recordings of the necessary sleep study type.

The diagnostic strategy can incorporate:

(a) An improved severity/stratification strategy, beyond AHI [6], based on day and night measurements and AI models,
(b) Determination of the OSA endotype, in terms of mechanisms underlying the condition (e.g., pharyngeal collapsibility, loop gain, arousal threshold) based on standard sleep measurements, wearable sensors, blood biomarkers, and AI methods [42], toward personalized medicine,
(c) Determination of OSA phenotypes or disease clusters [78] in terms of symptoms, comorbidities, lifestyle, and
(d) DSS to decide on treatment and secondary prevention based on "endophenotype," as CPAP is not the treatment for all. Data-driven prediction of treatment adherence can be considered.

As OSA patients can benefit in the long term from lifestyle changes, a patient DSS to support and guide patients in achieving and maintaining healthy habits (sleep, activity), as well as adhere to treatment, is also part of the solution. The DSS will also detect deteriorations of OSA symptoms and physiology, together with other interrelated comorbid conditions like COPD, including causal information, and will guide patients accordingly for self-management or transition to other health-care services and treatments.

Discussion: unmet needs and challenges for the future

Much attention has been drawn both to the promises and perils of CDSSs during the past decades. Despite the attention of both technical researchers and clinical practitioners, the adoption of such systems has been sparse. Among the main challenges identified in the CDSS in previous decades [3] was the technical complexity in accessing and integrating the data from diverse sources and in managing large datasets in a clinical context, the conceptual complexity and the wide range of problems to be dealt with (deterioration, optimization of intervention) in a real clinical environment, the lack of in-depth understanding of complex disease pathophysiology, which would enable building solid models, and the limitations related to regulatory issues [4].

Some of these challenges are present also in a respiratory DSS, due to: (a) the complex interaction of respiratory system with circulatory and other main systems, (b) the respiratory system decision-making spans in many directions and time scales, in acute and chronic care, and (c) a wealth of informative measurements and signals implicated.

In recent years, the concept of personalized medicine has gained popularity among respiratory clinicians, which implies an effort to address some of the fundamental challenges posed above, i.e., address the complexity and the multiscale nature of respiratory phenomena.

Emerging technologies in wearable sensors, -omics analysis, big data management, and computing, as well as AI methods, can leverage the RDSS research and development, but in acute and chronic care. In this direction, several issues must be considered:

Challenges for AI in RDSS. For AI methods to become trustworthy and adopted as part of routine RDSS, decision explainability, causability, and interpretability concepts must be incorporated [28].

Level of personalization in decisions. On the way to prediction medicine, the role of multiple sources of information in decision models must be recognized. This also has as a prerequisite the availability of rich and unbiased multiparametric data, which suggest the generation of integrated resources and data repositories as a necessary step. A synergetic use of clinical respiratory guidelines and AI, with the user in the loop, can be the basis for personalized decisions.

RDSS for new organizational schemes and services. RDSS ideas have been developing for critical care, or clinical setting. As models of care are being reorganized toward more efficient schemes, new RDSS tools need to be conceived developed, or integrated, to cover chronic care self-management services, tele-ICU, and transition care.

RDSS as part of a learning health system. The ability of an RDSS to learn from its mistakes [79] and improve its function, as well as improve its experience from ongoing use and data collection (i.e., via retraining), would be a valuable direction for development. This aspect can improve robust use in a real-life environment and can place an RDSS as a part of a learning health system, in which new knowledge gets embedded in daily practice and continuously improves systems and care.

References

[1] A. Badnjevic, L. Gurbeta, E. Custovic, An expert diagnostic system to automatically identify asthma and chronic obstructive pulmonary disease in clinical settings, Sci. Rep. 8 (1) (2018).

[2] T.L. Barken, E. Thygesen, U. Söderhamn, Advancing beyond the system: telemedicine nurses' clinical reasoning using a computerised decision support system for patients with COPD - an ethnographic study, BMC Med. Inf. Decis. Making 17 (1) (2017).

[3] P.E. Beeler, D. Westfall Bates, B.L. Hug, Clinical decision support systems, Swiss Med. Wkly. 144 (2014) w14073.

[4] A. Belard, T. Buchman, J. Forsberg, B.K. Potter, C.J. Dente, A. Kirk, E. Elster, Precision diagnosis: a view of the clinical decision support systems (CDSS) landscape through the lens of critical care, J. Clin. Monit. Comput. 31 (2) (2017) 261−271.

[5] L.M. Boer, M. van der Heijden, N.M.E. van Kuijk, P.J.F. Lucas, J.H. Vercoulen, W.J.J. Assendelft, E.W. Bischoff, T.R. Schermer, Validation of ACCESS: an automated tool to support self-management of COPD exacerbations, Int. J. Chron. Obstruct. Pulmon. Dis. 13 (2018) 3255−3267.

[6] M.R. Bonsignore, M.C. Suarez Giron, O. Marrone, A. Castrogiovanni, J.M. Montserrat, Personalised medicine in sleep respiratory disorders: focus on obstructive sleep apnoea diagnosis and treatment, Eur. Respir. Rev. 26 (2017).

[7] F. Braido, P. Santus, A. Guido Corsico, F. Di Marco, G. Melioli, N. Scichilone, P. Solidoro, Chronic obstructive lung disease 'expert system': validation of a predictive tool for assisting diagnosis, Int. J. Chron. Obstruct. Pulmon. Dis. 13 (2018) 1747−1753.

[8] K.K. Bressem, L.C. Adams, C. Erxleben, B. Hamm, S.M. Niehues, J.L. Vahldiek, Comparing different deep learning architectures for classification of chest radiographs, Sci. Rep. 10 (1) (2020).

[9] P. Cardinal-Fernández, N. Nin, J. Ruíz-Cabello, J.A. Lorente, Systems medicine: a new approach to clinical practice, Arch. Bronconeumol. 50 (10) (2014) 444−451.

[10] J. Chen, K.C. See, Artificial intelligence for COVID-19: rapid review, J. Med. Internet Res. 22 (10) (2020) e21476.

[11] J. Chen, D. Sun, W. Yang, M. Liu, S. Zhang, J. Peng, C. Ren, Clinical and economic outcomes of telemedicine programs in the intensive care unit: a systematic review and meta-analysis, J. Intensive Care Med. 33 (7) (2018) 383−393.

[12] I.G. Chouvarda, D.G. Goulis, I. Lambrinoudaki, N. Maglaveras, Connected health and integrated care: toward new models for chronic disease management, Maturitas 82 (1) (2015) 22−27.

[13] I. Chouvarda, N.Y. Philip, P. Natsiavas, V. Kilintzis, D. Sobnath, R. Kayyali, J. Henriques, R.P. Paiva, A. Raptopoulos, O. Chetelat, N. Maglaveras, Welcome — innovative integrated care platform using wearable sensing and smart cloud computing for COPD patients with comorbidities, in: 2014 36th Annual International Conference of the IEEE Engineering in Medicine and Biology Society, 2014.

[14] M. Chu, T. Nguyen, V. Pandey, Y. Zhou, H.N. Pham, R. Bar-Yoseph, S. Radom-Aizik, R. Jain, D.M. Cooper, M. Khine, Respiration rate and volume measurements using wearable strain sensors, Npj Digit. Med. 2 (1) (2019).

[15] G.D. Clifford, I. Silva, B. Moody, Q. Li, D. Kella, A. Chahin, T. Kooistra, D. Perry, R.G. Mark, False alarm reduction in critical care, Physiol. Meas. 37 (8) (2016) E5−E23.

[16] A. Dekker, S. Vinod, L. Holloway, C. Oberije, A. George, G. Goozee, G.P. Delaney, P. Lambin, D. Thwaites, Rapid learning in practice: a lung cancer survival decision support system in routine patient care data, Radiother. Oncol. 113 (1) (2014) 47−53.

[17] S. Dragonieri, G. Pennazza, P. Carratu, O. Resta, Electronic nose technology in respiratory diseases, Lung 195 (2) (2017) 157−165.

[18] P. Essay, T.B. Shahin, B. Balkan, J. Mosier, V. Subbian, The connected intensive care unit patient: exploratory analyses and cohort discovery from a critical care telemedicine database, JMIR Med. Inform. 7 (1) (2019) e13006.

[19] F.M.E. Franssen, P. Alter, N. Bar, B.J. Benedikter, S. Iurato, D. Maier, M. Maxheim, F.K. Roessler, M.A. Spruit, C.F. Vogelmeier, E.F.M. Wouters, B. Schmeck, Personalized medicine for patients with COPD: where are we? Int. J. Chron. Obstruct. Pulmon. Dis 14 (2019) 1465−1484.

[20] I. Frerichs, M.B.P. Amato, A.H. Van Kaam, D.G. Tingay, Z. Zhao, B. Grychtol, M. Bodenstein, H. Gagnon, S.H. Böhm, E. Teschner, O. Stenqvist, T. Mauri, V. Torsani, L. Camporota, A. Schibler, G.K. Wolf, D. Gommers, S. Leonhardt, A. Adler, E. Fan, W.R.B. Lionheart, T. Riedel, P.C. Rimensberger, F. Suarez Sipmann, N. Weiler, H. Wrigge, Chest electrical impedance tomography examination, data analysis, terminology, clinical use and recommendations: consensus statement of the TRanslational EIT DevelopmeNt StuDy group, Thorax 72 (1) (2017) 83−93.

[21] S. Fröhlich, N. Murphy, A. Doolan, O. Ryan, J. Boylan, Acute respiratory distress syndrome: underrecognition by clinicians, J. Crit. Care 28 (5) (2013) 663−668.

[22] M.T. García-Ordás, J. Alberto Benítez-Andrades, I. García-Rodríguez, C. Benavides, H. Alaiz-Moretón, Detecting respiratory pathologies using convolutional neural networks and variational autoencoders for unbalancing data, Sensors 20 (4) (2020).

[23] S. Gonem, W. Janssens, N. Das, M. Topalovic, Applications of artificial intelligence and machine learning in respiratory medicine, Thorax 75 (8) (2020) 695−701.

[24] S.M. Hartmann, R.W.D. Farris, O. Yanay, R.M. Diblasi, C.N. Kearney, J.D. Zimmerman, K. Carlin, J.J. Zimmerman, Interaction of critical care practitioners with a decision support tool for weaning mechanical ventilation in children, Respir. Care 65 (3) (2020) 333−340.

[25] V. Hartmann, H. Liu, F. Chen, W. Hong, S. Hughes, D. Zheng, Toward accurate extraction of respiratory frequency from the photoplethysmogram: effect of measurement site, Front. Physiol. 10 (June 2019) 732.

[26] D. Hashoul, H. Haick, Sensors for detecting pulmonary diseases from exhaled breath, Eur. Respir. Rev. 28 (152) (2019).

[27] Y. He, H. Yu, E. Ong, Y. Wang, Y. Liu, A. Huffman, H.H. Huang, J. Beverley, J. Hur, X. Yang, L. Chen, G.S. Omenn, B. Athey, B. Smith, CIDO, a community-based ontology for coronavirus disease knowledge and data integration, sharing, and analysis, Sci. Data 7 (1) (2020) 181.

[28] A. Holzinger, G. Langs, H. Denk, K. Zatloukal, H. Müller, Causability and explainability of artificial intelligence in medicine, Wiley Interdiscip. Rev. Data Min. Knowl. Discov. 9 (2019) e1312.

[29] S.M. Honaker, A. Street, A.S. Daftary, S.M. Downs, The use of computer decision support for pediatric obstructive sleep apnea detection in primary care, J. Clin. Sleep Med. 15 (3) (2019) 453−462.

[30] E. Iadanza, V. Mudura, P. Melillo, M. Gherardelli, An automatic system supporting clinical decision for chronic obstructive pulmonary disease, Health Technol. 10 (2) (2020) 487−498.

[31] I. Jankovic, J.H. Chen, Clinical decision support and implications for the clinician burnout crisis, Yearb. Med. Inform. 29 (1) (2020) 145−154.

[32] D.S. Karbing, C. Allerød, L.P. Thomsen, K. Espersen, P. Thorgaard, S. Andreassen, S. Kjaergaard, S.E. Rees, Retrospective evaluation of a decision support system for controlled mechanical ventilation, Med. Biol. Eng. Comput. 50 (1) (2012) 43−51.

[33] H. Kaul, Respiratory healthcare by design: computational approaches bringing respiratory precision and personalised medicine closer to bedside, Morphologie 103 (343) (2019) 194−202.

[34] R.G. Khemani, J.C. Hotz, K.A. Sward, C.J.L. Newth, The role of computer-based clinical decision support systems to deliver protective mechanical ventilation, Curr. Opin. Crit. Care 26 (1) (2020) 73−81.

[35] R.J. Khusial, P.J. Honkoop, O. Usmani, M. Soares, A. Simpson, M. Biddiscombe, S. Meah, M. Bonini, A. Lalas, E. Polychronidou, J.G. Koopmans, K. Moustakas, J.B. Snoeck-Stroband, S. Ortmann, K. Votis, D. Tzovaras, K. Fan Chung, S. Fowler, J.K. Sont, Effectiveness of myAirCoach: a MHealth self-management system in asthma, J. Allergy Clin. Immunol. Pract. 8 (6) (2020) 1972−1979, e8.

[36] V. Kilintzis, I. Chouvarda, N. Beredimas, P. Natsiavas, N. Maglaveras, Supporting integrated care with a flexible data management framework built upon linked data, HL7 FHIR and ontologies, J. Biomed. Inf. 94 (2019) 103179.

[37] R.D. Kindle, O. Badawi, L. Anthony Celi, S. Sturland, Intensive care unit telemedicine in the era of big data, artificial intelligence, and computer clinical decision support systems, Crit. Care Clin. 35 (3) (2019) 483−495.

[38] O. Kocsis, G. Arvanitis, A. Lalos, K. Moustakas, J.K. Sont, P.J. Honkoop, K.F. Chung, M. Bonini, O.S. Usmani, S. Fowler, A. Simpson, Assessing machine learning algorithms for self-management of asthma, in: 2017 E-Health and Bioengineering Conference, EHB, 2017, pp. 571−574.

[39] E. Koutsiana, A. Chytas, K. Vaporidi, I. Chouvarda, Smart alarms towards optimizing patient ventilation in intensive care: the driving pressure case, Physiol. Meas. 40 (9) (2019).

[40] S. Le, E. Pellegrini, A. Green-Saxena, C. Summers, J. Hoffman, J. Calvert, R. Das, Supervised machine learning for the early prediction of acute respiratory distress syndrome (ARDS), J. Crit. Care 60 (2020) 96−102.

[41] S. Ley, Lung imaging, Eur. Respir. Rev. 24 (136) (2015) 240−245.

[42] M. Light, R.L. Owens, C.N. Schmickl, A. Malhotra, Precision medicine for obstructive sleep apnea, Sleep Med. Clin. 14 (3) (2019) 391−398.

[43] S. Liu, K. Choong See, K. Yuan Ngiam, L. Anthony Celi, X. Sun, M. Feng, Reinforcement learning for clinical decision support in critical care: comprehensive review, J. Med. Internet Res. 22 (2020) e18477.

[44] F. Lococo, A. Cesario, A. Bufalo, A. Ciarrocchi, G. Prinzi, M. Mina, S. Bonassi, P. Russo, Novel therapeutic strategy in the management of COPD: a systems medicine approach, Curr. Med. Chem. 22 (32) (2015) 3655−3675.

[45] T. Loncar-Turukalo, E. Zdravevski, J.M. da Silva, I. Chouvarda, V. Trajkovik, Literature on wearable technology for connected health: scoping review of research trends, advances, and barriers, J. Med. Internet Res. 21 (9) (2019).

[46] G.F. Luger, W.A. Stubblefield, in: George F. Luger, William A. Stubblefield (Eds.), Artificial Intelligence and the Design of Expert Systems, Benjamin/Cummings Pub. Co, Redwood City, CA, 1989.

[47] A. Masood, P. Yang, B. Sheng, H. Li, P. Li, J. Qin, V. Lanfranchi, J. Kim, D.D. Feng, Cloud-based automated clinical decision support system for detection and diagnosis of lung cancer in chest CT, IEEE J. Transl. Eng. Health Med. 8 (2020).

[48] M.P. McRae, G.W. Simmons, N.J. Christodoulides, Z. Lu, S.K. Kang, D. Fenyo, T. Alcorn, I.P. Dapkins, I. Sharif, D. Vurmaz, S.S. Modak, K. Srinivasan, S. Warhadpande, R. Shrivastav, J.T. McDevitt, Clinical decision support tool and rapid point-of-care platform for determining disease severity in patients with COVID-19, Lab Chip 20 (12) (2020) 2075—2085.

[49] E. Mekov, M. Miravitlles, R. Petkov, Artificial intelligence and machine learning in respiratory medicine, Expet Rev. Respir. Med. 14 (6) (2020) 559—564.

[50] L. Mendes, I.M. Vogiatzis, E. Perantoni, E. Kaimakamis, I. Chouvarda, N. Maglaveras, C. Teixeira, J. Henriques, P. Carvalho, R.P. Paiva, Detection of wheezes and crackles using a multi-feature approach, Proc. Book 16 (2017) 31.

[51] S. Nikkonen, I.O. Afara, T. Leppänen, J. Töyräs, Artificial neural network analysis of the oxygen saturation signal enables accurate diagnostics of sleep apnea, Sci. Rep. 9 (1) (2019).

[52] A. Oliveira, J. Rodrigues, A. Marques, Enhancing our understanding of computerised adventitious respiratory sounds in different COPD phases and healthy people, Respir. Med. 138 (2018) 57—63.

[53] V.E. Papaioannou, I. Chouvarda, N. Maglaveras, C. Dragoumanis, I. Pneumatikos, Changes of heart and respiratory rate dynamics during weaning from mechanical ventilation: a study of physiologic complexity in surgical critically ill patients, J. Crit. Care 26 (3) (2011) 262—272.

[54] T. Penzel, I. Fietze, M. Glos, Alternative algorithms and devices in sleep apnoea diagnosis: what we know and what we expect, Curr. Opin. Pulm. Med. 26 (6) (2020) 650—656.

[55] E. Perantoni, D. Filos, K. Archontogeorgis, P. Steiropoulos, I.C. Chouvarda, Pre-diabetic patients with severe obstructive sleep apnea: novel parameters of hypoxia during sleep correlate with insulin resistance, in: in Proceedings of the Annual International Conference of the IEEE Engineering in Medicine and Biology Society, EMBS, 2019, pp. 5002—5005.

[56] T.J. Pollard, A.E.W. Johnson, J.D. Raffa, L.A. Celi, R.G. Mark, O. Badawi, The EICU collaborative research database, a freely available multi-center database for critical care research, Sci. Data 5 (2018) 180178.

[57] R.X.A. Pramono, S. Bowyer, E. Rodriguez-Villegas, Automatic adventitious respiratory sound analysis: a systematic review, PLoS One 12 (5) (2017).

[58] V. Quaresima, M. Ferrari, COVID-19: efficacy of prehospital pulse oximetry for early detection of silent hypoxemia, Crit. Care 24 (1) (2020).

[59] N. Reamaroon, M.W. Sjoding, K. Lin, T.J. Iwashyna, K. Najarian, Accounting for label uncertainty in machine learning for detection of acute respiratory distress syndrome, IEEE J. Biomed. Health Inform. 23 (1) (2019) 407—415.

[60] J.J. Reeves, H.M. Hollandsworth, F.J. Torriani, R. Taplitz, S. Abeles, M. Tai-Seale, M. Millen, B.J. Clay, C.A. Longhurst, Rapid response to COVID-19: health informatics support for outbreak management in an academic health system, J. Am. Med. Inf. Assoc. 27 (6) (2020) 853—859.

[61] D. Révész, E.G. Engelhardt, J.J. Tamminga, F.M.N.H. Schramel, B.D. Onwuteaka-Philipsen, E.M.W. Van De Garde, E.W. Steyerberg, E.P. Jansma, H.C.W. De Vet, V.M.H. Coupé, Decision support systems for incurable non-small cell lung cancer: a systematic review, BMC Med. Inf. Decis. Mak. 17 (1) (2017).

[62] B.M. Rocha, D. Filos, L. Mendes, G. Serbes, S. Ulukaya, Y.P. Kahya, N. Jakovljevic, T.L. Turukalo, I.M. Vogiatzis, E. Perantoni, E. Kaimakamis, P. Natsiavas, A. Oliveira, C. Jácome, A.S.P.D. Marques, N. Maglaveras, R. Pedro Paiva, I. Chouvarda, P. de Carvalho, An open access database for the evaluation of respiratory sound classification algorithms, Physiol. Meas. 40 (2019) 035001.

[63] C.J. Ryerson, T.J. Corte, J.S. Lee, L. Richeldi, S.L.F. Walsh, J.L. Myers, J. Behr, V. Cottin, S.K. Danoff, K.R. Flaherty, D.J. Lederer, D.A. Lynch, F.J. Martinez, G. Raghu, W.D. Travis, Z. Udwadia, A.U. Wells, H.R. Collard, A standardized diagnostic ontology for fibrotic interstitial lung disease an international working group perspective, Am. J. Respir. Crit. Care Med. 196 (10) (2017) 1249—1254.

[64] M.B. Sesen, M.D. Peake, R. Banares-Alcantara, D. Tse, T. Kadir, R. Stanley, F. Gleeson, M. Brady, Lung cancer assistant: a hybrid clinical decision support application for lung cancer care, J. R. Soc. Interface 11 (98) (2014) 20140534.

[65] O.P. Singh, R. Palaniappan, M. Malarvili, Automatic quantitative analysis of human respired carbon dioxide waveform for asthma and non-asthma classification using support vector machine, IEEE Access 6 (2018) 55245—55256.

[66] M.W. Sjoding, Translating evidence into practice in acute respiratory distress syndrome: teamwork, clinical decision support, and behavioral economic interventions, Curr. Opin. Crit. Care 23 (5) (2017) 406—411.

[67] C. Subirà, C. de Haro, R. Magrans, R. Fernández, L. Blanch, Minimizing asynchronies in mechanical ventilation: current and future trends, Respir. Care 63 (4) (2018) 464—478.

[68] R.T. Sutton, D. Pincock, D.C. Baumgart, D.C. Sadowski, R.N. Fedorak, K.I. Kroeker, An overview of clinical decision support systems: benefits, risks, and strategies for success, Npj Digit. Med. 3 (2020) 1—10.

[69] K. Sward, C. Newth, Computerized decision support systems for mechanical ventilation in children, J. Pediatr. Intensive Care 05 (03) (2015) 095—100.

[70] A. Teramoto, A. Yamada, T. Tsukamoto, K. Imaizumi, H. Toyama, K. Saito, H. Fujita, Decision support system for lung cancer using PET/CT and microscopic images, in: Advances in Experimental Medicine and Biology, vol. 1213, 2020, pp. 73—94.

[71] P.I. Terrill, A review of approaches for analysing obstructive sleep apnoea-related patterns in pulse oximetry data, Respirology 25 (5) (2020) 475—485.

[72] V. Tomicic, R. Cornejo, Lung monitoring with electrical impedance tomography: technical considerations and clinical applications, J. Thorac. Dis. 11 (7) (2019) 3122—3135.

[73] H.H. Tseng, Y. Luo, S. Cui, J.T. Chien, R.K. Ten Haken, I.E. Naqa, Deep reinforcement learning for automated radiation adaptation in lung cancer, Med. Phys. 44 (12) (2017) 6690—6705.

[74] N. Verstraete, G. Jurman, G. Bertagnolli, A. Ghavasieh, V. Pancaldi, M. De Domenico, CovMulNet19, integrating proteins, diseases, drugs, and symptoms: a network medicine approach to COVID-19, Netw. Syst. Med. 3 (1) (2020) 130—141.

[75] M.T. Wayne, T.S. Valley, C.R. Cooke, M.W. Sjoding, Electronic sniffer systems to identify the acute respiratory distress syndrome, Ann. Am. Thorac. Soc. 16 (4) (2019) 488—495.

[76] B.D. Winters, M.M. Cvach, C.P. Bonafide, X. Hu, A. Konkani, M.F. O'Connor, J.M. Rothschild, N.M. Selby, M.M. Pelter, B. McLean, S.L. Kane-Gill, Technological distractions (Part 2): a summary of approaches to manage clinical alarms with intent to reduce alarm fatigue, Crit. Care Med. 46 (1) (2018) 130—137.

[77] L. Wynants, B. Van Calster, G.S. Collins, R.D. Riley, G. Heinze, E. Schuit, M.M.J. Bonten, J.A.A. Damen, T.P.A. Debray, M. De Vos, P. Dhiman, M.C. Haller, M.O. Harhay, L. Henckaerts, N. Kreuzberger, A. Lohmann, K. Luijken, J. Ma, C.L. Andaur Navarro, J.B. Reitsma, J.C. Sergeant, C. Shi, N. Skoetz, L.J.M. Smits, K.I.E. Snell, M. Sperrin, R. Spijker, E.W. Steyerberg, T. Takada, S.M.J. Van Kuijk, F.S. Van Royen, C. Wallisch, L. Hooft, K.G.M. Moons, M.Van Smeden, Prediction models for diagnosis and prognosis of covid-19: systematic review and critical appraisal, BMJ 369 (2020) 18.

[78] L. Ye, G.W. Pien, S.J. Ratcliffe, E. Björnsdottir, E.S. Arnardottir, A.I. Pack, B. Benediktsdottir, T. Gislason, The different clinical faces of obstructive sleep apnoea: a cluster Analysis, Eur. Respir. J. 44 (6) (2014) 1600–1607.

[79] E. Yoshida, S. Fei, K. Bavuso, C. Lagor, S. Maviglia, The value of monitoring clinical decision support interventions, Appl. Clin. Inf. 9 (1) (2018) 163–173.

[80] J. Zhu, B. Shen, A. Abbasi, M. Hoshmand-Kochi, H. Li, T.Q. Duong, Deep transfer learning artificial intelligence accurately stages COVID-19 lung disease severity on portable chest radiographs, PLoS One 15 (July 7, 2020).

Chapter 11

Integrated care in respiratory function management

Iman Hesso, Reem Kayyali and Shereen Nabhani-Gebara
Kingston University London, KT, United Kingdom

An overview of integrated care

Understanding integrated care versus fragmented care

Fragmentation of care is a well-documented problem/issue across health-care systems worldwide [1]. According to the fragmentation hypothesis, care fragmentation occurs when delivery of care involves multiple providers and organizations with no single entity effectively coordinating different aspects of care [2]. This poor coordination across the different entities within the system potentially leads to suboptimal care, including poor patient outcomes, important health-care issues being inadequately addressed, and unnecessary or even harmful services that ultimately both increase costs and degrade quality [2].

Hence, fragmentation of care can undermine the value of health care especially for patients with long-term conditions [2,3] and complex bio-psychosocial needs [1]. In fact, it is a critical problem with detrimental effect on quality and costs of care [2] leading to unsustainable health-care systems and inequality in health-care services [4]. In the United States, fragmentation of care has been considered as an important source of inefficiency [4,5], ineffectiveness, inequality, commoditization, commercialization, depro-fessionalism, depersonalization, despair, and discord within the health-care system [4]. In a previous research in the United States, fragmented care was positively associated with increased costs of care, higher chance of deviation from clinical best practice, and higher rates of preventable hospitalizations among patients with long-term conditions [2]. Another research in the United States also reported that fragmented care was associated with higher resources utilization irrespective of the patient's clinical condition and preferences [5]. Similarly, in the United Kingdom, several reports have starkly underlined the fragmented and complex picture of services many patients still encounter when being treated by the National Health Service (NHS), emphasizing the

Wearable Sensing and Intelligent Data Analysis for Respiratory Management
https://doi.org/10.1016/B978-0-12-823447-1.00007-5
323

need for a more sustainable and a more patient-centered NHS [6,3]. Additionally, several articles have highlighted the need to overcome fragmentation in health care across many countries worldwide [7], including Canada [8], Austria, the Netherlands, and Germany [9], just to mention a few.

Thus, efforts have been directed toward fostering integration of care as an approach in the face of fragmentation experienced across health-care systems globally [1,4,10−12,3]. Integrated care has been increasingly advocated as an approach to promote better coordination of services and quality of care in the face of three main global and interlocking challenges including an aging demographic, increasing prevalence of chronic conditions and unmet needs for more personalized and connected care [11,3].

However, there has been an abundance of definitions for integrated care across the published literature with more than 175 definitions, indicating a wide variation in the concept understanding and definition [11,13]. Moreover, many synonyms have been used interchangeably such as coordinated care, managed care, seamless care, guided care, shared care, disease management, care management, disease management programs, and comprehensive care programs [11,14]. For some authors, integrated care is an organizational process for achieving continuity of care based on patients' holistic needs and views [11]. Others define integrated care as the coordination of the care delivery system, involving multiple interventions targeting patients, health-care professionals, and organizations as well [11]. However, the absence of a unifying definition or a common conceptual understanding of integrated care can be mostly attributed to the polymorphous nature of integrated care itself [14]. It can also be attributed to the perspective of the various stakeholders within the health-care systems [10]. Box 11.1 provides some of the most commonly used definitions based on these perspectives [10].

Given the polymorphous nature of integrated care, it is difficult to find a single model that can fit all settings, circumstances, and contexts [11]. This in return renders the evidence for consistent and reproducible benefits to be elusive [18]. Hence, although key lessons can be learnt from different successful integrated care models and experiences, transferring or replicating these experiences across different countries might not always be successful [11].

What becomes clear is that integrated care is not a concept that can be narrowly defined. It should be viewed as an overarching term for a broad and multicomponent set of principles, ideas, and practices intrinsically shaped by contextual factors that aim for better coordination of care around people's holistic needs [10,18].

Rationale for integrated care

As indicated above, the provision of integrated care becomes of paramount importance due to three important and interrelated challenges explained below.

BOX 11.1 Four commonly used definitions of integrated care

A health system—based definition

"Integrated health services: health services that are managed and delivered so that people receive a continuum of health promotion, disease prevention, diagnosis, treatment, disease-management, rehabilitation and palliative care services, coordinated across the different levels and sites of care within and beyond the health sector, and according to their needs throughout the life course" (Contadndriapoulos et al. [15] cited in Ref. [10]; p. 1).

A manager's definition

"The process that involves creating and maintaining, over time, a common structure between independent stakeholders ... for the purpose of coordinating their interdependence in order to enable them to work together on a collective project" (Kodner and Spreeuwenberg, [16] cited in Ref. [10]; p. 1).

A social science—based definition

"Integration is a coherent set of methods and models on the funding, administrative, organisational, service delivery and clinical levels designed to create connectivity, alignment and collaboration within and between the cure and care sectors. The goal of these methods and models is to enhance quality of care and quality of life, consumer satisfaction and system efficiency for people by cutting across multiple services, providers and settings. Where the result of such multipronged efforts to promote integration lead to benefits for people the outcome can be called 'integrated care'."

A patient's definition (patient-centered coordinated care)

"I can plan my care with people who work together to understand me and my carer(s), allow me control, and bring together services to achieve the outcomes important to me" (Lewis et al. [17] cited in Ref. [10]; p. 1).

Source: N. Goodwin, *Understanding integrated care*, Int. J. Integr. Care 16 (2016).

Aging society

The pace of population aging is increasing dramatically worldwide [14]. Population aging is currently considered a serious challenge across western and industrialized countries, with estimates indicating a drastic increase in the number of older people aged 60 and above from 900 million in 2015 to 1.4 billion by 2030 and 2.1 billion by 2050 worldwide [19,20]. This in turn poses a growing concern about the ways to maintain the health and well-being of this growing cohort of population [19].

Increased chronic disease prevalence

Multimorbidity constitutes another significant challenge for health-care systems. There has been a rapid increase in the prevalence of multiple long-term conditions worldwide, which comes in line with the aging nature of population demographics as an important factor driving chronic disease prevalence [11,21,22]. Currently, estimates indicate one in three adults to be living with one chronic condition or multiple chronic conditions and the figure becomes closer to three out of four in older adults living in developed countries [21]. Pertaining to this, comes the economic burden of chronic conditions which has been estimated to be in the span of $47 trillion from 2010 through 2030 worldwide [23]. The management of such chronic conditions necessitates the presence of multifaceted and multiinstitutional levels of care, which again underscores the importance of care integration in this context [11].

Need for more personalized and connected care

The active aging paradigm coupled with multimorbidity contribute significantly to the need for more personalized and connected care. Integrated care is seen as a potential solution to the growing demand for improved patient experience and health outcomes of patients living with multimorbid conditions and needing long-term care [14]. In addition, integrated care has the potential to leverage personalized health [24] especially with the deployment of information and communication technologies (ICTs).

Types, levels, and forms of integrated care

Several conceptual frameworks and taxonomies were proposed to help in understanding the concept [10]; these include:

- The type of integration: professional integration, service integration, functional integration, organizational integration, and system integration [10,11].
- The level of integration: at the macro level (system level), meso level (organizational level, professional level), and micro level (service level, personal level) [10,11].
- The process of integration: related to the organization and management of integrated care delivery [10].
- The breadth of integration: for example, integration across a whole population group or specific client group [10].
- The degree or intensity of integration: integration linking or coordinating both formal and informal care provision and with fully integrated teams or organizations [10].

There are several forms of integration, including:

- Horizontal integration: Integration of care between health services, social services, and other care providers. This form of integration usually depends on the development of multidisciplinary teams and/or care networks to support a specific group, for example, older people with specific complex needs [10].

- Vertical integration: Integration of care across primary, secondary, and tertiary care services. Integration is usually driven through the establishment of protocol (best practice), care pathways for people with specific conditions (such as chronic obstructive pulmonary disease (COPD) and diabetes), and/or care transitions between hospitals to intermediate and community-based care providers [10].
- Sectoral integration: Integration of care within the one sector. An example of this would be combining horizontal and vertical programs of integrated care within mental health services through multidisciplinary teams and networks of primary, secondary, and tertiary care providers [10].
- People-centered integration: Integration of care between patients and other service users and health-care providers to engage and empower people through shared decision-making, health education, supported self-management, and community engagement [10].
- Whole-system integration: This is the most ambitious form of integrated care since it focuses on the multiple needs of whole populations, not just those related to specific care groups or diseases. It is related to integration of care that encompasses public health to support both a population-based and person-centered approach to care [10].

Importance and benefits of integrated care in chronic respiratory conditions: Chronic Obstructive Pulmonary Disease

Asthma and COPD are among the most prevalent respiratory conditions [25], with more than half a billion people affected worldwide [26,27].

COPD is a common preventable and treatable respiratory condition that is characterized by persistent airflow limitation that is partially/not fully reversible, and progressive over the years. Chronic dyspnea, cough, and sputum production are the main clinical manifestations of the disease [27]. COPD is associated with a high burden in terms of prevalence, mortality, morbidity, and economic costs worldwide [28−30]. COPD affects more than 384 million people worldwide with a global prevalence of 11.7% [27], albeit the condition is well documented to be underdiagnosed in many countries [31−41]. According to World Health Organization (WHO), COPD is currently the fourth leading cause of mortality worldwide and it is projected to become the third by 2030 [42,43]. In 2015, more than 3 million deaths were attributed to COPD, equating to 5% of all deaths worldwide [44].

Furthermore, the economic burden of COPD is high on health-care systems across the world. Such high costs are usually attributed to drug expenditure, costly hospital admissions due to exacerbations, and productivity losses [27,45,46]. In the European Union (EU), the total direct cost of COPD is estimated at €38.6 billion representing 6% of the total health-care expenditure and 56% of disease expenditure on respiratory conditions, whereas productivity losses amount to a total of €28.5 billion annually [27]. Among respiratory conditions, COPD is considered the leading cause of lost work days. In the EU, approximately 41,300 lost work days per 100,000 population are due to COPD. The costs of COPD in the United States accounted for $32 billion as direct costs and $20.4 billion as indirect costs in 2010 [27]. In the United Kingdom, the total direct costs of COPD were estimated at £1.9 billion in 2014 [47]. In Australia, the health-care expenditure on COPD was estimated at $977 million in 2015–16, representing 24% of disease expenditure on respiratory conditions [48].

Most importantly, COPD is a multimorbid condition. The prevalence of comorbidities is common among COPD patients [45,46,49,50]. The most common associated comorbidities include: hypertension, chronic heart failure, diabetes, anxiety and depression, osteoporosis and osteoarthritis [49,50]. The presence of comorbid conditions has important consequences for COPD assessment and management. These comorbidities significantly contribute to the disease severity and complexity of disease management in individual patients [30,49,50]. Additionally important, these comorbidities have been identified as a major driver of excess costs of COPD patients [30,45].

As such, COPD is currently seen as a complex and heterogonous condition, with some authors arguing for it to be considered as a syndrome rather than a disease [51]. The heterogeneity of the condition is manifested in its clinical symptoms, natural history, prognosis, treatment response, and associated comorbidities. As a consequence of this heterogeneity, health-care professionals usually face several challenges when caring for COPD patients [51]. The successful management of COPD patients requires a multidisciplinary and more holistic approach including the assessment and appropriate treatment of the condition and associated comorbidities [49,50]. This in turn underscores the importance of the concept of integrated care for COPD patients with associated comorbidities. Hence, the current chapter focuses in particular on COPD, being a major chronic multimorbid respiratory condition.

Evidence supporting benefits of integrated care

Integrated care programs have been widely researched and evaluated for their effect on patient-specific outcomes and health-care provision costs. Pilot studies and programs of integrated care to improve care for people with long-term conditions (including those with chronic multimorbidity) have been established across high-income countries, including Australia, Canada, Italy, the Netherlands, New Zealand, Sweden, the United Kingdom, and the United States [18]. A review of systematic reviews highlighted the positive effect of integrated care programs on the quality of care for patients living with long-term conditions [12]. Furthermore, according to a more recent systematic review of United Kingdom and international evidence, there is strong evidence that new models of integrated care can enhance perceived quality of care, patient satisfaction, and improve access to services [13].

The positive effects of integrated care programs in respiratory chronic conditions has been also evident in several studies [12,25]. In COPD, the evidence from a systematic review of integrated disease management programs published in 2009 reported improvements in quality of life and reductions in hospitalizations in triple integration interventions (patient-related, professional-directed, and organizational interventions) [25]. In one study, the effect of integrated care on COPD patients recently discharged from hospital after being admitted with exacerbations was evaluated and compared to usual care. The integrated care model was conducted by a specialized respiratory nurse and included a comprehensive assessment of the patient (including adherence, comorbidities, and severity of the disease), patient education, individualized care with follow-up, and a 24-hour access to care via an ICT platform. A total of 113 patients were recruited and randomized to two groups: integrated care and usual care. After 12 months, the integrated care group had improved nutritional status and self-management scores which included knowledge, exacerbation identification, inhaler adherence, and correctness. These factors translated into a significant reduction of hospitalization and hospital readmission rates and increase in the percentage of patients without admission [28]. Other studies have also demonstrated positive effects associated with integrated care in COPD. One study has demonstrated the effectiveness of a multidisciplinary comprehensive COPD program in decreasing hospital readmissions and length of hospital stay among the enrolled patients in comparison to 1 year before joining the program [52]. The program consisted of input from respiratory physicians, nurses, and physiotherapists and involved optimization of COPD treatment, patient education, management of comorbidities, telephone call by nurses, telephone hotline for advice, and physiotherapy training [52]. A Cochrane systematic review evaluated the effects of integrated disease management programs in COPD. The review included 26 studies involving 2997 subjects from 11 countries and demonstrated significant improvements associated with these programs in relation to health-related quality of life, exercise capacity, in addition to reduction in hospital admissions and hospitalization days per person [53]. In a second related review, these programs were found to have favorable effects on both health outcomes and costs, but with considerable heterogeneity in results depending on patient, intervention, and study characteristics [54]. This review reported annual average savings in health-care costs of €898 per person (95% CI: €231 to €1566), hospitalization costs of €1060 per person (95% CI: €80 to €2040), and a decreased rate ratio of hospitalizations (0.75, 95% CI, 0.54−1.03). However, caution should be exercised when interpreting the results given that the costs of developing, implementing, and operating the programs were not included in these estimates. Interestingly, the effects and cost savings increased with severity of COPD [54].

Respiratory disease management: current state of care and care pathways

COPD care pathways

Different countries often tend to have different care delivery pathways when it comes to chronic disease management [55], and COPD is no exception. A previous research by Ref. [55] compared COPD care delivery pathways across five EU countries: the United Kingdom, Greece, Ireland, the Netherlands, and Germany. The research aimed to map COPD care pathways and underscored the similarities and differences in the pathways across the different investigated countries. The current section provides an overview about the results of this research. In essence, the research found that COPD care pathways were fragmented across the five countries.

Diagnostic parameters and guidelines

There was a slight difference with respect to the diagnostic parameters, normal ranges, and therapeutic targets, for COPD patients between the five countries. The guidelines used differed between countries. England and Ireland follow the National Institute for Health and Care Excellence (NICE) guidelines for COPD management, the Netherlands follows the American Thoracic Society (ATS)/European Respiratory Society (ERS) guidelines, Greece follows the Global Initiative for Chronic Obstructive Lung Disease (GOLD) guidelines, and Germany follows the National Guidelines for the Diagnosis and Therapy of COPD issued by "Deutsche Atemwegsliga" and "Deutsche Gesellschaft für Pneumologie und Beatmungsmedizin" [55].

Follow-up services

A marked variability in the follow-up care services within the pathways was observed between the five countries. In England, provision of follow-up services is done via community services, general practitioner (GP) visits, hospital chest clinics, and special clinical teams. Health-care professionals involved in the follow-up care services are GPs, consultants, physiotherapists, and nurses. England is the only country to provide community services through hospital specialist respiratory teams working in the community to provide respiratory clinics, pulmonary rehabilitation, and smoking cessation.

In the Netherlands, patients' follow-up occurs in primary or secondary care. GPs are involved in the diagnosis, treatment, prevention, and follow-up care of patients with COPD, and they have a gate-keeping role within the pathway. The gate-keeping principle determines what hospital care and specialist care to be provided to each patient, except for emergency care. Specialist care involves referral to special centers such as CIRO (Centre of Expertise for Chronic Organ Failure) where services such as pulmonary rehabilitation are provided to patients.

In Greece, there are no special services provided for outpatients with no role of primary care. In Ireland, inpatients are provided with an early supported discharge, assisted discharge, pulmonary rehabilitation, and transitional care program via an outreach team. The outreach team consists of a respiratory consultant, a practice physiotherapist, and nurses. For outpatients, a service in a respiratory COPD clinic and a pulmonary rehabilitation program are available.

The follow-up parameters are similar across the five countries. These include spirometry-derived pulmonary function measures, blood gases, oxygen saturation, and body mass index. The frequency of the follow-up was found to depend on patients' needs, but in general, it is performed in face-to-face clinics on an annual basis in England and biannual basis in Greece and Ireland.

England and Ireland are the only two countries that currently have special schemes to support COPD patients at home. In England, a telehealth service is provided to COPD patients in some districts to facilitate discharge or to provide rescue treatment. In Ireland, several programs are used to support patients who meet certain criteria at home. These include early supported and assisted discharge aiming to accelerate discharge from hospital and provide support in the community. Another service is the transitional care program to ensure the coordination/continuity of health care when patients transfer between different settings or levels of care [55].

COPD and comorbidities

The management of COPD did not differ even if the patient is diagnosed with comorbidities in the five countries. So even if the patient suffered from multiple comorbidities, each comorbidity is managed independently following its own national guideline and through separate clinical teams. Absence of direct communication between the different diagnosed condition services was also reported across the five countries, reflecting fragmentation of care at its heart/core. In all countries, the GP is the main responsible health-care professional for liaising and referring patients between these different services and clinical teams. The only exception is Greece where this is done via the pulmonologists [55].

Adherence and nonpharmacological interventions

The services provided for lifestyle management are similar between the five countries, with no specific tools in place to enhance patients' adherence, and no specified role/training for patients' carers (family, partners, and friends). Examples of these services include: advice on annual flu vaccination, smoking cessation, advice on importance of pulmonary rehabilitation, and referral to a dietician. Differences were noted across the five countries with respect to monitoring patients' adherence to therapy. In England, adherence to therapy and nonpharmacological interventions are monitored mainly by the appropriate teams (e.g., pulmonary rehabilitation and smoking cessation). In the Netherlands, patients' adherence is assessed during group sessions (training, education, smoking cessation). In Greece, the pulmonologists monitor adherence to therapy with questionnaires and device demonstration, whereas in Ireland, no specific services are available for monitoring adherence to therapy [55]. Fig. 11.1 illustrates the COPD care pathways across the five European countries.

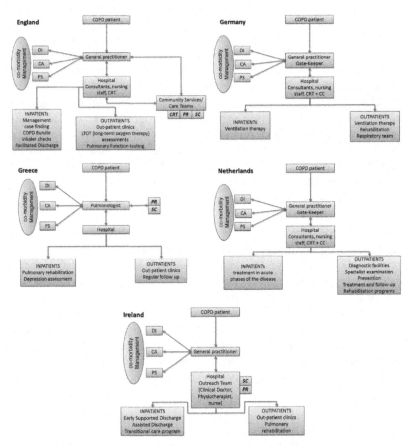

FIGURE 11.1   COPD care pathways across England, Germany, Greece, the Netherlands, and Ireland. Abbreviations: *CA*, cardiologist; *CC*, chest clinic; *DI*, diabetologists; *PR*, pulmonary rehabilitation; *PS*, psychiatrist; *RT*, respiratory team; *SC*, smoking cessation. *Source: From R. Kayyali, B. Odeh, B. et al. COPD care delivery pathways in five European Union countries: mapping and health care professionals' perceptions, Int. J. Chronic Obstr. Pulm. Dis. 11 (2016) 2831–2838. Original publisher: Dove Medical Press.*

Current state of COPD care

An important way to understand the current care pathway is by understanding the perceptions of the different stakeholders involved in these pathways. Hence, this section will draw on the perceptions of different stakeholders including health-care professionals, patients, and informal carers involved in COPD care pathway.

Health-care professionals' perceptions

A previous research by [11,55] investigated health-care professionals perceptions about state of COPD care in five European counties: the United Kingdom, Greece, Ireland, the Netherlands, and Germany. In this research, health-care professionals reported several issues regarding their view of the current state of care including lack of communication between health-care professionals, limited resources, and poor patient engagement. Lack of communication between different health-care professionals was a common and significant issue within the pathways across the five countries. Health-care professionals reported different forms of fragmentation: between the different health-care professionals involved in the care of a patient; between primary care and secondary care; and between health and social care. According to health-care professionals, lack of communication between the different health-care teams in primary and secondary care creates several challenges especially in terms of continuity of care to COPD patients and knowing what happens to patients on a day-to-day basis. In addition, health-care professionals highlighted the limited involvement of pharmacists in COPD patients care pathways. Lack of resources in terms of staff, time constraints, and training was another issue noted within current state of care for COPD patients. Lack of these resources was expressed as an important concern impacting quality of care especially considering the documented high hospitalization rates associated with COPD. In addition, lack of resources available for patients to contact doctors when they feel sick and knowing how patients' social and economic environment influences their physical condition were also reported as challenges by health-care professionals.

Poor patient engagement was also reported as an issue, as health-care professionals were not sure how to encourage patients to better manage their medication and treatment of COPD. This is particularly an issue taking into consideration the complications from complex treatment regimens in COPD. Information provision, enabling patients to self-manage their condition and other comorbidities, and recognizing problems in good time were reported as challenges by health-care professionals [11,55].

In another research in France, hospital practitioners across 25 main hospitals reported a global lack of coordination between ambulatory care and in-hospital care of COPD patients before and after hospitalization [56]. This was also echoed in another research [57]; albeit in primary care, where fragmented care and lack of communication among health-care professionals were perceived by community pharmacists to be negatively affecting care provision for COPD and asthma patients. Poor integration of community pharmacies into the health-care system was also a crucial weakness identified by community pharmacists in this research.

Patients' perceptions

According to the research conducted by [11,50], fragmentation of care and lack of communication between different health-care professionals was identified by patients as a common and main challenge in the current health-care system across four countries: the United Kingdom, Ireland, the Netherlands, and Greece. This resulted in patients facing long waiting times before getting appointments or referrals to specialists. In the Netherlands, patients elaborated on the differences in pulmonary care between the different care settings (primary and secondary/tertiary) stating that the process prior to referral to a pulmonologist was not adequate as referrals to pulmonologists are unduly late with GPs- not being able to address some of the COPD-related problems independently. Therefore, the majority of patients contact the pulmonary nurse or pulmonologist directly if they need any help. The pulmonary nurse was seen/depicted as a good intermediary between them and the pulmonologist [11,50].

Carers' perceptions

Carers echoed similar perceptions regarding the fragmented state of care when it comes to treating COPD patients. Carers across the four countries indicated that lack of communication between health-care professionals makes medicines arrangement very hard, which in turn causes delays in getting the medications. Additionally, carers expressed their frustration with having to make a judgment call about their relative's health status in case of an exacerbation as they do not always know when to call and seek help [11,50].

Interestingly, the carers in the UK cohort appreciated the respiratory HOT clinic at the Croydon University Hospital, which is a primary health clinic run by a multidisciplinary team from secondary care including respiratory consultant, nurse, and physiotherapist. The clinic provides rapid access to help with the management of COPD patients at risk of hospitalization in order to avoid potential hospital admissions [11,50].

Information and communication technology to support integrated care

What is technology enabled care?

The history of technology use in health care can be dated back to as early as the 1900s, which started with the use of radio communications to provide medical services to Antarctica [58]. The advent of ICT has revolutionized various aspects of health-care services [59].

Technology enabled care (TEC) has been recognized as an approach to promote the management of people with long-term complex health and social care needs and to maintain their independence using of modern ICTs. Telecare, tele-coaching, telemedicine, telehealth, wearable technology, and assistive technology are examples of TEC services [60]. These technologies are designed to enable the remote monitoring of health status and collect information that can inform treatment plans, and act as a powerful tool in the coordination and delivery of care. TEC is also seen as a component in health and social care pathways [60]. Additionally, TEC has been increasingly seen as an integral part of the solution to many of the challenges facing the health sector [61].

Nevertheless, the role of technology in health care has been described using different terminologies. The presence of several definitions for each of these technology based services indicates lack of consensus about these terms in the international literature [62]. For example, the terms "e-health," "telemedicine," and "telehealth" are often used interchangeably in the literature [63]. A bibliometric analysis which examined the occurrence of these terms in the Scopus database found 11,644 documents containing one of the three terms in the abstract or title. Interestingly, out of the 11,644 documents found, telemedicine was the most frequent term with 8028 documents thus accounting for 69%, followed by e-health with 2573 documents (22%), and telehealth with 1679 documents (14%). Hence, the variations found in the adoption level of these three terms within the health-care literature highlights a lack of clarity in terms of their definitions and the concepts they refer to [63]. In the below section, we provide an overview about some of the definitions pertaining to TEC services.

E-health

E-health is the broad umbrella term that encompasses all the previous mentioned technologies, referring to all forms of electronic health-care delivery. And while the terms telehealth and e-health are most often used interchangeably, yet the distinction between the two concepts lies in the fact that e-health applications are not limited to health care over a distance, as it is the case with telehealth [64].

One of the most widely used and cited definitions of e-health was published in 2001 [65], which defined e-health as "the intersection between medical informatics, public health, and business referring to services and information delivered or enhanced through the internet and related technologies." Interestingly, a previous systematic review compared 51 unique definitions of e-health. The review concluded that e-health as a term encompasses a set of disparate concepts including technology, health, and commerce, with varying degree of emphasis on these concepts among the retrieved definitions. In 2005, the WHO defined e-health as "the cost-effective and secure use of ICTs in support of health and health-related fields, including health-care

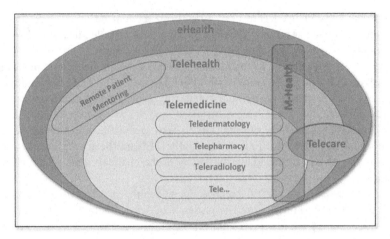

FIGURE 11.2 The relatioship between e-health, m-health, telecare, telemedicine, telehealth and remote patient monitoring. *Source: Modified from L. Van Dyk, A review of telehealth service implementation frameworks, Int. J. Environ. Res. Publ. Health, 11 (2014) 1279–1298.*

services, health surveillance, health literature, and health education, knowledge and research." Fig. 11.2 provides an illustration about the relationship between the different TEC approaches.

Telehealth

Telehealth involves the use of remote exchange of data between the patient and the health-care provider using technological devices to support the diagnosis and management of patients with long-term conditions [66,67]. The WHO defines telehealth as "delivery of health care services, where patients and providers are separated by distance. Telehealth uses ICT for the exchange of information for the diagnosis and treatment of diseases and injuries, research and evaluation, and for the continuing education of health professionals. Telehealth can contribute to achieving universal health coverage by improving access for patients to quality, cost-effective, health services wherever they may be. It is particularly valuable for those in remote areas, vulnerable groups and ageing populations" [68].

Interestingly, some scholars relate telehealth to telemedicine, in the same manner that health relates to medicine. In 1978, some scholars coined the term telehealth to extend the scope of telemedicine by incorporating a broader set of activities. As such, telehealth is envisaged as an expansion to telemedicine because it incorporates the curative, preventative, and promotive aspects of the health field, hence it should be distinguished from telemedicine which has a narrower focus on the curative aspect only, thus making telemedicine a subset of telehealth [64].

Telemedicine/teleconsultations

Telemedicine involves the use of advanced ICTs with the purpose of exchanging health information/provision of consultations between patients and health-care providers. It can allow patients and health-care professionals to communicate 24/7, using smartphones or webcam-enabled computers. It offers the opportunity to provide health-care services across time, geographic, social, and cultural barriers. This includes subsections such as teledermatology (to provide dermatological services) and telepharmacy (to deliver pharmaceutical care). Telemedicine is related to curative aspects in the sense that it is restricted to delivery of services with curative purposes only [60,64,69].

Telecare

Telecare involves the management of the risks associated with independent living through the continuous, automatic, and remote monitoring of real-time emergencies and lifestyle changes over time. This approach is quite important particularly for the elderly population and those with complex health and social care needs. Telecare is a preventative approach, hence it comes under the scope/umbrella of telehealth but not telemedicine [64].

Tele-coaching

Tele-coaching involves provision of advice via telephone from a coach to empower and support people to build knowledge, skills, and confidence to change behaviors [60]. It offers the opportunity to coach patients from a distance in an automated or semiautomated way, hence reducing the burden of face-to-face interactions for both health-care professionals and patients [70].

Remote patient monitoring/telemonitoring

Remote patient monitoring (RPM) stands for a variety of technologies that are used by patients at homes, outside the clinical environment (clinics, pharmacies, hospitals) in order to monitor their health conditions. A typical example of such monitoring devices include: weight scales, blood pressure monitors, glucometers, and pulse oximeter [11].

A significant evolution in telemonitoring came about with wearable technology. The latter is a form of mobile technology whereby sensors are embedded in the users clothing or as an accessory to collect data over periods of time [71,72]. There has been a growing interest in the development, design, and implementation of wearable technologies. Example of these technologies include: smart shirts for ubiquitous health, activity, and vital signs monitoring; smart vests with integrated sensors to monitor physiological parameters; armband wearable body monitors; and instrumented wearable belts for wireless health monitoring [73–77]; these devices can collect a large range of data from blood sugar, oxygen level, and exercise routines to sleep and mood. The market for wearable technologies in the health-care sector is expected to rise from $20 billion in 2015 to around $70 billion by 2025. This is because smart wearable technologies, when continuously monitoring vital signs, have the potential to become significantly effective in timely prevention, diagnosis, treatment, and control of long-term conditions [78].

mHealth

According to WHO, mobile health or mHealth is considered as a component of e-health [79], which makes it consequently a subcategory of telecare, tele-medicine, and telehealth [64]. It is defined as the "medical and public health practice supported by mobile devices, such as mobile phones, patient moni-toring devices, personal digital assistants (PDAs), and other wireless devices". It "involves the use of mobile phone's core utility such as voice and short messaging service (SMS) as well as more complex functionalities and mobile application software including third and fourth generation mobile telecom-munications (3G and 4G systems), global positioning system (GPS) and Bluetooth technology."

This brought about the explosion of mobile applications which along with wearables are considered two of the pillars of mHealth [80].

How can technology enabled care support integrated care?

In fact, the employment of ICT systems is considered as an important enabler for integrated care. The use of modern ICT systems can be considered as an aspect of functional integration [11,81]. Remarkably, functional integration is an important type of integrated care because it supports all the other types of integration (clinical, professional, organizational, and system integration) through linking financial, management, and information systems across the health system in order to add the greatest overall value to the system [11,82].

Perceptions of different stakeholders toward technology enabled care

The successful implementation of technologies in health care is largely depen-dent on the perceptions and acceptability of the different stakeholders and po-tential users of these technologies [83,84]. As such, the current sections provide an overview of the perceptions of health-care professionals, patients, and carers regarding some TEC services implemented for the management of COPD.

Health-care professionals' perceptions

A previous research by [11,50] aimed to explore perceptions of different stakeholders including health-care professionals, patients, and carers about a telehealth-based system called WELCOME in four European countries: the United Kingdom, Greece, the Netherlands, and Ireland. The system was designed to integrate care for COPD patients with comorbidities using a wearable vest with sensors for monitoring lung and heart function through electroimpedence tomography, analyzing chest sounds, ECG, and blood oxygen saturation. In addition, the system used other smart computing technologies including mobile measuring devices such as weight scale, blood pressure monitor, blood glucose meter, and an inhaler monitoring device. The system was also accompanied by support applications for patients/carers to address disease/medicine education, mental health assessment, lifestyle management, and physical well-being (Fig. 11.3).

Health-care professionals [GPs, specialist doctors (pulmonologists, cardiologist, diabetologist, physiatrist), nurses, and physiotherapists] perceived the system to be useful for COPD patients given the complexity of their profile in particular housebound patients, and for avoiding potential hospital admissions through early detection of exacerbations. Furthermore, they felt that the system can reduce the delay currently observed with patient seeking medical advice [11,50]. Another research [85] explored the views of health-care professionals from different disciplines and sectors about the role that telehealth can play in integrated care. Main themes derived from the interviews were: awareness and understanding of telehealth, experiences and benefits of telehealth, barriers and facilitators of telehealth, and misconceptions about telehealth. What was obvious from the views elicited is that the concepts of telehealth and integrated care were perceived as complementary and facilitators to each other. Health-care professionals envisaged telehealth as a facilitator of integrated care by acting as a tool for integrating secondary and primary care sectors. Some health-care professionals recognized that the main benefit of telehealth integrated care is the extended capacity it offers which allows many health-care professionals to come together to discuss clinical issues about patient care. This can occur without the need for the health-care

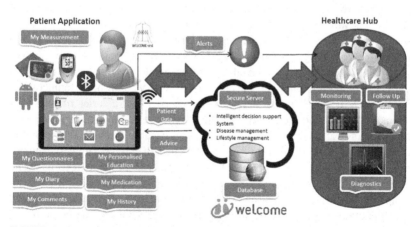

FIGURE 11.3   The WELCOME system. *Source: From D. Sobnath, et al. Mobile Self-Management Application for COPD Patients with Comorbidities: a Usability Study.* IEEE 18th International Conference on e-Health Networking, Applications and Services (Healthcom) (2016), *1−6, doi: 10.1109/HealthCom.2016.7749502.*

professional to travel to physically assess a patient, which consequently would save resources for both the health-care professional and the patient in terms of time and money. They though highlighted their skepticism about the ability and confidence of their patients in adopting and operating technological services, with some of them expressing that elderly patients would prefer the more traditional way of health-care provision and delivery. On the other hand, fear of losing face-to-face contact and missing vital care information were voiced as concerns among health-care professionals irrespective of the discipline. Nurses, in particular, perceived technological services such as telehealth to be a threat to their livelihood/profession [11,85].

Another research [86] explored nurses' perceptions toward an RPM telehealth service targeting patients with congestive heart failure (CHF) and COPD. The RPM system consisted of a set of peripheral devices to be used by patients at home, including pulse oximeter, thermometer, weighing scales, blood pressure monitor, and peak-flow monitor/spirometer. Patients were required to measure their vital signs using the provided devices and answer a series of questions about symptoms. Monitoring data were then transmitted to a monitoring center for triage. If the data fall outside the parameters set for the patient, a trigger was generated and forwarded to the nurse. The nurse can also access the data remotely, enabling monitoring trends over time and making informed decisions accordingly. The nurses described their experience with telehealth to be positive and reported the telehealth system to be beneficial to patients. Nevertheless, lack of resources, organizational support, and technical support were major issues impacting the service implementation. The nurses also highlighted patients' positive engagement and attitudes toward the service [86].

In another research in Australia [87], perceptions of nurses and COPD patients were elicited regarding a respiratory ambulatory care service with RPM telehealth service as an adjunct. The adjunct telehealth service consisted of a complete monitoring and care management unit provided through a fixed portal for recording patients' clinical parameters at home. Patients were required to measure physiological parameters related to lung function, pulse oximeter, blood pressure, electrocardiogram, body weight, body temperature, and blood glucose. Patients were also requested to fill a series of questions about symptoms and well-being and maintain an electronic diary of consultations with health-care professionals. An alert capability was available within the system in case of abnormal measurements. In this research, the nurses thought that patients were finding the telehealth system difficult to use. The nurses also felt that it would be better and more accurate for a nurse to manually take the vital measurements. Interestingly, the authors highlighted how the nurses' comments regarding general impressions of the home monitoring experience were in distinct contrast to comments provided by patients, which were overall much more positive. As such, the research concluded that utilizing home telemonitoring to manage high-care COPD patients in the community resulted in high patient acceptance and lower staff acceptance [87].

Additionally, another research [57] explored community pharmacists' perceptions and need for a novel mobile technology; an electronic monitoring device, to aid in the objective assessment of inhaler technique; and adherence among COPD and asthma patients. In this context, community pharmacists were receptive and open to the idea of incorporating such a technology into practice and were able to envisage the benefits of the investigated technology in terms of providing objective feedback and counseling about inhaler technique and adherence, given that both issues are historically reported in the literature to be problematic among respiratory patients. Nevertheless, the acceptance and willingness for this technology was conditioned by the provision of resources including training and human resources, remuneration, in addition to ease of use of the technology.

Patients and carers perceptions

As highlighted earlier, the research by [11,50] also aimed to explore perceptions of stakeholders like patients and carers about the WELCOME system in four European countries: the United Kingdom, Greece, the Netherlands, and Ireland. The results of the research showed that patients were receptive to all the elements of the proposed system and to the idea of telehealth and integration of their care, although reporting limited daily interaction with technology (limited computer and smartphone use). The majority of patients advocated the concept of self-monitoring while at home and underscored their willingness and acceptance to share all their monitoring results and parameters with all health-care professionals involved in their care pathway and their carers as well. However, the Dutch cohort of patients preferred to share monitoring results with health-care professionals only and not their carers to avoid additional burden. The patients also wanted to have access to their own measurements to be reassured that they are fine.

In a similar manner to patients, informal carers were receptive to the components of the system and the concept of telehealth without due concern, despite their limited interaction with technology on a daily basis. Carers indicated their willingness to access the parameters measured by patients (monitoring data) and to act upon any recommendations provided to patients such as taking antibiotics and/or steroids. Like patients, carers also like these monitoring results to be shared with all the health-care professionals involved in the care of the patient. Interestingly, there was little concern over security and confidentiality from the perspective of patients and carers as they felt that the benefit of such a system outweighs any concerns [11,50].

In the aforementioned research by [87], although, as indicated, the nurses were skeptical about patients' ability to use the system, this was not the case with patients (n = 21) who were more positive and satisfied with the telehealth service. In fact, the patients highlighted that the telehealth service provided them with a sense of empowerment and control over their condition. This was evident through patients' adherence with the measurement regimes and system usage which was reported as excellent in the study. Furthermore, some patients were distressed when reminded that the telehealth service will be removed at the end of the study [87].

In the follow-up research by [88], COPD patients were very satisfied with the RPM telehealth service (as described earlier in [86]) provided to them. The provision of emotional and medical support underpinned patients' satisfaction with the RPM telehealth service. According to the patients, the service had improved their health and it was a convenient form of health-care delivery for them. Interestingly, patients felt they were empowered and more involved in decision-making about their treatment and care. Moreover, patients had no major concerns regarding confidentiality, loss of direct contact with a health-care professional during a telehealth consultation, or the use of the necessary equipment for the service [88]. Another research confirmed how COPD patients valued the relationship continuity because of the telemonitoring service offered to them, which underpinned service satisfaction. Additionally, patients expressed gratitude for the personalized advice and support they received from the service [89].

Another research by [90] explored the perceptions and experiences of COPD patients who received telehealth supported discharge service versus those receiving standard community-based discharge service provided by nurses. Interestingly, although patients receiving the telehealth service had 50% fewer home visits from the clinicians compared to those having the traditional community-based nursing intervention, the patients were enthusiastic about the telehealth service, with some describing it as the best service they had ever received.

Additionally another research reported that a 12-week physical activity tele-coaching intervention was well accepted and perceived as feasible by 159 COPD patients with the majority (89.3%) indicating that they enjoyed taking part in the intervention. The majority reported positive experience with the tele-coaching intervention while describing it as motivational. In fact, most of the patients highlighted that being monitored provided them with an external motivational cue to be physically active [70].

In summary, what is obvious from the literature is the disconnection in perceptions and beliefs among health-care professionals and patients when it comes to the implementation and use of TEC services. As mentioned earlier, health-care professionals often appear to be skeptical about patients' ability and confidence in adopting and operating technological services. However, this was not the case as shown in the abovementioned studies which demonstrated how patients were open, receptive, and even grateful for such technological services. Interestingly, this leaves health-care professionals with a distorted perception about the reality, with the assumption that patients are often not receptive or ready for technological interventions.

Barriers for technology enabled care

Effective implementation of TEC hinges on the acceptance of different stakeholders [60]. In fact, acceptability is regarded as a key concept in the design, evaluation, and implementation of health-care interventions including those involving technology. Furthermore, acceptability can have a significant impact on the effectiveness of the intervention [70,91]. Hence, understanding the challenges underpinning the acceptance and adoption of technologies are key to promote successful implementation of TEC [60]. This section provides an overview about some of the barriers in relation to the adoption of technology enabled services.

Factors related to the technology itself

Perceived usefulness and perceived ease of use are two pivotal factors in modeling the process of acceptance of new technologies. This is underpinned by many theories in relation to technology acceptance, most notably the Technology Acceptance Model (TAM), its updated version (TAM2), and the Unified Theory of Acceptance and Use of Technology (UTAUT) [92]. Perceived usefulness is defined as the extent to which the person perceives that using the proposed technology will enhance their job performance, whereas perceived ease of use is defined as the extent to which the person perceives that using the proposed technology will be free of effort [92]. As highlighted in the literature, it is important for health-care professionals and patients to perceive the ease of use and usefulness of technologies, otherwise there would be less incentive to use them [62]. Design and technical concerns including but not limited to complexity of using features, limited features, and screen size of employed devices were frequently identified as barriers to TEC services. In addition, the interoperability of the technology (integration with other systems), cost, privacy, and security concerns and technology reliability were among the barriers cited in several studies [62,85].

Pertaining to this comes the characteristics of the innovation itself which are particularly important to determine its rate of adoption. Rogers diffusion of innovation (DOI) theory is a well-established theory in the literature to explain the prerequisites and processes for successful adoption and implementation of innovations/new technologies [93]. The theory highlights five attributes that are important in determining the adoption rate of any innovation/new technology; these include: relative advantage, complexity, compatibility, trialability, and observability. Relative advantage is defined as the degree to which an innovation is seen better than the current available practices. This attribute relates positively to the rate of adoption of an innovation, hence the more advantageous the innovation, the more rapid its rate of adoption will be. The relative advantage can be economic, social, or associated with relative effectiveness [93].

Complexity stands for the degree to which an innovation/new technology is seen difficult to understand and use. This attribute relates negatively to the rate of adoption of an innovation, hence innovations that are simpler to understand and use are more rapidly adopted [93].

Compatibility stands for the degree of consistency of an innovation with the already existing values, needs, beliefs, and experiences of the potential adopters. An innovation that is congruent with their preexisting values, needs, beliefs, and experiences will be adopted more rapidly because adopters will be less uncertain and feel comfortable with it. Hence, this attribute relates positively to the rate of adoption of an innovation [93].

Trialability stands for the degree to which an innovation can be experimented or tested before any commitment to adoption is made. This attribute relates positively to the rate of adoption of an innovation, by giving the chance for an adopter to examine how an innovation works under one's own environment, thus eliminating any uncertainties still associated with the innovation [93].

Observability stands for the extent to which an innovation provides tangible results to others. This attribute relates positively to the rate of adoption of an innovation. Observing the impact of the innovation/new technology can reduce any uncertainties and facilitate adoption [93].

Hence, according to the theory, these five attributes determine between 49% and 87% of the variation in the adoption of innovations. Moreover, it is argued that innovations/new technologies offering more relative advantage, simplicity, compatibility, trialability, and observability will be adopted faster than other innovations/new technologies [93].

Individual factors: knowledge, attitude, and sociodemographic characteristics

Some of the barriers falling under this category include: familiarity with the technology itself, self-efficacy, time-related issues, awareness of the objectives, and/or existence of technologies, in addition to experience and impact on professional security for health-care professionals [62,85]. With regard to impact on professional security, a previous research by [85] reported that although telehealth was perceived by health-care professionals as beneficial, yet skepticism was prevalent. The fear of losing face-to-face contact emerged as a major limitation to adoption of telehealth; the issue was prevalent among pharmacists, nurses, and other health-care professionals irrespective of discipline, experience, and sector. Nurses perceived telehealth as a threat to their profession/livelihood. Interestingly, this was also reflected in other studies throughout the literature [94–98]. Interestingly, the research by [85] also highlighted lack of awareness about telehealth among health-care professionals, particularly those in primary care, which was reflective of the lack of experience reported, given that most of the study sample who worked in primary care did not use telehealth in their work. This picture may have now changed considering the increased virtual consultations that took place in primary care during the Coronavirus disease

2019 (COVID-19) pandemic. With the emergence of this pandemic and cities being in lockdown, the use of virtual care solutions became inevitable to provide quality health care to the population, ensure continuity of care while maintaining risk-mitigation strategies [99]. The use of digital technologies has been considered as a way to remediate the COVID-19 pandemic [100]. In fact, several studies have reported on the increased uptake and success of TEC services such as telemedicine/teleconsultations across many developed countries during the pandemic [99,101−103].

Pertaining to this comes the necessity of raising awareness and education among all health-care professionals regarding digital health including TEC. Currently, topics related to TEC, health technology, or digital health are not included or considered as core topics within the curriculum for health-care students. However, in an era of digital revolution, it becomes imperative to integrate digital health into the curriculum in order to educate future health-care professionals on how to work in an era of digital tools [104]. This is particularly pertinent in light of the current COVID-19 pandemic and shift to virtual consultations.

External factors: human environment

Barriers in this category were based on the perceptions of the health-care professionals in terms of the applicability of the technology to the characteristics of the patients, patients' attitudes and preferences toward the technology, and patient−human interaction. For example, the evidence base regarding staff skepticism about patients' beliefs toward technology and patients' ability and confidence level in adopting and operating technology has been reported as a key challenge for telehealth implementation in the literature [85]. This reported skepticism is mostly underpinned by age. From health-care professionals' point of view, most of the potential users of TEC services are elderly who are most likely not familiar with technology use, or can be skeptical about the service as they are used to the traditional way of health-care delivery resembled in seeing health-care professionals face-to-face [85].

External factors: organizational environment

Lack of resources and health-care professionals' status and reimbursement are among the factors identified under this category. Problems related to cost and funding and lack of resources in particular time and workload were frequently reported in the literature as key barriers to implementation of technologies in the health-care sector [62,85,86]. Patient and health-care professionals' education and training were also reported as key barriers pertaining to resources [85]. In one research, health-care professionals raised the need for education and training for both patients and health-care professionals to overcome some of the barriers encountered with telehealth implementation, especially in terms of the service functionality [85]. Pertaining to this comes another important issue which is the level of information provided to patients prior to the introduction of any technology enabled service. A previous research aimed to explore how telehealth is portrayed in telehealth leaflets provided to users via discourse analysis [105]. The discourse analysis highlighted certain key gaps and variations within the content and presentation of the screened leaflets in relation to the following aspects: cost of the telehealth service, confidentiality, patients' choices to withdraw from the service, in addition to assurances about the technology used particularly equipment use and technical support. The same research also interviewed patients to explore their perceptions about the screened leaflets and their engagement with telehealth as a concept. Analysis of the interviews revealed patients' need for clear and sufficient information about the telehealth service within the leaflets. Patients wanted to have simplified terminologies for telehealth description and clear simple texts with pictorial presentations. The interviews also revealed certain limitations against adoption of telehealth by the patients, including lack of privacy and confidentiality of information, fear of technology breakdown and equipment failure, loss of face-to-face contact with health-care professionals, and being too dependent on the telehealth service [105]. Lack of investment in resources related to the technology infrastructure whether for patients or health-care professionals or other stakeholders and lack of human resources for technical support are other crucial barriers [85,86].

The introduction of technological innovations into health-care organizations is often disruptive and requires a comprehensive change management plan to be in place, given the complexity associated with these technological interventions/innovations [85,86]. Several studies reported organizational and management barriers to affect the implementation of TEC services [85–87]. Hence, it is paramount to carefully plan the organizational change when introducing such interventions/innovations through: integrating and engaging potential stakeholders such as health-care professionals and patients at an early stage, careful integration of the technology/innovation into the established care pathway, redesigning job descriptions, and allocating sufficient resources. As highlighted in the literature, uncertainty on the part of health-care professionals is considered as a potential barrier to the implementation of TEC services [96], hence advocacy and support from frontline health-care staff is a key to the success of these innovations/interventions [86,106]. Low acceptance of TEC services among health-care professionals was reported as a key issue affecting telehealth implementation in the literature [85–87]. In the aforementioned research by [87], there was a low level of acceptance to the telehealth service among nurses compared to patients. This was attributed to the fact that the staff were regarding the technology as an undefined adjunct to their current service model and not well integrated into their clinical routine. Hence, the nurses continued to take patients' vital measurements manually rather than relying on those already recorded through the telehealth service when conducting home visits to the patients. In the aforementioned research by [86] which explored nurses' perceptions regarding a telehealth service for COPD patients, nurses reported lack of organizational support which led to negative feelings about the success and future of the service. Almost all nurses had no say in the decision about joining the telehealth service, given that their names were allocated by the GP. Hence, the nurses felt that the service was imposed on them without soliciting their views [86]. Maintaining the sustainability of technological interventions/innovations is also another key aspect, which can be only achieved with the identification and engagement of what we term as "champions" within the organization who are willing and prepared to promote these technological interventions/innovations on the long run [86,107].

Changing health service delivery requires a long process of negotiation as well as a complete business strategy. Among the components of this strategy, standard policies and clinical protocols for the implementation of TEC services will be essential for acceptance of these technologies into mainstream clinical practice/care. Additionally, such strategies should also consider the incorporation of incentives for health-care staff to change existing practices [87].

Conclusion

With many perspectives and definitions in the literature, it can be concluded that integrated care is not a simple or straightforward concept to adopt and implement. In essence, it can be best envisaged as an approach to overcome care fragmentations through better coordination in order to improve patient care, while taking contextual factors into consideration. Integrated care becomes crucial in the context of respiratory care management, due to several factors, most importantly, the complexity of patients' profiles, the huge burden imposed by chronic respiratory conditions on health-care systems worldwide, and the reported fragmentation in current care pathways to date. And while this chapter focused on COPD as an example of a long-term respiratory condition, yet the same can be argued for the other chronic conditions as well. Equally important, is the reported benefits of integrated care in respiratory care management as evident in several studies documented in this chapter. Moreover, the utilization of ICTs in health care, also termed as TEC, can be envisaged as an enabler of integrated care through promoting functional integration. Lack of consensus about the definitions of the various TEC services was also evident throughout the literature. However, the current chapter aimed to resolve some of the ambiguity associated with these services, through clarifying these definitions. Incorporation of TEC services is complex with many barriers at the individual, organizational, and contextual levels, including: characteristics of the technology itself, perceived ease of use and usefulness, familiarity with technology, time issues, resources availability, and change management. Understanding these barriers is key to promote integrated care if TEC services are to support such integration.

References

[1] H.M. Lloyd, et al., Collaborative action for person-centred coordinated care (P3C): an approach to support the development of a comprehensive system-wide solution to fragmented care, Health Res. Pol. Syst. 15 (1) (2017) 98.

[2] B.R. Frandsen, et al., Care fragmentation, quality, and costs among chronically ill patients, Am. J. Manag. Care 21 (5) (2015) 355—362.

[3] Royal College of Physicians (RCP), Putting the Pieces Together: Removing the Barriers to Excellent Patient Care, 2015. Available at: file:///C:/Users/ku70283/Downloads/Putting% 20patients%20first%20removing%20the%20barriers%20to%20excellent%20patient%20care_ 0.pdf. (Accessed 12 December 2020).

[4] K.C. Stange, The problem of fragmentation and the need for integrative solutions, Ann. Fam. Med. 7 (2009) 100—103, https://doi.org/10.1370/afm.971.

[5] L. Agha, B. Frandsen, J.B. Rebitzer, Causes and Consequences of Fragmented Care Delivery: Theory, Evidence, and Public Policy, National Bureau of Economic Research, Cambridge, MA, 2017.

[6] British Medical Association, Growing Older in the UK: A Series of Expert-Authored Briefing Papers on Ageing and Health, 2016. Available at: https://www.bma.org.uk/media/ 2105/supporting-healthy-ageing-briefings-final.pdf.

[7] T. Manser, Fragmentation of patient safety research: a critical reflection of current human factors approaches to patient handover, J. Public Health Res. 2 (3) (2013).

[8] D. Martin, A.P. Miller, A. Quesnel-Vallee, N.R. Caron, B. Vissandjee, G.P. Marchildon, Canada's global leadership on health 1. Canada's universal healthcare system: achieving its potential, Lancet 391 (2018) 1718—1735.

[9] E. Nolte, et al., Overcoming fragmentation in health care: chronic care in Austria, German and The Netherlands, Health Econ. Policy Law 7 (2012) 125—146.

[10] N. Goodwin, Understanding integrated care, Int. J. Integr. Care 16 (4) (2016).

[11] R. Kayyali, S. Nabhani-Gebara, et al., User Profiling for Coordinated and Integrated Care, IEEE, 2016, pp. 473—476.

[12] M. Ouwens, et al., Integrated care programmes for chronically ill patients: a review of systematic reviews, Int. J. Qual. Health Care 17 (2) (2005) 141—146.

[13] S. Baxter, et al., The effects of integrated care: a systematic review of UK and international evidence, BMC Health Serv. Res. 18 (1) (2018) 1—13.

[14] World Health Organization-WHO, Integrated Care Models: An Overview, 2016. Available at: https://www.euro.who.int/__data/assets/pdf_file/0005/322475/Integrated-care-models-overview.pdf. (Accessed 9 December 2020).

[15] A.P. Contandriapoulos, J.L. Denis, N. Touati, C. Rodriguez, Groupe de recherche interdisciplinaire en santé. Working Paper N04—01. Montréal: Université de Montréal, The integration of health care: dimensions and implementation (2003 Jun). Available from, http://www.irspum.umontreal.ca/rapportpdf/n04-01.pdf.

[16] D. Kodner, C. Spreeuwenberg, Integrated care: meaning, logic, applications and implications — a discussion paper, Int. J. Integr. Care 2 (14) (2002), https://doi.org/10.5334/ijic.67.

[17] R. Lewis, R. Rosen, N. Goodwin, J. Dixon, Where next for integrated care organisations in the English NHS? The Nuffield Trust, London, 2010.

[18] G. Hughes, S.E. Shaw, T. Greenhalgh, Rethinking integrated care: a systematic hermeneutic review of the literature on integrated care strategies and concepts, Milbank Q. 98 (2) (2020) 446—492.

[19] L. Liu, et al., Smart homes and home health monitoring technologies for older adults: a systematic review, Int. J. Med. Inf. 91 (2016) 44–59.

[20] J.A. Botia, A. Villa, J. Palma, Ambient assisted living system for in-home monitoring of healthy independent elders, Expert Syst. Appl. 39 (9) (2012) 8136–8148.

[21] C. Hajat, E. Stein, The global burden of multiple chronic conditions: a narrative review, Prev. Med. Rep. 12 (2018) 284–293.

[22] N. Zonneveld, et al., Values of integrated care: a systematic review, Int. J. Integr. Care 18 (4) (2018).

[23] D.E. Bloom, S. Chen, M. Kuhn, M.E. McGovern, L. Oxley, K. Prettner, The economic burden of chronic diseases: estimates and projections for China, Japan, and South Korea, National Bureau of Economic Research Working Paper Series, Cambridge, MA, 2017. No. 23601. 10.3386/w23601.

[24] N. Maglaveras, et al., Integrated Care and Connected Health Approaches Leveraging Personalised Health through Big Data Analytics, 2016, p. 117.

[25] K.M. Lemmens, A.P. Nieboer, R. Huijsman, A systematic review of integrated use of disease-management interventions in asthma and COPD, Respir. Med. 103 (5) (2009) 670–691.

[26] R. Ahmed, R. Robinson, K. Mortimer, The epidemiology of noncommunicable respiratory disease in sub-Saharan Africa, the Middle East, and North Africa, Malawi Med. J. 29 (2) (2017) 203–211.

[27] GOLD, Global Strategy for the Diagnosis, Management and Prevention of Chronic Obstructive Pulmonary Disease, 2020. Available at: https://goldcopd.org/wp-content/uploads/2019/12/GOLD-2020-FINAL-ver1.2-03Dec19_WMV.pdf. (Accessed 12 December 2020).

[28] A. Casas, et al., Integrated care prevents hospitalisations for exacerbations in COPD patients, Eur. Respir. J. 28 (1) (2006) 123–130.

[29] J. Garcia-Aymerich, et al., Effects of an integrated care intervention on risk factors of COPD readmission, Respir. Med. 101 (7) (2007) 1462–1469.

[30] M.E. Wacker, et al., Direct and indirect costs of COPD and its comorbidities: results from the German COSYCONET study, Respir. Med. 111 (2016) 39–46.

[31] A. Casas Herrera, et al., COPD underdiagnosis and misdiagnosis in a high-risk primary care population in four Latin American countries. A key to enhance disease diagnosis: the PUMA study, PLoS One 11 (4) (2016) e0152266.

[32] S. Carlone, et al., Health and social impacts of COPD and the problem of under-diagnosis, Multidiscip. Respir. Med. 9 (1) (2014) 1–6.

[33] J. Ancochea, et al., Underdiagnosis of chronic obstructive pulmonary disease in women: quantification of the problem, determinants and proposed actions, Arch. Bronconeumol. 49 (6) (2013) 223–229.

[34] K. Hill, et al., Prevalence and underdiagnosis of chronic obstructive pulmonary disease among patients at risk in primary care, Can. Med. Assoc. J. 182 (7) (2010) 673–678.

[35] D.S. Kim, et al., Prevalence of chronic obstructive pulmonary disease in Korea: a population-based spirometry survey, Am. J. Respir. Crit. Care Med. 172 (7) (2005) 842–847.

[36] A. Lindberg, et al., Prevalence and underdiagnosis of COPD by disease severity and the attributable fraction of smoking: report from the Obstructive Lung Disease in Northern Sweden Studies, Respir. Med. 100 (2) (2006) 264–272.

[37] D.M. Mannino, et al., Chronic obstructive pulmonary disease surveillance-United States, 1971–2000, Respir. Care 47 (10) (2002) 1184–1199.

[38] D.M. Mannino, 'No Title', Underdiagnosed Chronic Obstructive Pulmonary Disease in England: New Country, Same Story, 2006.

[39] M.C. Queiroz, M.A.C. Moreira, M.F. Rabahi, Underdiagnosis of COPD at primary health care clinics in the city of Aparecida de Goiânia, Brazil, J. Bras. Pneumol. 38 (2012) 692–699.

[40] T. Takahashi, et al., Underdiagnosis and undertreatment of COPD in primary care settings, Respirology 8 (4) (2003) 504–508.

[41] J.A. Walters, et al., Under-diagnosis of chronic obstructive pulmonary disease: a qualitative study in primary care, Respir. Med. 102 (5) (2008) 738–743.

[42] S. Ehteshami-Afshar, et al., The global economic burden of asthma and chronic obstructive pulmonary disease, Int. J. Tubercul. Lung Dis. 20 (1) (2016) 11–23.

[43] P. Rajkumar, et al., A cross-sectional study on prevalence of chronic obstructive pulmonary disease (COPD) in India: rationale and methods, BMJ Open 7 (5) (2017).

[44] World Health Organization (WHO)., Chronic obstructive pulmonary disease (COPD). Key facts, 2017. Available at: https://www.who.int/news-room/fact-sheets/detail/chronic-obstructive-pulmonary-disease-(copd). (Accessed 9 December 2020).

[45] W. Chen, et al., Excess economic burden of comorbidities in COPD: a 15-year population-based study, Eur. Respir. J. 50 (1) (2017).

[46] A. Lenferink, et al., A self-management approach using self-initiated action plans for symptoms with ongoing nurse support in patients with Chronic Obstructive Pulmonary Disease (COPD) and comorbidities: the COPE-III study protocol, Contemp. Clin. Trials 36 (1) (2013) 81–89.

[47] British Lung Foundation, Estimating the Economic Burden of Respiratory Illness in the UK, 2016. Available at: https://cdn.shopify.com/s/files/1/0221/4446/files/PC-1601_-_Economic_burden_report_FINAL_8cdaba2a-589a-4a49-bd14-f45d66167795.pdf?1309 501094450848169&_ga=2.18521209.1753974424.1608231327-881901112.1608231327. (Accessed 12 December 2020).

[48] Australian institute of Health and Welfare, Chronic Obstructive Pulmonary Disease (COPD), 2020. Available at: https://www.aihw.gov.au/getmedia/6a614e0d-f0ac-48 f5-ad5b-e1e50d7a430d/Chronic-obstructive-pulmonary-disease-COPD.pdf.aspx?inline= true. (Accessed 12 December 2020).

[49] F.M. Franssen, C.L. Rochester, 'No Title', Comorbidities in Patients with COPD and Pulmonary Rehabilitation: Do They Matter?, 2014.

[50] S. Nabhani-Gebara, et al., WELCOME project: what do stakeholders want?, in: Depth Analysis of COPD Patients, Carers and Healthcare Professional Views IEEE, 2014, p. 365.

[51] N. Roche, 'No Title', Adding Biological Markers to COPD Categorisation Schemes: A Way towards More Personalised Care?, 2016.

[52] F.W. Ko, et al., COPD care programme can reduce readmissions and in-patient bed days, Respir. Med. 108 (12) (2014) 1771–1778.

[53] A.L. Kruis, N. Smidt, W.J. Assendelft, et al., Integrated disease management interventions for patients with chronic obstructive pulmonary disease (Review), Cochrane Database Syst. Rev. (10 (2013) CD009437.

[54] M.R. Boland, et al., The health economic impact of disease management programs for COPD: a systematic literature review and meta-analysis, BMC Pulm. Med. 13 (1) (2013) 40.

[55] R. Kayyali, B. Odeh, et al., COPD care delivery pathways in five European Union countries: mapping and health care professionals' perceptions, Int. J. Chronic Obstr. Pulm. Dis. 11 (2016) 2831–2838. Original publisher: Dove Medical Press.

[56] J. Cittée, B. Sauteron, S. Brossier, E. Ferrat, C. Attali, C. Chouaïd, B. Housset, COPD Patient Care Pathways: Points of View of Hospital Personnel, Santé Publique (2015) 177–187, https://doi.org/10.3917/spub.150.0177.

[57] I. Hesso, R. Kayyali, S. Nabhani-Gebara, Supporting respiratory patients in primary care: a qualitative insight from independent community pharmacists in London, BMC Health Serv. Res. 19 (1) (2019) 5.

[58] P. Sullivan, D.J. Lugg, Telemedicine between Australia and Antarctica: 1911−1995, 1995, https://doi.org/10.4271/951616. SAE Technical Paper 951616.

[59] A.Z. Woldaregay, S. Walderhaug, G. Hartvigsen, Telemedicine services for the arctic: a systematic review, JMIR Med. Inform. 5 (2) (2017) e16.

[60] National Health Service, Technology Enabled Care Services. Resource for Commissioners, 2015. Available at: https://www.england.nhs.uk/wp-content/uploads/2014/12/TECS_FinalDraft_0901.pdf. (Accessed 12 December 2020).

[61] A.L. Leonardsen, et al., Patient experiences with technology enabled care across healthcare settings-a systematic review, BMC Health Serv. Res. 20 (1) (2020) 1−17.

[62] M. Gagnon, et al., m-Health adoption by healthcare professionals: a systematic review, J. Am. Med. Inf. Assoc. 23 (1) (2016) 212−220.

[63] F. Fatehi, R. Wootton, Telemedicine, telehealth or e-health? A bibliometric analysis of the trends in the use of these terms, J. Telemed. Telecare 18 (8) (2012) 460−464.

[64] L. Van Dyk, A review of telehealth service implementation frameworks, Int. J. Environ. Res. Publ. Health 11 (2) (2014) 1279−1298.

[65] G. Eysenbach, What is e-health? J. Med. Internet Res. 3 (2) (2001) e20.

[66] A. Steventon, M. Bardsley, S. Newman, Effect of telehealth on use of secondary care and mortality: findings from the Whole System Demonstrator cluster randomised trial, BMJ 344 (2012) e3874.

[67] C. Sanders, A. Rogers, R. Bowen, P. Bower, S. Hirani, M. Cartwright, et al., Exploring barriers to participation and adoption of telehealth and telecare within the Whole System Demonstrator trial: a qualitative study, BMC Health Serv. Res. 12 (2012).

[68] World Health Organization, Telehealth, 2020. Available at: https://www.who.int/gho/goe/telehealth/en/. (Accessed 9 December 2020).

[69] J.R. Reid, A Telemedicine Primer: Understanding the Issues, Artcraft Printers, Billings, MT, 1996.

[70] M. Loeckx, et al., Smartphone-based physical activity telecoaching in chronic obstructive pulmonary disease: mixed-methods study on patient experiences and lessons for implementation, JMIR Mhealth Uhealth 6 (12) (2018) e200.

[71] E. Papi, A. Belsi, A.H. McGregor, A knee monitoring device and the preferences of patients living with osteoarthritis: a qualitative study, BMJ Open 5 (9) (2015).

[72] H.P. Profita, et al., Don't Mind Me Touching My Wrist: A Case Study of Interacting with On-Body Technology in Public, 2013, p. 89.

[73] Y. Lee, W. Chung, Wireless sensor network based wearable smart shirt for ubiquitous health and activity monitoring, Sensor. Actuator. B Chem. 140 (2) (2009) 390−395.

[74] P.S. Pandian, et al., Smart Vest: wearable multi-parameter remote physiological monitoring system, Med. Eng. Phys. 30 (4) (2008) 466−477.

[75] F. Pitta, et al., Quantifying physical activity in daily life with questionnaires and motion sensors in COPD, Eur. Respir. J. 27 (5) (2006) 1040−1055.

[76] V. Rocha, et al., Wearable Computing for Patients with Coronary Diseases: Gathering Efforts by Comparing Methods, IEEE, 2010, p. 1.

[77] E. Sardini, M. Serpelloni, Instrumented wearable belt for wireless health monitoring, Procedia Eng. 5 (2010) 580−583.

[78] S. Ajami, F. Teimouri, Features and application of wearable biosensors in medical care, J. Res. Med. Sci. 20 (12) (2015) 1208.

[79] World Health Organization, mHealth: New Horizons for Health Through Mobile Technologies: Second Global Survey on eHealth, 2011. Available at: https://www.who.int/goe/publications/goe_mhealth_web.pdf. (Accessed 9 December 2020).

[80] I. García-Magariño, D. Sarkar, R. Lacuesta, 'No Title', Wearable Technology and Mobile Applications for Healthcare, 2019.

[81] N. Goodwin, et al., Providing Integrated Care for Older People with Complex Needs: Lessons from Seven International Case Studies, King's Fund London, 2014.

[82] P.P. Valentijn, et al., Understanding integrated care: a comprehensive conceptual framework based on the integrative functions of primary care, Int. J. Integr. Care 13 (2013).

[83] A.H.Y. Chan, et al., Electronic adherence monitoring device performance and patient acceptability: a randomized control trial, Expet Rev. Med. Dev. 14 (5) (2017) 401–411.

[84] J.M. Foster, et al., Patient-perceived acceptability and behaviour change benefits of inhaler reminders and adherence feedback: a qualitative study, Respir. Med. 129 (2017) 39–45.

[85] R. Kayyali, I. Hesso, A. Mahdi, et al., Telehealth: misconceptions and experiences of healthcare professionals in England, Int. J. Pharm. Pract. 25 (3) (2017) 203–209.

[86] B. Odeh, et al., Implementing a telehealth service: nurses' perceptions and experiences, Br. J. Nurs. 23 (21) (2014) 1133–1137.

[87] T. Shany, et al., Home telecare study for patients with chronic lung disease in the Sydney West Area Health Service, Stud. Health Technol. Inf. 161 (2010) 139–148.

[88] B. Odeh, et al., Evaluation of a Telehealth Service for COPD and HF patients: clinical outcome and patients' perceptions, J. Telemed. Telecare 21 (5) (2015) 292–297.

[89] P. Fairbrother, et al., Continuity, but at what cost? The impact of telemonitoring COPD on continuities of care: a qualitative study, Prim. Care Respir. J. 21 (3) (2012) 322–328.

[90] D.A. Fitzsimmons, et al., Comparison of patient perceptions of Telehealth-supported and specialist nursing interventions for early stage COPD: a qualitative study, BMC Health Serv. Res. 16 (1) (2016) 420.

[91] M. Sekhon, M. Cartwright, J.J. Francis, Acceptability of healthcare interventions: an overview of reviews and development of a theoretical framework, BMC Health Serv. Res. 17 (1) (2017) 1–13.

[92] N.Ö. Gücin, Ö.S. Berk, Technology Acceptance in Health Care: An Integrative Review of Predictive Factors and Intervention Programs, vol. 195, Procedia-Social and Behavioral Sciences, 2015, pp. 1698–1704.

[93] E.M. Rogers, Diffusion of Innovations, Simon and Schuster, 2010.

[94] M. Gagnon, et al., Implementing telehealth to support medical practice in rural/remote regions: what are the conditions for success? Implement. Sci. 1 (1) (2006) 18.

[95] R. Giordano, M. Clark, N. Goodwin, Perspectives on Telehealth and Telecare: Learning from the 12 Whole System Demonstrator Action Network (WSDAN) Sites, King's Fund, 2011.

[96] V. Joseph, et al., Key challenges in the development and implementation of telehealth projects, J. Telemed. Telecare 17 (2) (2011) 71–77.

[97] N. Oudshoorn, Physical and digital proximity: emerging ways of health care in face-to-face and telemonitoring of heart-failure patients, Sociol. Health Illn. 31 (3) (2009) 390–405.

[98] U. Sharma, J. Barnett, M. Clarke, Clinical users' perspective on telemonitoring of patients with long term conditions: understood through concepts of giddens's structuration theory & consequence of modernity, in: MEDINFO, IOS Press, 2010, pp. 545–549.

[99] B.A. Jnr, Use of telemedicine and virtual care for remote treatment in response to COVID-19 pandemic, J. Med. Syst. 44 (132) (2020).

[100] D.S.W. Ting, et al., Digital technology and COVID-19, Nat. Med. 26 (2020) 458–461.

[101] E.Z. Barsom, et al., Coping with COVID-19: scaling up virtual care to standard practice, Nat. Med. 26 (2020) 632−634.

[102] R.P. Murphy, et al., Virtual geriatric clinics and the COVID-19 catalyst: a rapid review, Age Ageing 49 (2020) 907−914.

[103] L.M. Quinn, et al., Virtual consultations and the role of technology during the COVID-19 pandemic for people with type 2 diabetes: the UK perspective, J. Med. Internet Res. 22 (8) (2020) e21609, https://doi.org/10.2196/21609.

[104] T.D. Aungst, R. Patel, Integrating digital health into the curriculum- considerations on the current landscape and future developments, J. Med. Educ. Curric. Dev. 7 (2020) 1−7.

[105] R. Kayyali, I. Hesso, E. Ejiko, et al., A qualitative study of Telehealth patient information leaflets (TILs): are we giving patients enough information? BMC Health Serv. Res. 17 (1) (2017) 362.

[106] J. Hendy, T. Chrysanthaki, J. Barlow, et al., An organisational analysis of the implementation of telecare and telehealth: the whole systems demonstrator, BMC Health Serv. Res. 12 (2012) 403, https://doi.org/10.1186/1472-6963-12-403.

[107] A. Hockey, P. Yellowlees, S. Murphy, Evaluation of a pilot second-opinion child telepsychiatry service, J. Telemed. Telecare 10 (Suppl. 1) (2004) 48−50.

Index

sociodemographic characteristics, 344–345
stakeholders toward technology enabled care, 338–342
technology enabled care, 334–335
 barriers for, 343–347
telecare, 337
tele-coaching, 337
telehealth, 336
telemedicine/teleconsultations, 337
types, 326–327
Integrated tools, applications/websites with, 280
Intensive care, 301–303
Interfaces, inefficiency of, 216
Interstitial lung disease (ILD), 23
Interstitium, 23
Inverse model, 173–175

L
Learning health system, 316
Level of personalization, 316
Linear predictive coding (LPC), 151
Long-term adherence, strategies for
activation, 283
adherence, 273
animals, 290
apps/web portals, 286
cockroach, 290
decision aid, 283
dust mites, 289–290
elderly person, 292–295
inclusion, 277–279
indoor mold, 290
informational videos/text, 285
informative videos/text, 281
initial search, 276
integrated tools, applications/websites with, 280
materials and methods, 276–279
medication taking schedule, 281, 285
messaging services, 281
outdoor seasonal allergens, 290
participation, 284
patient diaries/reports, 285
patient participation, 283
personalization, 283
screening, 276
SMS sending, 284–285
social media, 281–283
steps
 activation, 294

participation, 294–295
take over, 294
strategy, 289–290, 292–294
suitability, 276–277
take over, 283–284
wearables/EMD, 286
Lung
cancer diagnosis, 303
diseases
 asthma, 16–18
 blood vessels, 23–25
 chest wall, 26–27
 diagnosis, 17
 interstitial lung disease (ILD), 23
 interstitium, 23
 pathology, 16–27
 pleura, 25–26
 pneumonia, 23
 pulmonary edema, 23
 pulmonary embolism (PE), 23–25
 pulmonary hypertension, 25
 treatment, 16–27
function, 5
function and structure, 306–307
monitoring
 abdomen, 60–62
 capacity, 61–62
 chest, 60–62
 electrical impedance tomography
  bioimpedance measurements, technological challenges of, 66–68
 electrical impedance tomography, cooperative sensors for, 70
 impedance-based methods, 63–70
 inductance, 62
 measurement principle, 63
 multichannel, electrical impedance tomography, 64–66
 optical methods, 60
 resistivity, 60–61
 respiratory functions assessment through monitoring of the airways, 70–76
 safety considerations, 63
 single-channel, impedance pneumography, 64
 single vs. multichannel bioimpedance measurement modalities, 64–66
 wearable electrical impedance tomography systems, 68–70
perfusion, 199
sounds, 126
vesicular sounds, 126–127

Printed in the United States
by Baker & Taylor Publisher Services